Micro and Nano Energy Harvesting Technologies

For a complete listing of titles in the
Artech House Microelectromechanical Systems library,
turn to the back of this book.

Micro and Nano Energy Harvesting Technologies

Bin Yang

Huicong Liu

Jingquan Liu

Chengkuo Lee

ARTECH HOUSE

BOSTON | LONDON
artechhouse.com

Library of Congress Cataloging-in-Publication Data
A catalog record for this book is available from the U.S. Library of Congress.

British Library Cataloguing in Publication Data
A catalogue record for this book is available from the British Library.

Cover design by John Gomes

ISBN 13: 978-1-60807-814-1

685 Canton Street
Norwood, MA 02062

10 9 8 7 6 5 4 3 2 1

Contents

Preface

Energy shortages and the influence of global warming attributed to the emission of greenhouse gases have become global issues. Renewable and clean energies are essential for reducing heavy reliance on mineral-based energy and remedying the threat of global warming to our environment. Solar energy, mechanical energy (mostly wind), and bioenergy are common substitutions and have been more widely used. Solar energy is abundant but not accessible in dark areas or under water. Mechanical energy is ubiquitous but mostly at a lower energy level, such as friction, body movement, heat, wave, and acoustic. Bioenergy research examines bioenergy crops with a higher energy yield, an easy method of conversion to fuels, and agriculture. However, in the last decade, explosive growth in research and development efforts devoted to microelectromechanical systems (MEMS) technology and nanowire-related nanotechnology has paved a great foundation for new mechanisms of harvesting mechanical energy at the micrometer and nanometer scales. Friction, heat, inertial force, and movement are physical phenomena that are strongly subject to the interface and surface properties at the micrometer and nanometer scales. Leveraging the microtechnology and nanotechnology, we can scavenge the small amount of mechanical energy that is not used on the basis of on macroscale mechanisms, and we can also properly design micromechanisms and nanostructures for the energy conversion of micrometer and nanometer scale displacement, deformation, movement, vibration, shock, and heat, which are introduced by friction, body movement, wave and acoustic-induced vibrations, and heat. Recent advances in low-power consumption integrated circuits (ICs) and intelligent wireless sensor networks (WSNs) have led to the rapid emergence of energy-harvesting technology as an alternative power source of traditional batteries. With respect to small mobile or portable electronics or advanced WSN nodes that do not have a high energy requirement, mechanical energy would be a perfect power source for them to operate in a self-sustained or self-powered mode. Considering the projected technology progress in the low-power consumption ICs and *wireless sensor network* (WSN) nodes, in the next decade, the MEMS and nanotechnology-based, energy-harvesting technology and its devices will become indispensible in mobile electronics, wearable electronics, and sensor nodes for diversified applications ranging from information technology, automotive, point of care, home health care, and implantable medical devices. We have contributed to MEMS and nanotechnology-based, energy-harvesting mechanisms in the past few years and witnessed the technology progress in terms of significantly increased output power and novel mechanisms. The strong demand of energy-harvesting technology has motivated the publication of this book to present the current state of knowledge in this field. The content of this book consists mainly of technical reviews, discussions, and basic knowledge in the design and fabrication of energy-harvesting materials and systems.

In order to address students, researchers, scientists, and application engineers with different backgrounds, this book starts with fundamental physics and theoretical design and the method of modeling for four mainstream energy-harvesting mechanisms (piezoelectric, electromagnetic, electrostatic, and triboelectric) in Chapters 1 through 4. A comprehensive technical review is undertaken in each chapter so that readers will quickly obtain a historical view and how the technology evolution has taken shape in the past few years. Chapter 5 is dedicated to the researchers who want to develop advanced energy-harvesting technology. The current major challenges in energy-harvesting technology are highlighted first. Then hybrid mechanisms and frequency-broadening approaches are provided as methods of realization of advanced energy harvesters. Chapter 6 reports the design requirement of energy-harvesting circuits and provides two case studies. Acoustic-based energy harvesters and wind-driven energy harvesters are described in Chapters 7 and 8. Chapter 9 introduces thermoelectric and pyroelectric energy harvesters. Chapter 10 is dedicated to the nanowire-based piezoelectric energy harvesters, and Chapter 11 reports unique and representative examples of energy harvester applications.

We hope that this book will serve as a valuable reference source for all those faced with the challenging problems created by the ever-increasing interest in MEMS and nanotechnology-based energy harvesters and their applications. We also hope that it will aid in stimulating further technology fusion between MEMS and nanotechnology. Nowadays, MEMS-based inertial sensors have been the enabler for numerous applications associated with smart phones, tablets, and mobile electronics. We believe that the advance of energy-harvesting technology will enable the next-generation WSN technology and wearable electronics. Thus, we hope that more people join in this exciting field and enjoy the new world created by energy-harvesting technology.

1

Piezoelectric MEMS Vibration Energy Harvesting

Over the last decade, energy-harvesting technology has gained growing attention due to the reduced power consumption of electronic components such as wireless sensor nodes. Vibration-based energy harvesting converts ambient available kinetic energy in the forms of structural or machines vibrations, human motions, and fluid flow into electricity. It could be a promising alternative for reducing the maintenance costs and chemical waste of battery and eventually achieving self-powered electronic devices and wireless sensor nodes. The *energy transduction mechanisms* (ETMs) such as piezoelectric, electromagnetic, and electrostatic are incorporated in the system to maximize the energy coupled from the ambient environment. This chapter provides an overview of vibration-based energy harvesting using piezoelectric mechanism. The working principle of piezoelectric energy conversion is introduced and equivalent circuit modeling is investigated for high-power efficiency. Typical MEMS-based piezoelectric energy harvesters will be reviewed.

1.1 Working Principle

1.1.1 Piezoelectricity

The piezoelectric effect was first discovered in 1880 by the brothers Pierre Curie and Jacques Curie. It is a unique property of certain crystalline materials, understood as the linear electromechanical interaction between the mechanical and the electrical state in crystalline materials with no inversion symmetry. As shown in Figure 1.1, when subjected to a mechanical force, the crystals became electrically polarized. The electric voltage of opposite polarity is generated by tension and compression and is in proportion to the applied force. Subsequently, the converse of this relationship was confirmed: if the crystal is exposed to an electric field, it experiences mechanical strain according to the polarity of the field, and is in proportion to the strength of the electric field. These phenomena were labeled as the direct piezoelectric effect and the inverse piezoelectric effect, respectively.

The direct and converse piezoelectric effects are governed by the piezoelectric constitutive (1.1) as [1]:

$$\begin{bmatrix} \delta \\ D \end{bmatrix} = \begin{bmatrix} s^E & d^t \\ d & \varepsilon^T \end{bmatrix} \begin{bmatrix} \sigma \\ E \end{bmatrix} \tag{1.1}$$

where δ and σ represent the strain and stress components; D and E refer to the electric displacement and electric field components; s, ε, and d are the elastic compliance, the dielectric constant, and the piezoelectric coefficient, respectively; the superscripts E and T denote that the respective constants are evaluated at constant electric field and constant stress, respectively; and the superscript t stands for the transpose. In applications, the direct piezoelectric effect is essential for sensing and energy harvesting where the applied stresses are used to generate surface charges on the piezoelectric materials.

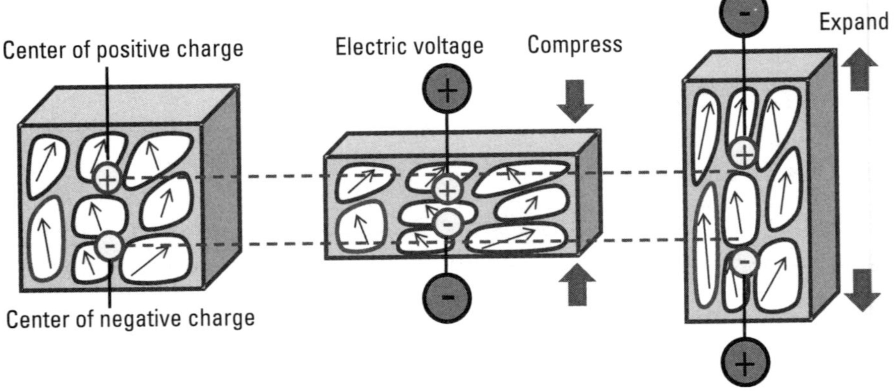

Figure 1.1 The direct piezoelectric effect causes crystal materials to generate an electric charge when compressed or pulled. The reverse effect causes the crystal materials to compress or expand when an electric voltage is applied.

1.1.2 Piezoelectric Materials

Piezoelectric materials possess the unique property of direct electromechanical coupling, either with the generation of an electric charge under an applied external stress or the induction of mechanical strain due to an applied electric field. According to the structure characteristics of the limited crystal materials (32-point groups) [2], only 20 noncentrosymmetric point groups exhibit piezoelectricity (Figure 1.2). Of these 20 groups, 10-point groups are classified as polar materials, which present electric dipole moments without an electric field being applied. This is known as spontaneous polarization that changes with temperature enabling pyroelectricity.

Among these polar materials, there is a special class of materials called ferroelectric, such as $BaTiO_3$ and lead zirconate titanate (PZT). They enable the permanent reorientation of spontaneous polarization between crystallographically defined stable states by applying an electric field (i.e., before dielectric breakdown occurs) [3]. A traditional ferroelectric is a mass of perovskite crystal structures, ABO_3, each consisting of a lattice of larger, divalent metal ions, A^{2+} (e.g., Pb or Ba), a small, tetravalent metal ion, B^{4+} (e.g., Ti or Zr), and O^{2-} ions. Above a critical temperature, Curie point (T_c), each element crystal has a simple cubic symmetry (isotropic) without dipole moment [Figure 1.3(a)]. As the temperature falls below T_c, each crystal becomes tetragonal or rhombohedral symmetry (anisotropic) with a dipole moment as shown in Figure 1.3(b).

The groups of dipoles with parallel orientation are Weiss domains. A net dipole moment to the domains gives a net polarization of the ceramic element. The polarization P can be permanently reoriented into different equilibrium states by applying an electric field E to the ferroelectric ceramic element. Figure 1.4 shows a typical hysteresis loop of the polarization versus electric field exhibited in a ferroelectric ceramic. Before the poling treatment at point O, the Weiss domains are random in a raw ferroelectric, while there is no overall polarization.

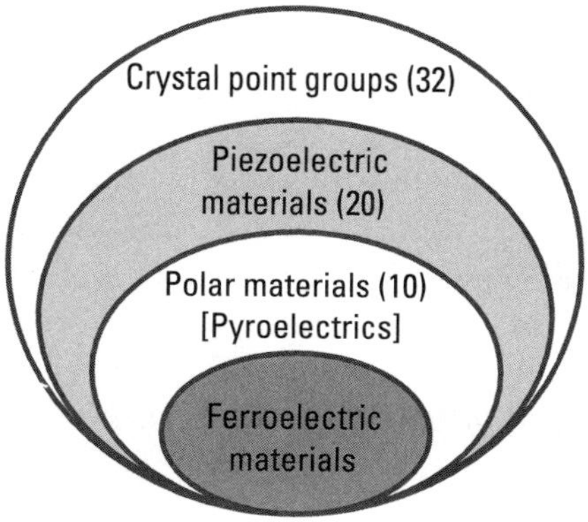

Figure 1.2 The relationship diagram of piezoelectric materials.

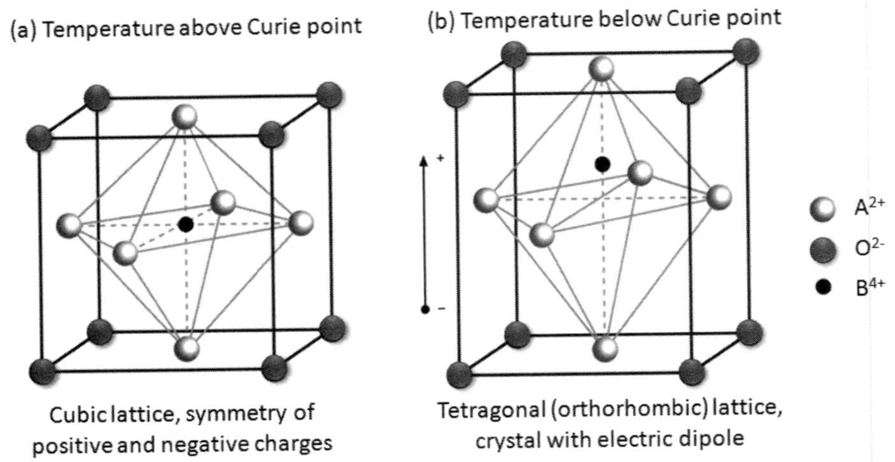

Figure 1.3 Crystal structure of a traditional ferroelectric ceramic.

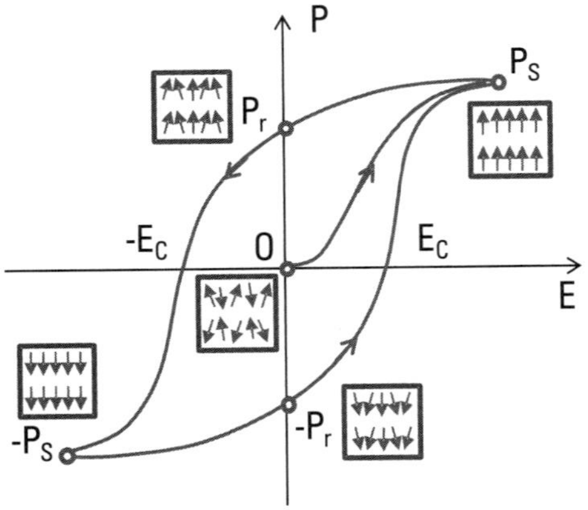

Figure 1.4 Effects of electric field (E) on polarization (P) and corresponding elongation/compression of a ceramic element.

The domains can be aligned by exposing the ferroelectric elements to a strong, direct electric field, normally at a temperature slightly below T_c. Through this poling (polarizing) treatment until the maximum polarization, P_s, the Weiss domains are mostly aligned with the electric field. When the electric field is removed, most of the dipoles are locked into an arrangement of near alignment, which determines the remanent polarization, P_r. As a result, there is a distortion that causes an elongation of the ferroelectric dimensions aligned with the field and a compression along the axes normal to the electric field. When a reverse electric field is applied, a negative maximum polarization, $-P_s$, and negative remanent polarization, $-P_s$, can be attained.

Most ferroelectric materials have a very high relative permittivity or dielectric constant (ε_r). The dielectric constant is defined as the ratio of the material permittivity (ε) to the permittivity of free space (ε_o). In addition, these ferroelectrics undergo structural phase transitions that lead to the large piezoelectric coefficient near the Curie temperature T_c. However, pure polar materials such as aluminum nitride (AlN) and zinc oxide (ZnO) do not allow the orientation change relative to the crystal lattice by polarization. The polar direction is crystallographically defined along the c-axis of the Wurtzite crystal structure [4]. The key advantages of the pure polar materials include the nearly constant ratio of the dielectric constant and piezoelectric coefficient, absence of depoling or aging of the piezoelectric coefficient, process compatibility with conventional complex metal-oxide semiconductor (CMOS) processing, and high acoustic velocities.

The efficiency of a piezoelectric energy harvester is fundamentally limited by the piezoelectric properties of the material. The material properties of the most common piezoelectric materials including quartz, AlN, ZnO, soft PZT piezoceramics (PZT-5H) and hard piezoceramics (PZT-5A), barium titanate ($BaTiO_3$), and polyvinylidene fluoride (PVDF) are given in Table 1.1 [5]. These properties typically vary with age, stress, and temperature. A few features to recognize are the piezoelectric coefficients d_{33}, d_{31}, and d_{51} of the ferroelectric material. These features have an order of magnitude larger than those of the nonferroelectric materials, such as AlN and ZnO. Furthermore, the relative permittivity of PZT under constant stress and strain, $\varepsilon^T{}_{33}$ and $\varepsilon^S{}_{33}$, are typically two orders of magnitude larger than AlN and ZnO. Some researchers have studied a range of piezoelectric crystals from *hard materials* with low losses but small piezoelectric coupling to *soft materials* with higher piezoelectric coupling but also higher losses [6]. It showed that the hard materials with smaller piezoelectric coefficient can produce larger power output than soft materials. However, this depends on the magnitude of the electrical power harvested compared to other sources of loss (i.e., the efficiency). For the systems with nonharvested losses dominated, the material with strong coupling would be most effective.

1.1.3 Piezoelectric 33-Mode and 31-Mode

To produce electrical energy, a piezoelectric material must be able to generate both electric charge and voltage by applying mechanical stress. It should be noted that most piezoelectric materials for energy harvesting exhibit a well-defined polar axis, and the direction of the applied stress relative to the polar axis would affect the energy harvesting performance. For a ferroelectric ceramic or polymer, the polar axis is dependent on the poling direction. However, for the nonferroelectric crystalline materials (e.g., AlN or ZnO), the polar axis is defined by the crystal orientation (along the c-axis of the Wurtzite crystal struc-

Table 1.1
Material Properties of the Most Common Piezpelectric Materials

Material	GaN	AIN	Cds	ZnO	z-quartz	$BaTiO_3$	PZT-1 hard PZT	PZT-5H soft PZT	PMN-PT	LiNb03	PVDF
Piezoelectric	✓	✓	✓	✓	✓	✓	✓	✓	✓	✓	✓
Pyroelectric	✓	✓	✓	✓	✖	✓	✓	✓	✓	✓	✓
Ferroelectric	✖	✖	✖	✖	✖	✓	✓	✓	✓	✓	✓
Const. strain rel. perm. (E^T_{33})	11.2 (ref 188)	10.0 (ref 189)	9.53 (ref 190)	8.81 (ref 191)	4.63 (ref 192)	910 (ref 193)	635 (ref 193)	1470 (ref 193)	680 (ref 194)	27.9 (ref 195)	(5–13) (ref 196)
Const. stress rel. perm. (E^T_{33})	—	11.9 (ref 197)	10.33 (ref 190)	11.0 (ref 191)	1.63 (ref 192)	1200 (ref 193)	1300 (ref 193)	3400 (ref 193)	8200 (ref 194)	28.7 (ref 195)	7.6 (ref 198)
d_{33}pCN^{-1}	3.7 (ref. 199), 13.2 (NW)[51]	5 (ref. 199)	10.3 (ref. 190)	12.4 (ref. 200)	d_{11} – 2.3 (ref. 192)	149 (ref. 193)	289 (ref. 193)	593 (ref. 193)	2820 (ref. 194)	6 (ref. 195)	–33 (ref. 198)
d_{21}pCN^{-1}	1.9 (ref. 199) –9.4 (NW)[51]	2 (ref. 199)	5.18 (ref. 190)	–5.0 (ref. 200)	—	58 (ref. 193)	123 (ref. 193)	274 (ref. 193)	–1330 (ref. 194)	–1.0 (ref. 195)	21 (ref. 198)
d_{18}pCN^{-1}	3.1 (ref. 201)	3.6 (ref. 201)	–13.98 (ref. 190)	–8.3 (ref. 200)	d_{14} – 0.67 (ref. 192)	242 (ref. 193)	495 (ref. 193)	741 (ref. 193)	146 (ref. 194)	69 (ref. 202)	–27 (ref. 198)
Mechanical quality (Q_m)	2800 (ref. 203) (NW)	2490 (ref. 204)	–1000 (ref. 205) (NW)	1770 (ref. 204)	10^5 –10^6 (ref. 202)	400 (ref. 193)	500 (ref. 193)	65 (ref. 193)	43–2050 (ref. 204, 206)	10^4 (ref. 207)	3–10 (ref. 208)
Electro-mechanical coupling (k_{21})	—	0.23 (ref. 204)	0.26 (ref. 190)	0.18 (ref. 200)	0.1 (ref. 209)	0.49 (ref. 210)	0.7 (ref. 211)	0.75 (ref. 211)	0.94 (ref. 194)	0.23 (ref. 202)	0.19 (ref. 209)
Pyro. coeff. p ($\mu Cm^{-2}K^{-1}$)	4.8 (ref. 212)	6–8 (ref. 213)	4 (ref. 162)	9.4 (ref. 161)	—	200 (ref. 161)	260 (ref. 161) –533 (ref. 157) (PZT)	260 (ref. 161) –533 (ref. 157) (PZT)	1790 (ref. 214)	83 (ref. 161)	33 (ref. 157)
s^E_{11}(pPa^{-1})	3.326 (ref. 215)	2.854 (ref. 215)	20.69 (ref. 216)	7.86 (ref. 216)	12.77 (ref. 192)	8.6 (ref. 157)	12.3 (ref. 193)	16.4 (ref. 157)	69.0 (ref. 193)	5.83 (ref. 194)	365 (ref. 198)
s^E_{13}(pPa^{-1})	2.915 (ref. 215)	2.824 (ref. 215)	16.97 (ref. 216)	6.94 (ref.216)	9.73 (192)	9.1 (ref. 193)	15.5 (ref. 193)	20.8 (ref. 193)	119.6 (ref. 194)	5.02 (ref. 195)	472 (ref. 198)

Source: [5]. (Copyright: RSC, 2013.)

ture). The polar axis is referred to the "3" direction. Due to the symmetry, other directions at right angles to the polar axis are equivalent and can be referred to the "1" directions. The direction of the applied stress can be either along the polar axis (3-direction) or at right angles to it (1-direction), resulting in two common piezoelectric energy harvesting configurations, 33-mode and 31-mode, as illustrated in Figure 1.5. A piezoelectric material used in 33-mode means the compressive stress/strain is applied parallel to the 3-direction, while the voltage generated along the same axis. In the 31-mode, the stress/strain is applied perpendicular to the polar axis and the direction of the generated voltage is at right angle of the applied force. The piezoelectric coefficient (d_{3i}) is used to quantify the piezoelectric material performance, which is the ratio of the open circuit charge density to the applied stress (in unit of C/N). As referred to in Table 1.1, typically a d_{33} coefficient is higher than the d_{31} coefficient. However, operation in 31-mode leads to the use of large strain in the 1-direction and thus is commonly implemented in vibration energy harvesting.

Piezoelectric energy harvesting usually takes the forms of bimorph or unimorph cantilevers [7, 8]. In Figure 1.6, the most common 31-mode bimorph cantilever contains two separate piezoelectric sheet bonded together, sometimes

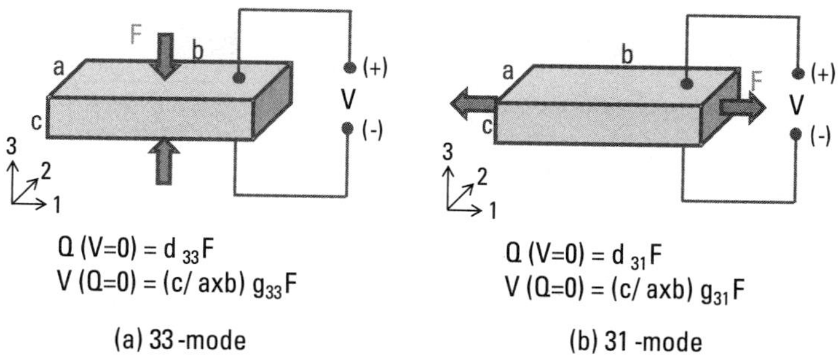

Figure 1.5 (a) The 33-mode and (b) 31-mode piezoelectric stress driven generator configurations. *Q* and *V* are the electric charge and voltage, respectively; *F* is the applied force; *P* is the polarization direction; d_{33} and d_{31} are the piezoelectric charge coefficients; g_{33} and g_{31} are the piezoelectric voltage coefficients.

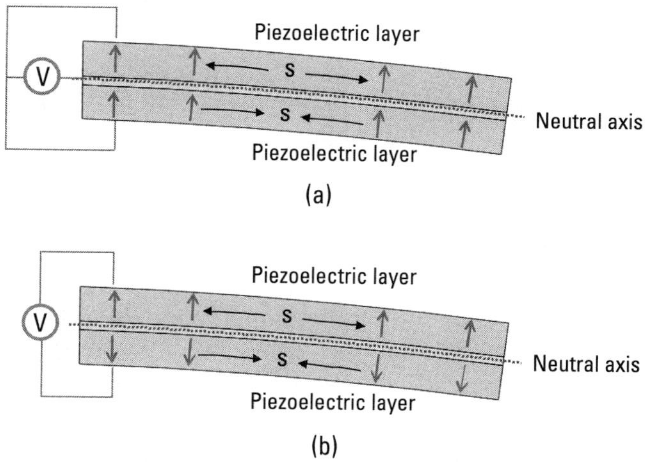

Figure 1.6 Piezoelectric bimorph cantilever constructions: (a) parallel poling and (b) series poling.

with a center shim in between. The structure is designed to operate in a bending mode thereby straining the top layer of the elements in tension and the bottom layer in compression or vice versa, and generating electric charge from the piezoelectric effect. If the top and bottom layers are poled in the same direction and electrodes are wired properly, the total electric current induced by each layer will be added. This is termed as parallel poling. Conversely, if the top and bottom layers are poled in the opposite directions, the induced electric voltage by each layer will be added. This is called series poling [9]. The bending cantilever can be made of multiple piezoelectric layers with proper electrodes and wiring in between each layer. In all cases, the power conversion potential is the same and theoretically the poling direction and number of layers only affect the voltage to current ratio.

The unimorph cantilever is another basic configuration for vibration-based piezoelectric energy harvesting, since the bending of cantilever can be easily excited. As demonstrated in Figure 1.7, in 31-mode, the piezoelectric layer is sandwiched by top and bottom electrodes, while in 33-mode, the electrodes are on top of the piezoelectric layer with interdigital finger pattern. In both modes, the piezoelectric thin film layer is coated on top of an elastic substrate (e.g., Si). The electric charge of the piezoelectric layer can be induced perpendicular or parallel to the direction of the applied strain in either 31-mode or 33-mode. The open circuit voltage of the piezoelectric layer V_{oc} is given as:

$$V_{oc} = \frac{d_{ij}}{\varepsilon_r \varepsilon_0} \sigma_{ij} g_e \tag{1.2}$$

From (1.2), the open circuit voltage V_{oc} is proportional to the applied stress σ_{ij}, piezoelectric coefficient d_{ij}, and the gap distance between electrodes g_e. ε_r and ε_0 are the relative dielectric constant and the permittivity of vacuum, respectively.

Clearly, the performance and merits of a piezoelectric unimorph cantilever are strongly dependent on the type of piezoelectric mode. Taking the most popular piezoelectric material PZT as an example, the piezoelectric coefficient d_{33} is approximately two times higher than that of d_{31}. Therefore, the generated voltage for 33-mode energy harvester is expected to be double than that of 31-mode device by assuming that both modes have the same configuration parameters. In addition, the voltage generation is proportional to the distance between the top and bottom electrodes for 31-mode device and to the distance between the electrode fingers for 33-mode device. Since the thickness of the PZT layer fabricated by microelectromechanical systems (MEMS) is normally

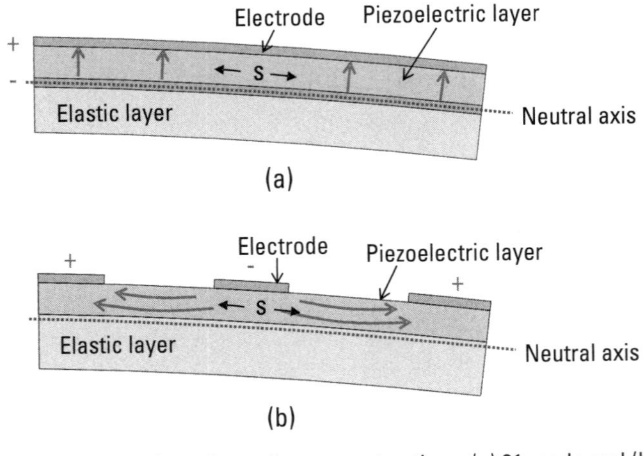

Figure 1.7 Piezoelectric unimorph cantilever constructions: (a) 31-mode and (b) 33-mode.

very thin, the electrode distance g_e in 31-mode is shorter than that in 33-mode. It follows that the 33-mode energy harvester has the advantage of producing a higher voltage output while the 31-mode device can be superior in larger current output. In the case of the output power obtained by the product of voltage and current, a better performance in 31-mode than in 33-mode was reported by Lee et al. [10]. Kim et al. [11] performed a similar comparison for MEMS PZT cantilevers based on 31-mode and 33-mode. They concluded that higher voltage and power can be obtained from the 33-mode mode device by optimizing the interdigital electrode (IDE) design.

Besides considering the output performance of 31-mode or 33-mode device, other factors such as the microfabrication and poling processes should be taken into account. It is hard to conduct the polarization of piezoelectric layer underneath IDEs in 33-mode due to the dead-layer effect [12, 13]. The electrode patterning process for 31-mode requires one more etching step or lithography mask than that for 33-mode, since the sandwiched electrodes are located on the two sides of the piezoelectric layer are compared to the IDE only at the top surface. The detailed microfabrication processes for piezoelectric energy harvesters will be discussed in the next section.

1.2 Mechanical and Electrical Modeling

1.2.1 Modeling of Vibration Energy Harvesting

A vibration-based energy harvester can be essentially considered as a single-degree-of-freedom (SDOF) lumped spring-mass-damper system as shown in Figure 1.8, which consists of an inertial mass (M) suspended by a spring with stiffness (K). The dynamic characteristic is unique to each system and can be simply described by two parameters: damping constant C and resonant frequency ω_n. The resonant frequency of the structure can be expressed as $\omega_n = \sqrt{K/M}$. The motion of the mass is damped by mechanical damping (C_m) and electric damping (C_e).

When the system is excited by an external displacement $y(t)$, the inertial mass moves out of phase with the system base, and the relative displacement is $z(t)$. The governing equation of motion of a lumped spring-mass-damper system can be written as:

$$M\ddot{z}(t)+C\dot{z}(t)+Kz(t)=M\ddot{y}(t) \tag{1.3}$$

Equation (1.3) can also be written in terms of damping constant and natural frequency. A damping factor, ζ, including mechanical damping ζ_m and electric damping ζ_e ($\zeta=\zeta_m+\zeta_e$) is a dimensionless number defined as

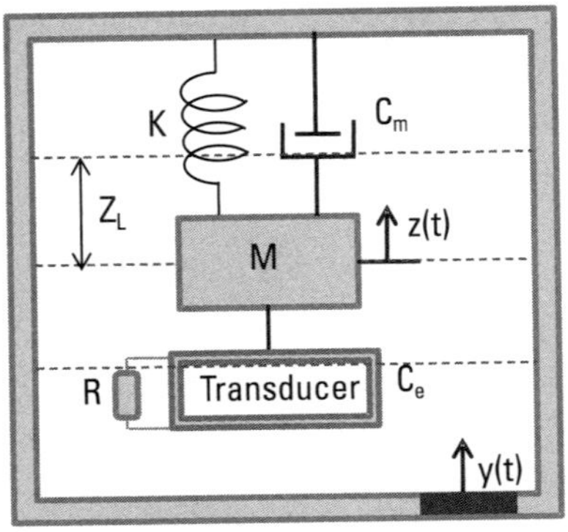

Figure 1.8 A spring-mass-damper model of the vibration energy-harvesting system.

$$\zeta = C / 2m\omega_n = C / 2\sqrt{MK} \tag{1.4}$$

In case of a cantilever beam, the stiffness K is given by $K = 3Y_c I / L^3$, where Y_c is the Young's modulus, I is the moment of inertia, and L is the length of the cantilever beam. The moment of inertia for a rectangular cross-section can be given as $I = (1/12)bh^3$, where b and h are the width and thickness of the cantilever beam, respectively. For the other cross-sectional area and stiffness, formulas are available in standard mechanical engineering handbook [14].

The ratio of the relative displacement $z(t)$ (i.e., output) and the external displacement $y(t)$ (i.e., input) can be obtained by applying Laplace transform with zero initial condition on (1.3) as:

$$\left|\frac{Z(s)}{Y(s)}\right| = \frac{s^2}{s^2 + 2\zeta\omega_n s + \omega_n^2} \tag{1.5}$$

By assuming the external excitation $y(t) = Y_0 \sin(\omega t)$, Y_0 and ω are the excitation amplitude and frequency, respectively, the time domain response $z(t)$ can be obtained by applying inverse Laplace transform on (1.5) given as:

$$z(t) = \frac{Y_0(\omega / \omega_n)^2}{\sqrt{\left(1-(\omega/\omega_n)^2\right)^2 + \left((2\zeta\omega/\omega_n)^2\right)}} \sin(\omega t - \phi) \tag{1.6}$$

where $z(t)$ is the net displacement $z(t) = Z_0 \sin(\omega t - \phi)$, the phase angle between the output and input can be expressed as:

$$\phi = \arctan\left(\frac{C\omega}{K - M\omega^2}\right) \tag{1.7}$$

Figure 1.9 shows a schematic diagram of the energy conversion for an energy-harvesting system. The input vibration energy U_{IN} is transformed into kinetic energy U_K and spring potential energy U_P of the spring-mass-damper system. Due to the electrical and mechanical damper, the energy dissipated U_C from the system is thus transformed into electric energy U_E and energy loss U_L. The energy dissipated per cycle U_C is simply the distance integral of the damping factor $C\dot{z}$ over a full cycle:

$$U_C = 2C\int_{-Z_0}^{Z_0} \dot{z}dz \tag{1.8}$$

Calculating the integral in (1.8) for the magnitude Z_0 given by (1.6) and multiplying by angular frequency give an expression for the dissipated power:

$$P = \frac{m\zeta Y_0^2 (\omega / \omega_n)^3 \omega^3}{\left(1-(\omega / \omega_n)^2\right)^2 + (2\zeta\omega / \omega_n)^2} \tag{1.9}$$

The maximum power can be obtained by setting the operating frequency as resonant frequency $(\omega = \omega_n)$ in (1.9) as:

$$P_{\max} = \frac{mY_0^2\omega_n^3}{4\zeta} \tag{1.10}$$

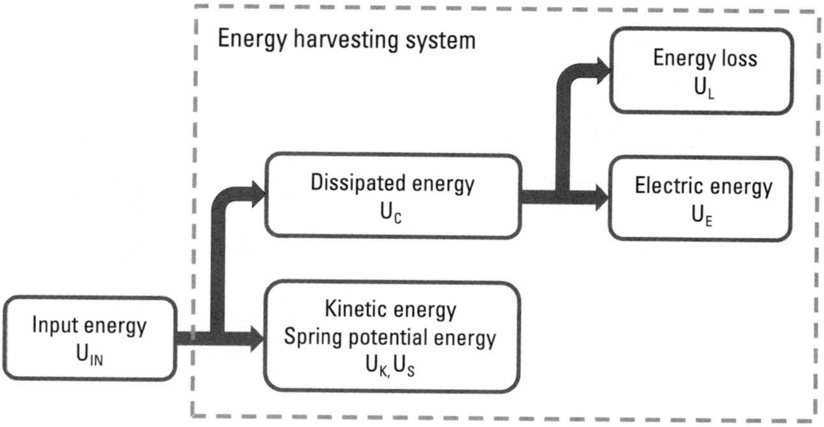

Figure 1.9 A schematic illustration of energy conversion for an energy harvesting system.

From (1.10), it can be seen that when the energy harvester operates at the matched frequency, the power generated can be maximized by lowering damping, increasing natural frequency, mass, and amplitude of excitation. In principle, a damping factor of zero would generate infinite power at resonance; however, in practice this is impossible. Decreasing the damping factor would increase the displacement of the mass. The available maximum displacement of the mass Z_L is limited by the size and geometry of the energy harvester as indicated in Figure 1.8. So the damping factor should be large enough to prevent the mass displacement exceeding the limitation. An optimal value of damping factor ς_{opt} exists when the mass displacement is reduced to just below the limit Z_L and an unclipped resonant cycle is achieved. Therefore, the constrained optimal power is achieved by operating as close to ς_{opt} as possible while observing the displacement constraint [15]. By the rearrangement of (1.6), the constrained optimal damping ratio ς_{opt} is:

$$\zeta_{opt} = \frac{1}{2\omega_c}\sqrt{\omega_c^4\left(\frac{Y_0}{Z_L}\right)^2 - \left(1-\omega_c^2\right)^2} \tag{1.11}$$

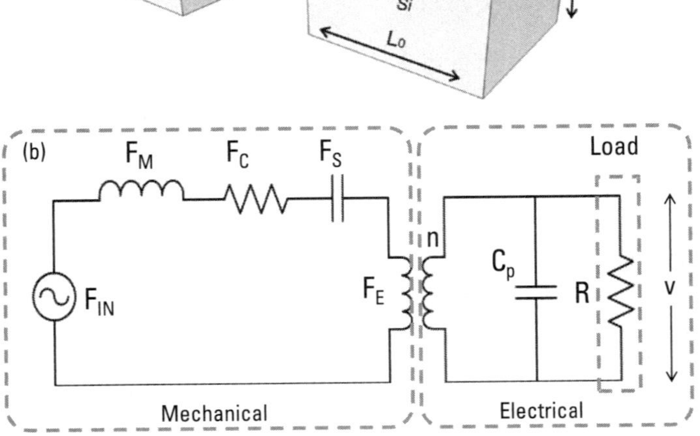

Figure 1.10 (a) A schematic drawing and (b) equivalent circuit model of a traditional piezoelectric MEMS energy harvester.

where ω_c is the ratio of excitation frequency and resonant frequency. The maximum dissipated power in the displacement constrained condition is obtained by substituting (1.11) into (1.10) as:

$$P_{C\max} = \frac{mY_0^2\omega^3}{2\omega_c^2}\left(\frac{Z_L}{Y_0}\right)^2\sqrt{\omega_c^4\left(\frac{Y_0}{Z_L}\right)^2 - \left(1-\omega_c^2\right)^2} \tag{1.12}$$

Thus the dissipated power at resonance can be given by setting $\omega_c = 1$ (i.e., $\omega = \omega_n$) in (1.12) as:

$$P_{Cres} = \frac{1}{2}mY_0\omega_n^3 Z_L \tag{1.13}$$

Equation (1.13) gives the maximum dissipated power at resonance condition for a limited mass displacement of Z_L. The dissipated power can be converted to electric power by using piezoelectric, electromagnetic, and electrostatic transduction mechanisms.

1.2.2 Equivalent Circuit

A cantilever structure with a tip mass is the most widely used configuration for piezoelectric MEMS energy harvesters. Figure 1.10(a) shows a schematic drawing of the piezoelectronic microcantilever with a proof mass at the end. The piezoelectronic thin film on the beam is sandwiched between the bottom and top electrode layer. Figure 1.10(b) describes its equivalent circuit. The motion of the mass is coupled to the piezoelectric transducer in the form of electric damping. Therefore, the equation of the piezoelectric harvester can be obtained by applying Kirchhoff's law (KCL):

$$F_M + F_C + F_K + F_E = F_{IN} \tag{1.14}$$

where F_{IN} is the force applied to the system, F_M is the kinetic energy of the inertial mass, F_C is the mechanical damping force, F_K is the elastic potential force of the spring, and F_E is the internal force induce by the voltage generation of the piezoelectric material. The electrical displacement and internal strain of the piezoelectric material of the microcantilever in 31-mode or 33-mode are given by the constitutive (1.1) as:

$$\delta = \frac{1}{Y_c}\sigma + d_{3i}E \tag{1.15}$$

$$D = d_{3i}\sigma + \varepsilon_{3i}E \tag{1.16}$$

where d_{3i} and ε_{3i} are the piezoelectric coefficient and dielectric constant in 31-mode or 33-mode, respectively. From (1.15), the induced stress due to the voltage generation of the piezoelectric film is given as:

$$\sigma_e = d_{3i}Y_cE \tag{1.17}$$

Therefore, the internal and external forces for the mechanical side of the equivalent circuit are redefined as:

$$M\ddot{z} + C_m\dot{z} + Kz + \frac{c_1Y_cd_{3i}}{g_e}v = -M\ddot{y} \tag{1.18}$$

where the v is the voltage across the load resistor and c_1 is the ratio of the internal force to the stress in the piezoelectric film (i.e., $c_1 = F_E/\sigma_e$). The electrical side can be expressed by using Kirchhoff's law:

$$\frac{c_2d_{3i}g_e}{\varepsilon_{3i}}\dot{z} - \frac{1}{RC_p}v = \dot{v} \tag{1.19}$$

where c_2 is the ratio of the stress of the piezoelectric film to the vertical displacement of the proof mass [16] (i.e., $c_2 = z/\delta$). R is the load resistor, and C_p is the total capacitance between the electrodes. For 33-mode harvester with n pairs of IDEs, the total capacitance is the sum of each individual capacitance C_i.

By taking the Laplace transform of (1.18) and (1.19), the output voltage v is given as

$$v = \frac{-j\omega\frac{c_2d_{3i}g_e}{\varepsilon_{3i}}\ddot{y}}{\left[\frac{1}{RC_p}\omega_n^2 - \left(\frac{1}{RC_p} + 2\zeta_m\omega_n\right)\omega^2\right] + j\omega\left[\omega_n^2\left(1+k_p^2\right) + \frac{2\zeta_m\omega_n}{RC_p} - \omega^2\right]} \tag{1.20}$$

where k_p is the electromechanical coupling coefficient represented by $k_p^2 = Y_cd_{3i}^2/\varepsilon_{3i}$. When the ambient vibration frequency matches with the resonant frequency of the piezoelectric cantilever ($\omega = \omega_n$), (1.20) can be simplified as:

$$v = \frac{-j\dfrac{c_2 d_{31} g_e}{\varepsilon_{31}}\ddot{y}}{2\zeta_m \omega_n^2 + j\left(\omega_n^2 k_p^2 + \dfrac{2\zeta_m \omega_n}{RC_p}\right)} \tag{1.21}$$

Therefore, the output power transferred to the load is calculated as:

$$P = \frac{1}{2}\frac{RC_p^2\left(\dfrac{c_2 d_{3i} g_e}{\varepsilon_{3i}}\right)^2}{\left(RC_p\omega_n\right)^2\left(4\zeta_m^2 + k_p^4\right) + RC_p 4\zeta_m \omega_n k_p^2 + 4\zeta_m^2}\frac{\ddot{y}^2}{\omega_n^2} \tag{1.22}$$

1.3 Fabrication of Piezoelectric MEMS Energy Harvesters

1.3.1 Deposition and Patterning of Piezoelectric Thin Films

To fabricate high-performance energy-harvesting devices, one of the key concerns is the deposition and patterning of the piezoelectric materials with high crystalline quality and controlled morphology. Various piezoelectric materials have been deployed widely in MEMS energy harvesters [17–20], such as AlN thin film, PZT ($PbZr_xTi_{(1-x)}O_3$) thin film or thick film, and ZnO and $BiTO_3$ nanowires. This section discusses the deposition and patterning techniques of the most common piezoelectric materials: AlN and PZT.

1.3.1.1 Deposition and Patterning of AlN Thin Films

To date, AlN deposition techniques have been reported including metal-organic chemical vapor deposition (MOCVD) [21–24], molecular beam epitaxy (MBE) [25, 26], ion beam assisted deposition [27, 28], and physical vapor deposition (PVD), also known as sputtering [29–32]. In general, AlN films are deposited via reactive magnetron sputtering with either RF or pulsed-dc sources. During the magnetron process, molecules of a reactive gas combine with the sputtered atoms from a metal target to form a compound thin film on a substrate. The growth of polycrystalline AlN films can be easily controlled at a low temperature (as low as 200°C or even at room temperature) with the similar properties as films grown by CVD and MBE. Table 1.2 lists several examples with successful deposition parameters of AlN thin films [33–36]. Achieving highly textured AlN films requires control of the nucleation and growth characteristics, substrate choice, surface roughness, and defect density. The quality

Table 1.2
Sputtering Deposition Parameters for AlN Thin Films

	Medjani [33]	Chu [34]	Dubois [35]	Abdallah [36]
System	RF mag	RF mag	DC Pulse	DC Pulse
Target	Al	Al	Al	Al
RF Power (W)	250	200	500	100
Target or substrate distance (mm)	65	17	80	40
Pressure (Pa)	0.5	0.27	<0.67	0.3
Ar (sccm or %)	15.5	3	0–10	90%
N_2 (sccm or %)	2.5	12	10–20	10%
Bias (V or W)	0–100V	0–320V	0–12W	0–250V
Substrate temperature (C°)	RT to 800	RT to 80	400	RT to 225

of the piezoelectric crystalline texture plays an important role in both the piezoelectric coefficient and the electromechanical coupling factor [37, 38].

One common design used in AlN-based energy harvesting is the 31-mode configuration in which the piezoelectric thin film is sandwiched between two conductive metal layers with an elastic substrate layer, as illustrated in Figure 1.10(a). Depending on the application, additional layers (i.e., elastic and metal electrode layers) are required beneath the AlN thin film. Normally, silicon (Si) is the well-known choice of elastic substrate for MEMS application, and AlN can be well deposited onto silicon. The choice of the electrode layer can directly influence the crystal texture of AlN, thereby affecting its piezoelectric properties. Besides the common electrode, materials include Ti, Pt, and Al electrodes [35, 39]; recently, there has been a push toward Mo electrodes [40–42]. Akiyama et al. [43] studied the influence of 15 metal electrodes (Ag, Al, Au/Ti, Co, Cr, Cu, Fe, Mo, Nb, Ni, Pt/Ti, Ti, W, Zn, Zr) on the crystal orientation of AlN thin film. The AlN films grown on Au/Ti and Pt/Ti electrodes show the highest crystallinity and orientation among them. This is because the crystallinity of the Au/Ti and Pt/Ti electrodes is high, the surface roughness of the two electrodes is low, and Au and Pt (111) planes match well with hexagonal AlN crystal structure. Lee et al. [44] deposited AlN films by the reactive RF sputtering method on various metal substrates including Al, Cu, Ti and Mo, to form an Al (top)/AlN/metal (bottom)/Si configuration. As seen in Figure 1.11, the AlN film deposited on the Mo electrode reveals a relatively dense and well-textured columnar structure with fairly uniform grains, while the films deposited on the other electrode metals exhibit a granular type of structure with mixed small and large grains.

The fabrication of AlN-based energy-harvesting devices requires the development of well-controlled AlN etch process at sufficient high etching rate.

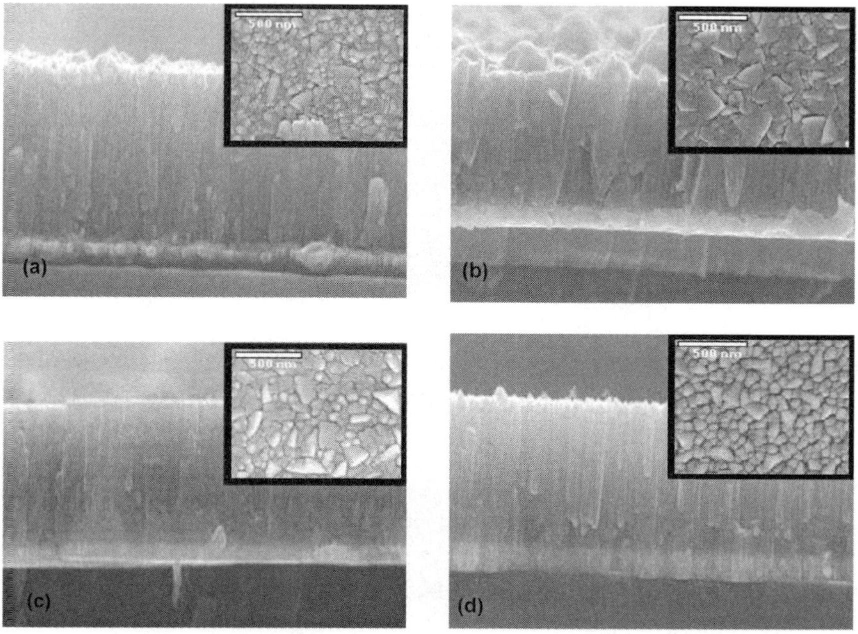

Figure 1.11 FE-SEM cross-sectional and surface morphologies for AlN films deposited on various bottom metals: (a) Al/Si, (b) Cu/Si, (c) Ti/Si, and (d) Mo/Si. (Copyright Elsevier. [44].)

Furthermore, the exact pattern transfer is also essential to obtain vertical sidewalls. For the AlN etch and patterning, both wet (chemical) etching and dry plasma (physical) etching processes are applicable. A list of common chemical etching techniques is given in Table 1.3. The wet etch process could induce a reasonably large lateral etching rate due to the isotropic etch tendency. However, it may prevent the physical damage and low selectivity of physical etch techniques in many conditions. It is noted that the crystalline quality dependent on the deposition parameters of AlN has a very strong relationship to the etching quality of the AlN thin film. In other words, the etching condition could be highly dependent on the magnetron sputtering parameters, such as RF power and substrate temperature (see Figure 1.12). In addition, the surface

Table 1.3
Summary of Wet Etching Techniques for AlN Thin Films

References	**[45, 46]**	**[46 ,47]**	**[48]**	**[49, 50]**
Chemicals	AZ400K (KOH)	KOH (10%)	TMAH	H_3PO_4 (85%)
Mask	Apiezon wax	Ti	Cr	60
Temperature (°C)	85	RT-85	RT	
Etch rate (nm/min.)	6-1000	6-1500	22	6-3000

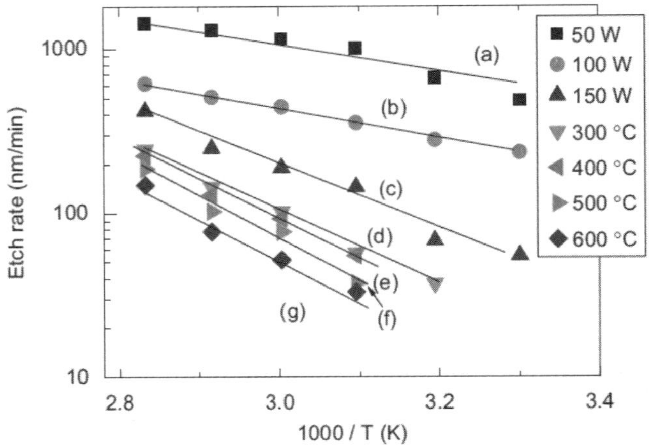

Figure 1.12 Wet etch rates of AlN films deposited at different RF power and substrate temperatures: (a) 50V and RT; (b) 100W and RT; (c) 150W and RT; (d) 300°C and 150W; (e) 400°C and 150W; (f) 500°C and 150W; and (g) 600°C and 150W. The etchant was the 10wt% KOH solution. (Copyright Elsevier. [47].)

morphology and lateral etching rate are also affected by the crystalline quality of the film [45].

In contrast to wet chemical etching, physical etching leads to nearly vertical sidewalls under the appropriate etch conditions. As summarized in Table 1.4, the dry etching of AlN was reported including ion milling [51], electron cyclotron resonance (ECR) [52, 53] and inductively coupled plasma (ICP) in $CH_4/H_2/Ar$ gases, and chlorine-based (Cl-based), and fluoride-based (F-based) plasma [54–56]. A high etching rate (>400 nm/min) of AlN can be achieved by using Cl-based plasma via ICP reactive ion etching (ICP-RIE) [57]. Unfortunately, the etching process in pure Cl-based plasma is usually isotropic; some methods have been reported by adding C, H, or O in the plasma to form etch-inhibiting layer on vertical sidewalls [54, 58]. The etching of AlN using F-based

Table 1.4
Summary of Physical Dry Etch Techniques for AlN Thin Films

References	**[51]**	**[52]**	**[53]**	**[54]**	**[55]**	**[56]**
System	Ion milling	ECR	ECR	ICP	ICP	RIE
Chemicals	Ar	$CH_4/H_2/Ar$	Cl_2/H_2	$BCl_3/Cl_2/O_2$	$Cl_2/H_2/Ar$	SF_6/Ar
Etching conditions	500 eV ion energy	–300 V, 200 W, 1 mTorr	–150 V, 1000 W, 1 mTorr	–83 V, 500 W, 2 mTorr	–280 V, 500 W, 2 mTorr	–1000 V, 20 mTorr
Etch rate (nm/min)	30	30	110	90	184	140
Film growth	MBE	—	—	Sputtering	MBE	RF sputtering

plasma was found to leave the AlN surface almost unaffected since F reacts with Al to form AlF_3, which has a low volatility and may block further etching [59].

1.3.1.2 Deposition and Patterning of PZT Thin Films

Deposition of lead zirconate titanate ($PbZr_xTi_{1-x}O_3$) or PZT thin films require strict control of the stoichiometry to prevent nucleation of nonferroelectric fluorite and pyrochlore structures [60, 61]. The early deposition of PZT thin films have been successfully carried out by a wide variety of PVD techniques, such as ion beam sputtering [62], RF planar magnetron sputtering [63, 64], or dc magnetron sputtering [65]. Current research efforts primarily utilize MOCVD [66–70], chemical solution deposition (CSD), including sol-gel routes [71–73] and metal-organic deposition (MOD) [74–76], and pulsed laser deposition (PLD) [77, 78] techniques. MOCVD has attracted great interest due to its high deposition rates, ease of conformal coating, and well-controlled film stoichiometry capabilities. One major drawback of vacuum-based deposition is the large capital investments; by contrast, CSD methods seem to meet the requirement of an easy implementation with standard Si wafer technology. The CSD process [79] can be generally defined as the chemical elaboration of ceramic materials by synthesis of precursor solution, gelation of the sol, drying, and organics pyrolysis followed by crystallization at high temperature. It offers advantages that include ease of control of chemical composition, good stoichiometry, high purity, and lower processing temperatures [80].

The most common researched CSD processes use 2-methoxyethanol (2-MOE) as a solvent and lead acetate trihydrate, titanium iso-propoxide, and zirconium n-propoxide as the precursors [71, 81, 82]. A general solution process of the 2-MOE-based PZT solution precursors is highlighted in Figure 1.13. This process has been used to produce PZT films with the highest piezoelectric coefficients to date. The PZT solution precursors were deposited on substrate by multiple spin coatings to achieve a uniform film thickness. It is followed by pyrolysis on a hot plate to remove the volatile organics. The spin and pyrolysis steps are repeated as required (i.e., one to four layers) prior to the crystallization anneal in oxygen. The process is continuously repeated until the desired film thickness is achieved. However, this CSD method requires an absolute planar surface, which brings the major inconvenience for the three-dimensional (3-D) structures usually prefabricated.

For 31-mode energy harvesting application, the choice of the bottom electrode will influence crystalline texture, quality, and properties of the piezoelectric PZT film, and thus is of primary importance [83–85]. PZT/Pt/Ti/SiO_2/Si is the most widely applied deposition sequence, where a homogeneous (111) texture of the Pt is commonly used to obtain a homogeneous nucleation of the same perovskite orientation [86]; the Ti is not only used as an adhesion layer, but also plays an important role in the diffusion phenomena. During the

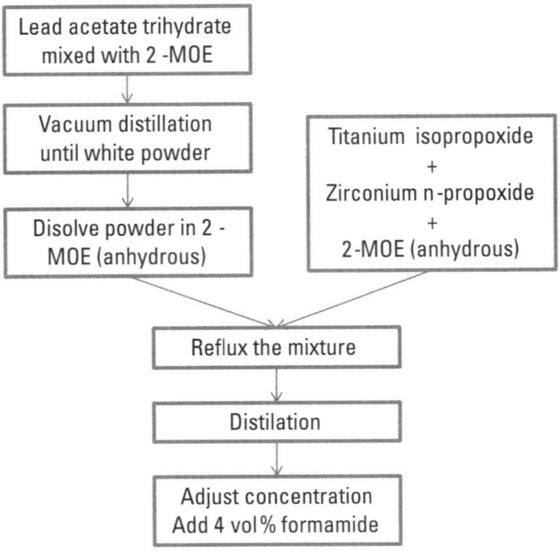

Figure 1.13 Preparation of the PZT solution precursors.

process, Ti reacts with oxygen and serves as nucleation center for PZT, since Pt does not inhibit the diffusion of Ti to PZT [87]. Furthermore, TiO_2/Ti adhesion layers have been utilized because the TiO_2 layer is used to partially prevent the interdiffusion of the Ti on top of the Pt electrode potentially modifying the nucleation conditions. Figure 1.14(a, b) shows the SEM images of a cross-section and a top view of a 4-μm PZT 53/47 film, which indicates columnar growth of the grains nucleated on the bottom electrode, and a dense and crack-free morphology. The XRD spectrum of the PZT 53/47 films with thickness ranging from 1 to 4 μm is shown in Figure 1.14(c).

Etching and patterning of PZT compounds remain a challenging aspect, accompanied by additional contamination problems of heavy metal elements (e.g., lead). From the literatures, PZT has been reported by various etching techniques including wet etching, the ECR or RIE process, and ICP as indicated in Table 1.5. In general it is difficult to achieve a high etching rate of PZT and an adequate selectivity to electrode materials and photoresist at the same time. Therefore, many efforts have been conducted to optimize the etching processes, which could meet the existing requirements for PZT and electrodes (micron scale, residue free, fencing free). Wet etching remains the easy solution, especially for a high selective PZT etching with a platinum electrode underneath. However, because of the lack of lateral etch control and the existence of photoresist undercut, the patterning dimension is limited to several microns to tens of microns. The typical etching chemistries contain the combinations of fluorine (F) and chlorine (Cl) by-products introduced by the mixtures of hydrochloric (HCl) and hydrofluoric (HF) acids [88–90]. The HCl acid is used

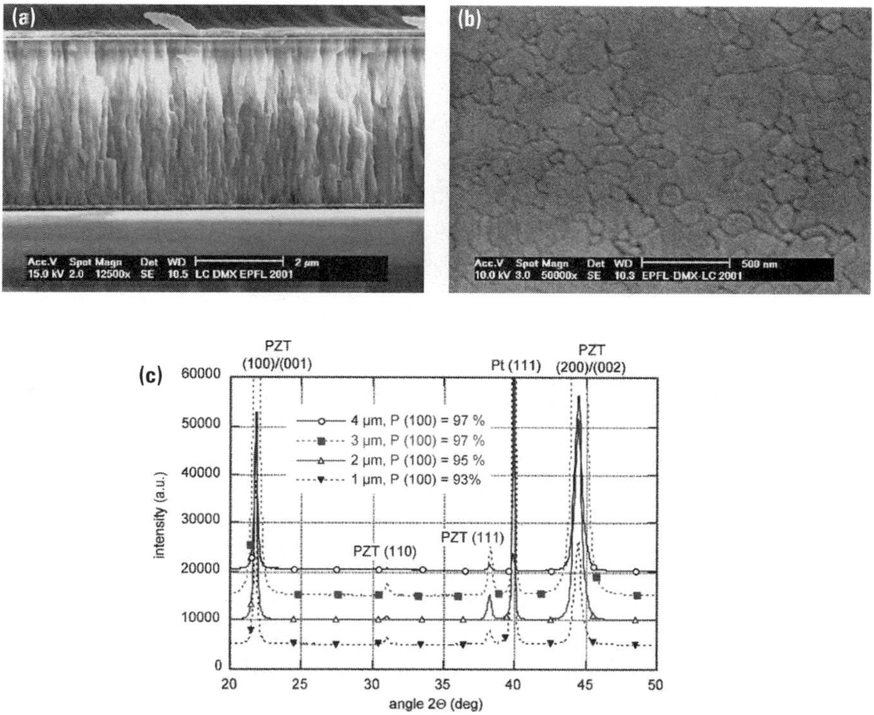

Figure 1.14 SEM (a) cross-section, (b) top view, and (c) θ–2θ X-ray diffractogram of the (100) textured PZT 53/47 film that is 4μm thick. (Copyright Elsevier [82].)

Table 1.5

Summary of Etch Techniques for PZT Thin Films

References	**[90]**	**[91]**	**[92]**	**[93]**	**[94]**
System	Wet etch	ECR/RF	ICP	ICP	Ae ion milling
Gas/chemical	BOE: 2HCl: $4NH_4Cl$: $4H_2O$ + $2HNO_3$: H_2O	CCl_4, CF_4, Ar	$Cl_2/C_2F_6/$ $ArCF_4$, Ar	BCl_3: Ar Cl_2: Ar	Ar
RF power (W)	—	0–150	400–800	500–800	150
Bias (V or W)	—	100–1000 V	100–400 V	200–350 V	36 V
Pressure (Pa or mT)	—	Se^{-4} to 1 Pa	1–10 mT	0.1–2.0 Pa	0.7 mT
Resist selectivity (PR/PZT)	1.5:1	3:1–13:1	1,5–3:1	3–6:1	1:1
Etch rate (nm/min)	960	20–100	100–180	63–193	25

to remove lead (Pb) and titanium (Ti), while the HF is used to remove Ti and zirconium (Zr). Miller et al. [89] made their efforts to develop an etch recipe that successfully minimized the undercut, providing a uniform etch and a controlled etch rate suitable for 1–10-μm thick films. Both RIE and ion-milling etch can provide the necessary dimensional control of microscale or nanoscale structures [91–94]. RIE is an etch process that combines physical ion bombardment of the PZT film with chemical etching in a plasma. Similar to the wet etching chemistries, RIE methods involve the combinations of Cl-based and F-based compounds (e.g., CCl_4 and CF_4). Baborowski et al. [91] used an electron cyclotron resonance/radio frequency (ECR/RF) reactor with 40% CCl_4, 40% CF_4, and 20% Ar gases to etch the PZT, which can achieve an etching rate of approaching 100 nm/min using 100-W RF operating at a pressure of 1 mPa.

1.3.2 Fabrication of Piezoelectric Microcantilevers

This section is intended to give the readers an understanding of the full process flow of microcantilever-based piezoelectric energy harvesters including AlN thin films in 31-mode, polycrystalline PZT thin films in 31-mode and 33-mode, epitaxial PZT thin film in 31-mode and bulk PZT thick films in 31-mode. Each of the main aspects covers the detailed fabrication descriptions of several highlighted energy-harvesting devices.

1.3.2.1 AlN Thin Film-Based Microcantilevers

AlN is a good choice for piezoelectric energy harvesters for its CMOS compatibility and relatively easy processing compared with the more commonly used piezoelectric material PZT. In addition, considerable power generation figure of merit (FOM) can be achieved. A power generation FOM of an energy harvester is defined as ${d_{31}}^2/\varepsilon_r$, where d_{31} is the piezoelectric constant and ε_r is the relative dielectric constant. The d_{31} and ε_r of AlN are –1.1 C/m^2 and 10.5, respectively [95]. For PZT these values range from –8 up to –12 C/m^2 for d_{31} and 300 up to 1,300 C/m^2 for ε_r [96]. It is seen although the piezoelectric coefficients of AlN are lower than PZT, the power generation FOM of AlN is comparable or higher due to the lower dielectric constant.

Piezoelectric energy harvesters using uniform AlN microcantilever have been successfully demonstrated by the research group from IMEC/Holst Centre [97]. An AlN-based microcantilever is comprised of an AlN capacitor located on top of a Si cantilever beam with an attached proof mass. The AlN capacitor consists of an AlN thin film sandwiched by a Pt bottom electrode and an Al top electrode. A standard micromachining process flow is schematically shown in Figure 1.15(a). It starts from the deposition of SiO_2 and Si_3N_4 isolation layers on a Si wafer (step 1). It is followed by the deposition and patterning of a Pt bottom electrode with a Ta adhesion layer (step 2). Then an

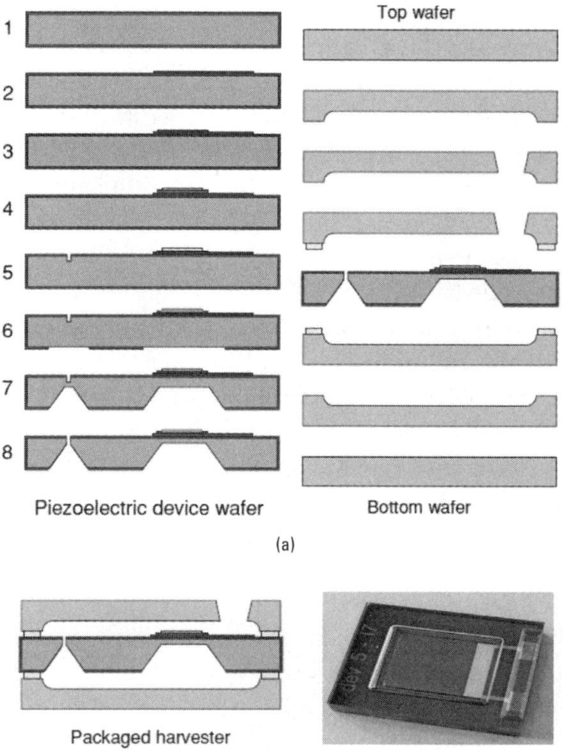

(b)

Figure 1.15 (a) Micromachining process flow and (b) the final packaged AlN-based energy harvester using bare Si wafer. (Copyright: IOP Publishing. [97].)

AlN film of 400 nm is deposited by reactive sputtering from an Al target and patterned (step 3). The AlN layer typically contains a thin amorphous layer at the AlN-Pt interface and a columnar structure on top. Step 4 is the deposition and patterning of an Al top electrode to form the AlN capacitor stack. On the front side of the wafer, a deep trench is etched to pattern the microcantilever in step 5. The depth of the trench should be deep enough to connect front to backside during the release etch. Then the SiO_2/Si_3N_4 is patterned as a hard mask for KOH etching on the backside of the wafer (step 6). A KOH etching is conducted to shape the proof mass of the microcantilever and is stopped before the complete release (step 7). The complete release is done either with dry etching or wet etching with a tetramethylammonium hydroxide (TMAH) solution due to the bad etch selectivity of KOH toward Al and AlN (step 8). Two glass wafers are etched by HF and patterned with cavity depth of 400 μm to form the top and bottom covers. The open holes on the top cover for wire bonding are fabricated by powder blasting. The harvester is finally packaged in between two glass substrates with 400-μm deep cavities to allow enough space for the mass displacement as shown in Figure 1.15(b). By controlling the dimension of the microcantilver beam and mass, a series of AlN-based MEMS energy harvesters have been fabricated with resonance frequencies ranging from 200 to 1,200 Hz. A maximum output power of 60 μW was achieved for the device type 5 with resonant frequency of 572 Hz at an acceleration of 2.0g.

Besides the use of a bare Si wafer, silicon-on-insulator (SOI) wafers are a good choice for microfabrication. An SOI wafer consists of two Si wafers fusion bonded together at approximately 1,100°C without interfacial voids with a thermal buried oxide (BOX) layer (SiO_2) of 1.0 ± 0.05 μm thick in between. The top wafer is ground and polished down to a desired thickness with a total-thickness variation (TTV) of less than 0.6 μm, called a device layer, while the bottom wafer is referred to as a handle layer. Andosca et al. [98] successfully fabricated AlN-based microcantilevers using SOI wafers at the Cornell NanoScale Science and Technology Facility in Ithaca, New York. It is a pure-play MEMS device completed by standard semiconductor fabrication equipment, which are CMOS compatible. The microfabrication process as shown in Figure 1.16 starts from a p-type double-side-polished (DSP) SOI wafer of a 100-mm diameter. The overall wafer thickness is 400 ± 5 μm with a device layer of 10 μm. Step 1 is the growth of a 1-μm wet thermal SiO_2 (TOX) on both sides of the SOI wafer. Then thin films including a 30-nm AlN adhesion layer, a 100-nm Mo bottom electrode layer, a 1-μm AlN piezoelectric layer, and a 100-nm Mo top electrode layer are deposited sequentially via a dual chamber of the Tegal sputtering system (step 2). The AlN film deposited on the Mo electrode reveals a relatively dense and well-textured columnar structure with fairly

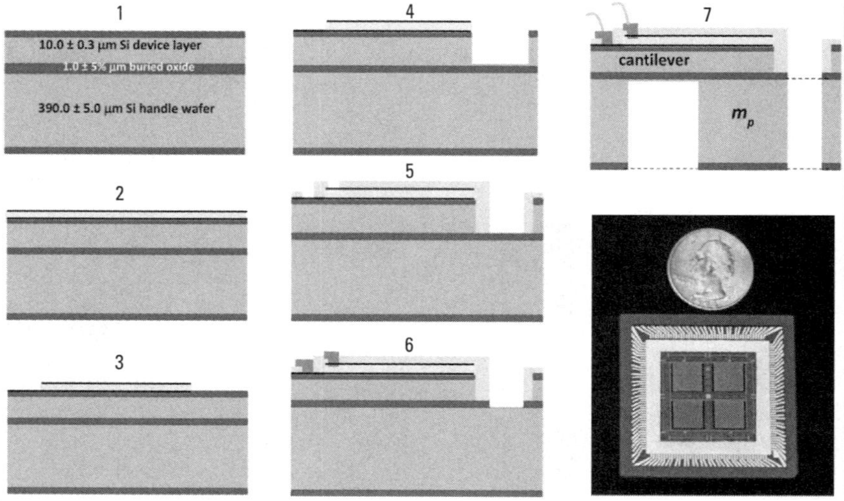

Figure 1.16 Micromachining process flow and the final packaged AlN-based energy harvester using SOI wafer. (Copyright: Elsevier. [98].)

uniform grains. This is followed by etch and patterning of top and bottom Mo electrodes via a RIE system and AlN piezoelectric and adhesion layer via hot phosphoric acid (step 3). Then a 500-μm-wide trench is formed by sequential RIE of a 1-μm TOX, deep-RIE (DRIE) of a 10-μm device layer and RIE of a 1-μm BOX (step 4). A passivation layer of a 1-μm SiOx is deposited via plasma-enhanced chemical vapor deposited (PECVD) to provide stress compensation to the tensile stressed AlN/Mo stack (step 5). Subsequently, square holes are formed via RIE to the top and bottom Mo electrodes for electrical connections and Al pads are evaporated with good step coverage down to the Mo electrodes (step 6). Finally, a DRIE process on the backside is conducted to pattern the Si proof mass and final release the microcantilever (step 7). The overall beam thickness is 13.9 μm and proof mass thickness is 390 μm. After die cleaning, the microcantilvers are packaged and wire-bonded as shown in the photo in Figure 1.16. One prototype resonating at 58 ± 2 Hz under 0.7g acceleration reaches a peak power of 63 μW.

1.3.2.2 PZT Thin Film-Based Microcantilevers

Polycrystalline PZT is the most popular piezoelectric material utilized for energy-harvesting applications due to its high piezoelectric coefficient. There are two configurations used for PZT microcantilevers, 31-mode, and 33-mode. A 31-mode cantilever has a PZT layer in between the top and bottom electrodes, while a 33-mode cantilever eliminates a bottom electrode by utilizing IDEs on the top of a PZT layer.

Different types of microcantilevers utilizing polycrystalline PZT films have been designed and fabricated using sol-gel methods [99–101]. In this case study, a unimorph 31-mode PZT microcantilver with optional proof nickel (Ni) mass reported by Fang et al. [99] was selected. As shown in Figure 1.17(a, b), the composite cantilever is made up of a piezoelectric PZT film sandwiched between a pair of Pt/Ti electrodes on SiO_2/Si substrate. A Ni proof mass at the free end of the cantilever is used to decrease its natural frequency for a low-frequency vibration application. The microfabrication techniques mainly involve functional film preparation and patterning, bulk Si micromachining, structure release, and mass assemblage as shown in Figure 1.14(c). First, a (100)-oriented Si wafer that was 500 μm thick was wet oxidized at 2 μm, which serves as a hard mask during the silicon wet etching process. A 30-nm-thick Ti adhesion layer and a 300-nm (111)-oriented thick Pt bottom electrode layer sputtered successively. It is followed by a sol-gel deposition of PZT films: a precursor solution is prepared from Pb acetate, Zr isopropoxide, and Ti tetrobutoxide, in 2-MOE solvent, where the Zr/Ti ratio is 52/48. A 0.25-M PZT precursor solution with 20 mol% excess Pb content was prepared and spin-coated on Pt/Ti/SiO_2/Si substrate at 3,000 rpm for 20 seconds, followed by pyrolysis at 300°C for 2 minutes. The spin-coating and pyrolysis processes were repeated for 15 times to

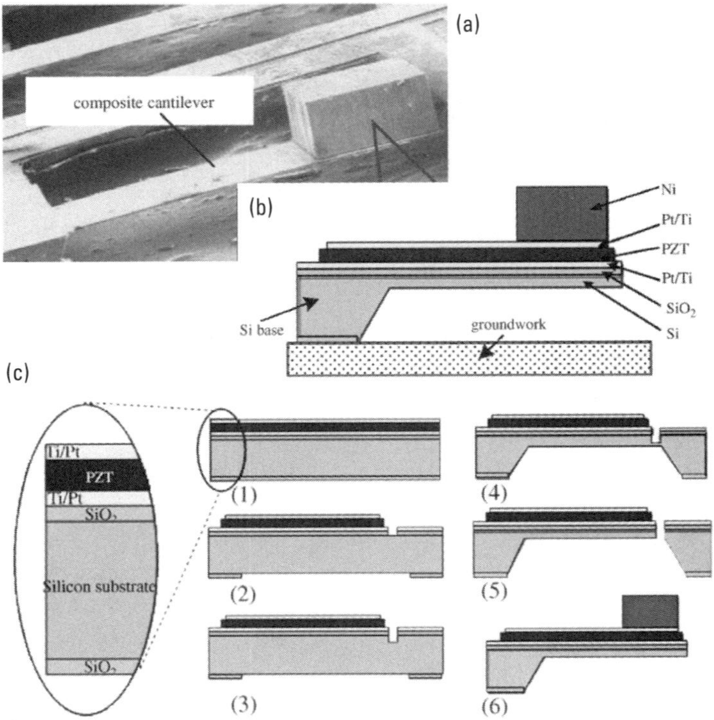

Figure 1.17 A unimorph 31-mode PZT microcantilver with optional proof nickel (Ni) mass using a bare Si wafer. (Copyright: Elsevier. [99].)

obtain a crackless PZT film that was 1.64 mm thick. Finally, the RTA process was performed at 600°C for 30 minutes to obtain the perovskite phase PZT film. Then a Ti/Pt top electrode layer sputtered on the top surface. Following this, the multilayer Si wafer was patterned using standard photolithography technique in step (2). The top electrodes, bottom electrodes, and oxide/Si layers were etched by RIE, while the PZT thin film was wet etched by ($HF:HCl:H_2O$ = 1:25:74), respectively. After the back oxide window opening by HF etching, backside micromachining of bulk Si was carried out via KOH chemistry etch. A thin Si layer was left before the KOH was etched through. Finally, an RIE process was utilized to release the cantilever structure. The Ni mass of 600 μm × 500 μm (width thickness) was then affixed at the end of the cantilever using glue. The PZT film was poled by applying 10-V dc voltage for 5 minutes. Thus, the PZT-based microcantilever was fabricated with a beam dimension of 2,000 μm in length, 600 µm in width, and 12 μm in thickness. The prototype produced the maximum power output of 2.16 μW at its resonant frequency of 608 Hz at 1.0g. Later on, this group demonstrated a PZT microcantilevers array for energy harvester using the aforementioned fabrication techniques [100].

More recently, the use of SOI wafer consisting of double polished device and handle layers to fabricate PZT microcantilevers has been widely illustrated [102–104]. Compared to the bare Si wafer, the beam thickness can be easily and accurately defined by the device layer, and a micromachined Si proof mass structure can be produced through the backside DRIE of the hand layer. Figure 1.15(a) shows a process flow of an S-shaped PZT microcantilever reported by Liu et al. [104]. The SOI wafer has a 5-μm-thick Si device layer, a 1-μm-thick BOX layer, and a 400-μm-thick Si handle layer. As shown in Figure 1.18, the process started from multilayer deposition of Pt/Ti/PZT/Pt/ Ti/SiO_2 on the SOI wafer. The thermal oxidation was conducted at 1,100°C, and the top and bottom electrodes of Pt (0.2 μm)/Ti (0.05 μm) were deposited by the dc magnetron sputtering. In between, a $Pb(Zr_{0.52},Ti_{0.48})O_3$ film that was 2.5 μm thick was deposited by sol-gel method. Step 2 shows the etch of top and bottom electrodes by Ar ions, and the etch of the PZT thin film by a mixture of HF, HNO_3, and HCl. Next, a SiO_2 thin film was deposited by RF-magnetron sputtering as an insulation layer. In step 4, the bonding pads were formed by etch and patterning of contact holes with Pt. To pattern and release the microcantilever structure, the SiO_2 layer and Si device layer at the front side surface were etched by RIE using feed gases of CHF_3 and SF_6, respectively, while the Si handle layer and BOX layer at backside were etched using DRIE. As shown in Figure 1.18(b), the successfully released S-shaped microcantilever was assembled onto a dual in-line package (DIP) with an extremely low resonant frequency of 27 Hz.

Epitaxial PZT thin films grown on Si or SOI wafers exhibit not only excellent piezoelectric coefficient but also low dielectric constant due to the high c-axis orientation [105]. Due to such enhancement of the energy-harvesting performance, there have been few reported microcantilever-based energy harvesters based on epitaxial PZT thin films [106–111]. Reilly et al. [106] first demonstrated an energy harvester based on epitaxial PZT thin films with a high quality by pulsed laser deposition (PLD). Morimoto et al. [107] successfully deposited epitaxial PZT film by RF magnetron sputtering on the substrate of (100) MgO single crystals with an epitaxial (001) Pt bottom electrode. The PZT film was later on transferred onto a stainless steel cantilever to enhance the output power efficiency and to improve the structural toughness. Isarakorn et al. [108] have magnetron sputtered $Pb(Zr_{0.2}Ti_{0.8})O_3$ layer on a Si wafer through proper intermediate layers, including a $SrTiO_3$ (STO) buffer layer of 6 nm and a $SrRuO_3$ (SRO) bottom electrode of 30 nm. As mentioned above, such PZT stoichiometry presents a good lattice match with the underneath SRO layer. The Au(100 nm)/Cr(10 nm) top electrodes (60 μm × 60 μm) are deposited by an e-beam evaporator. The epitaxial PZT thin films exhibit a relative dielectric constant ε_r of 70–100, and the piezoelectric coefficient gained is around 130–140 pm/V.

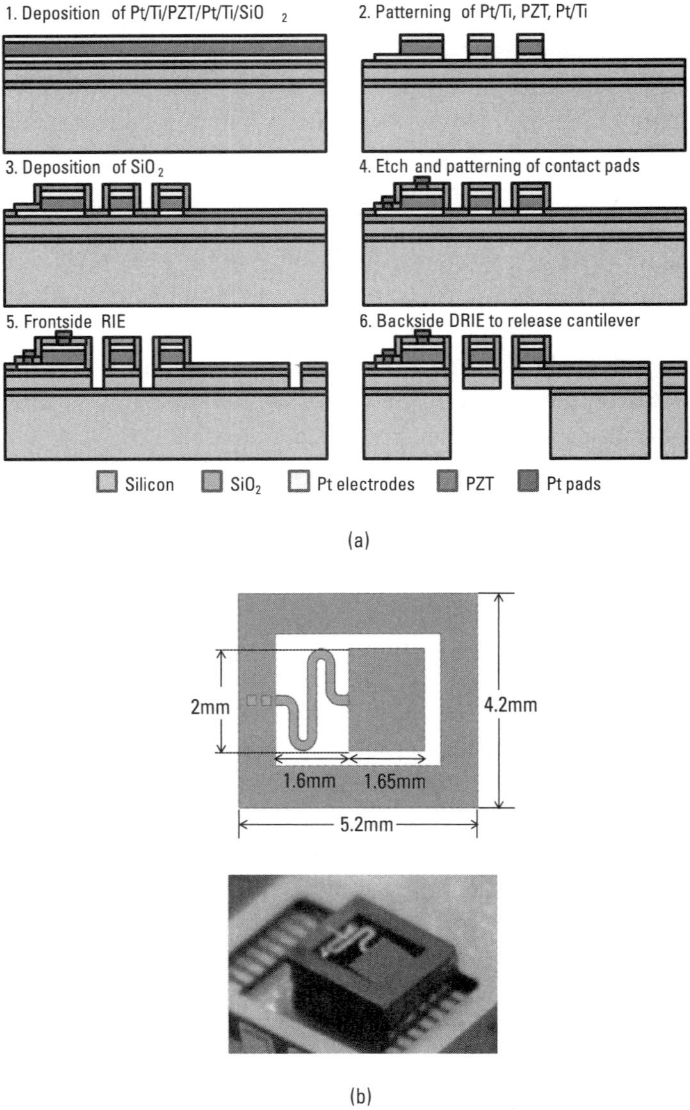

Figure 1.18 A unimorph 31-mode PZT microcantilver with Si mass using SOI wafer [104].

PZT material exhibits an approximately two times higher piezoelectric coefficient in the 33-mode than the 31-mode. The voltage constant is expected to be double for a 33-mode device compared to a 31-mode similar device for the same dielectric constant. Meanwhile, the voltage generation of a PZT energy harvester is proportional to the electrodes distance (i.e., distance between the top and bottom electrodes for a 31-mode device and distance between the electrode fingers for a 33-mode device). Since the PZT film thickness of a 31-mode MEMS device is relatively thin, resulting in a shorter electrode distance than that of a 33-mode device, it is expected that a higher voltage can be obtained for a 33-mode MEMS energy harvester. Jeon et al. [112] have demonstrated the MEMS-based PZT energy harvesters with IDEs working in the 33-mode. The microcantilever structure consists of a membrane layer (SiN_x) deposited on a (100) Si wafer, a diffusion barrier/buffer layer (ZrO_2) for preventing electrical charge diffusion from the piezoelectric layer, a piezoelectric thin film layer (PZT), a top IDEs (Pt/Ti) layer, and an optional proof mass layer (SU-8). The PZT solution with a composition of $Pb/(Zr_{0.52}Ti_{0.48})$ of 118/100 is deposited by sol-gel method. The PZT and membrane layers are patterned via RIE with BCl_3:Cl_2 (30:10) gas chemistry. The top IDEs layer is deposited via e-beam evaporation and patterned using a lift-off procedure. The thick photoresist, SU-8, is spin-coated and patterned on top of the existing layers to create a proof mass. After a final releasing step by KOH, the PZT microcantilever has a resonant frequency of 13.9 kHz. This 33-mode design generated 20 times higher voltage than that of the 31-mode design with the same beam dimension. The modeling, fabrication, and characterization of a similar 33-mode PZT microcantilever were reported by Park et al. [113]. A $PbTiO_3$ seed layer has been applied as an interlayer between the ZrO_2 and $Pb(Zr_{0.52}Ti_{0.48})O_3$ thin films to improve the piezoelectric property of the sol-gel spin-coated PZT thin film. The device generated an electrical power of 1.1 μW for a load of 2.2 MΩ with 4.4 V_{pp} at an acceleration of 0.39g and resonant frequency of 528 Hz.

Lee et al. [114] compared the output performance of the PZT microcantilevers in 31-mode and 33-mode by incorporating 3,000 × 1,500 μm^2 size cantilever structures that were 11 μm thick and comprised of 5-μm aerosol deposited PZT thin films. For the 33-mode PZT microcantilever, the IDEs were fabricated with 30 μm in width and 30 μm in gap. The microfabricated devices in two modes are shown in Figure 1.19(a, b). The size of the proof mass of the 31-mode device was 500 × 1,500 × 500 μm^3 and of the 33-mode device was 750 × 1,500 × 500 μm^3. Therefore, the resonance frequency of the 31-mode device (i.e., 255.9 Hz) was higher than that of the 33-mode device (i.e., 214 Hz). For the same beam dimension, it is obvious that the capacitance of the 31-mode device was larger than the 33-mode device, due to a relatively smaller electrodes distance of the 31-mode device. Therefore, the 31-mode device exhibited a smaller optimal load resistance than that of 33-mode. From the experimental results listed in Figure 1.19(c), it is seen that the output power of the 33-mode microcantilever was smaller than that of the 31-mode device for the same acceleration of 2.0g. This is because the PZT material of the 33-mode device was poled by the IDEs, which results in a nonuniform poling direction. The material under the electrodes was not poled properly. Moreover, the thicker the distance from the surface of the PZT material, the less effective the poling electric field strength will be. This caused a significant drop in the output power

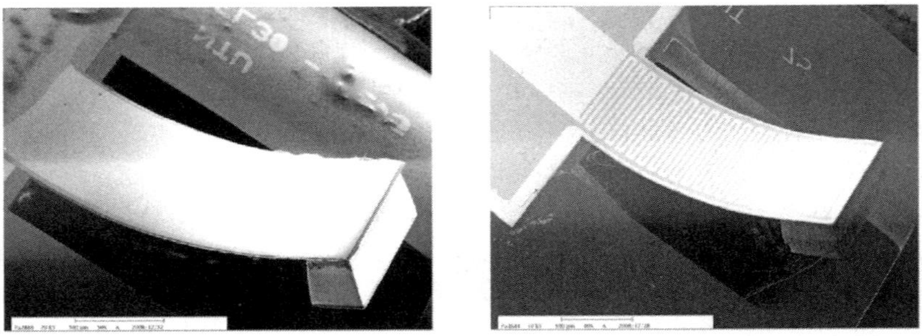

(a) 31-mode (b) 33-mode

Piezoelelectric mode	Resonant frequency	Optimal load	Power output	Voltage output (open circuit)	Voltage output (with load)
{3–1}	255.9 Hz	150 kΩ	2.099 μW	2.415 V	1.587 V
{3–3}	214.0 Hz	510 kΩ	1.288 μW	4.127 V	2.292 V

(c) Output performance comparison for 31-mode and 33-mode at 2g acceleration

Figure 1.19 The microfabricated PZT cantilevers in (a) 31-mode and (b) 33-mode, and (c) the output performance comparison for 31-mode and 33-mode at 2-g acceleration. (Copyright: IOP Publishing, [114].)

for the 33-mode piezoelectric MEMS energy harvester. Nevertheless, the 33-mode device possesses a higher output voltage than that of the 31-mode device and can be easily adjusted by the design gap of the IDEs for the same cantilever dimensions. Kim et al. [115] designed a series of 33-mode PZT microcantilevers with the width of the IDE ranging from 8 to 16 μm and finger spacing ranging from 4 to 16 μm. The results predict that the output voltage and power of the 33-mode device strongly depend on the dimensions of the IDE. For a predefined electrode thickness of 2 μm, an optimal power of 1.71 μW can be obtained when the electrode width is reduced to be the same as the thickness (i.e., 2 μm) with a finger spacing of 8 to 20 μm. In contrast, the 31-mode device with the same cantilever dimensions and PZT thickness generates a higher output power of 2.15 μW. However, it is also noted from the modeling results that a 33-mode device can favorable over a 31-mode device for application because higher voltage and power can be obtained with appropriate IDE design.

1.3.2.3 PZT Thick Film-Based Microcantilevers

The common deposition processes of PZT and AlN using sol-gel and sputtering methods limit the thin film thickness within 1~2 μm. Bulk PZT ceramics provide higher electromechanical coupling and power efficiency than piezoelectric thin films. The previous reported energy harvesters with bulk PZT [116, 117] possess relatively high-power generation but at the cost of large dimensions. Recently, researchers have utilized the screen-printing techniques [118, 119] and bulk-PZT microprocesses [120–125] to significantly increase the piezoelectric film thickness. As a result, the voltage and power output can be impressively improved. Lei et al. [119] combined the thick-film screen printing technique and micromachining process using the SOI wafer to fabricate a microcantilever with a unimorph PZT thick film. The fabricated harvester device produces 14.0 μW with an optimal resistive load of 100 kΩ from 1.0g input acceleration at its resonant frequency of 235 Hz. In recent years, a useful technique of bonding and thinning PZT thick films on Si has been investigated, which is the wafer bonding of bulk PZT and Si using an intermediate layer and then thinning the PZT. The basic fabrication processes for cantilever beam structures with thinned PZT films is outlined in Figure 1.20. In steps (1) and (2), the surfaces of bulk PZT and Si wafers are polished and bonded together. Following that, the bulk PZT wafer is thinned to a suitable thickness and the thinned PZT wafer is micromachined into the desired pattern by the dicing method in steps (3) and (4). Finally, the backside surface of the Si wafer is structured by wet or dry etching.

Tanaka et al. [120] successfully bonded the bulk PZT by using Au as the intermediate layer and thinned by mechanical lapping method. However, it is required special fabrication processing and bonding equipment, resulting in high cost. Meanwhile, since the bonding temperature was more than 500°C, the repolarization and residual stress issue would appear. The residual stress problem is caused by the thermal mismatch between the PZT ceramics and silicon substrate. Aktakka et al. [121] fabricated a unimorph piezoelectric cantilever with thinned-PZT layer that was 20 μm thick for vibration energy harvesting. The bulk-PZT pieces were bonded to an SOI wafer using Au/In bonding layer of 5-μm thickness and was later on thinned to about 20 μm. The

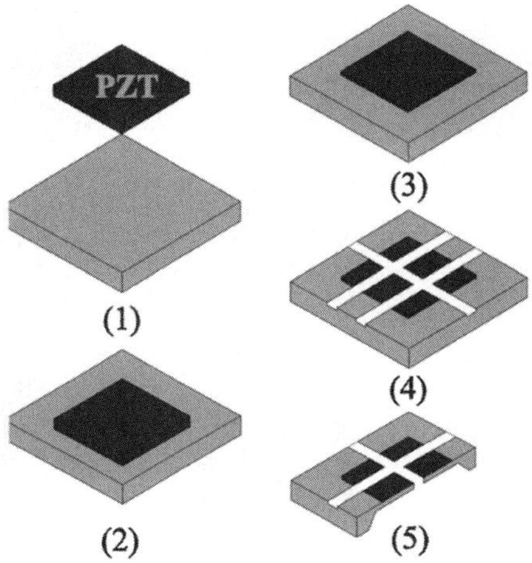

Figure 1.20 Basic fabrication processes of cantilever beam structures with thinned PZT films. (Copyright: IOP Publishing, [124].)

unpackaged harvester with a tungsten proof mass produces the maximum power of 2.74 μW at 0.1g (167 Hz), and 205 μW at 1.5g (154 Hz) at resonance. An intermediate layer of epoxy resin was used to reduce the fabrication cost of the intermediate layer of Au [122]. However, it is very difficult to expose the bottom electrodes for output signal by the micromachining process, because the epoxy resin is nonconductive. Wang et al. [123] developed the integration of bulk PZT and silicon by a patternable bond layer Cytop, while the thinning of the bulk PZT is done by chemical mechanical polishing (CMP).

Recently, a novel process to fabricate low-frequency piezoelectric MEMS energy harvesters by bonding bulk PZT on an SOI structure was proposed by Tang et al. [124]. The key techniques of the process include the low-temperature bonding technique using conductive epoxy resin, thinning of the bulk PZT using mechanical lapping and wet-etching combined method, and the micromachining of bulk ceramics by dicing. As shown in Figure 1.21(a), a unimorph PZT microcantilever array constructed with an Au/Cr/PZT/conductive epoxy/Cr/Au/SiO_2/Si/SiO_2 multilayer structure with Ni proof mass was designed and fabricated. For an excitation of 1.0g, the output voltage, power, and power density of this device are 2.72 Vpp, 11.56 μW, and 28 856.7 μW cm^{-3} at the resonant frequency of 514.1 Hz with a matched load resistance of 70 kΩ. The output performance of this device is higher than the other reported

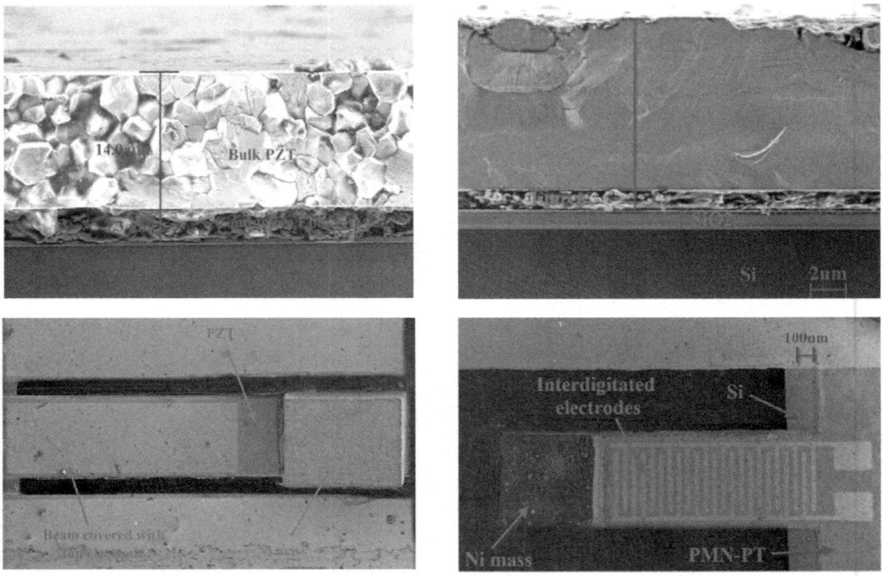

(a) 31 -mode thick PZT microcantilever (b) 33 -mode thick PMN-PT microcantilever

Figure 1.21 SEM cross-section view and top view of the fabricated (a) 31-mode thick PZT microcantilever (Copyright: IOP Publishing, [124]) and (b) 33-mode thick PMN-PT microcantilever. (Copyright: Elsevier, [125].)

MEMS devices using piezoelectric thin films. Therefore, this novel technique has great potential to fabricate high-performance piezoelectric MEMS energy harvesters. Using this fabrication technique, the same group [125] developed a high-performance piezoelectric 33-mode MEMS energy harvester based on PMN-PT single crystal thick film as shown in Figure 1.21(b). The PMN-PT piezoelectric thick film is deployed to improve the coupling coefficient and electromechanical coefficient. The 33-mode harvester based on IDEs contributes to improve its energy conversion efficiency. The experimental results show that the fabricated microcantilever produced the maximum output voltage and power of 5.36 Vpp and 7.182 μW, respectively, at the resonant frequency of 406 Hz. The calculated power density of 17,181.8 $\mu W/cm^3$ is also very promising compared to other reported devices.

1.3.3 Performance Summary of Piezoelectric Microcantilevers for Vibration Energy Harvesting

A variety of piezoelectric MEMS energy harvesters have been developed and their form factors are all different. Useful metrics in comparing the vibration energy-harvesting devices are their input acceleration, resonant frequency, effective volume, generated power and power densities in volume. In Table 1.6, piezoelectric MEMS energy harvesters with relatively higher power densities are selectively shown. As can be seen, most reported devices show orders of magnitude smaller normalized power density than required. To address this challenge, one direct solution is to improve the electromechanical coupling coefficient of the piezoelectric thin films. Another effective method is to utilize the aforementioned bulk PZT bonding and thinning technique to increase the piezoelectric film thickness and thus improve the output performance. Meanwhile, energy harvesters exhibiting high resonant frequencies or requiring high accelerations are less favorable because most of the ambient vibrations are of low frequency, and low *g* characteristics.

Additionally, bandwidth is an important characteristic for evaluating the performance of piezoelectric harvesters under unpredictable or uncontrollable spectra of ambient vibrations. Most piezoelectric energy harvesters have been designed based on a linear resonator using a cantilever configuration. The generated power of such device is proportional to quality factor of the resonator, which is inversely proportional to the bandwidth. Hence, there is a tradeoff between the output power and the bandwidth. Considering the fact that the frequency of ambient vibrations is unpredictable and uncontrollable, a linear energy harvester with a narrow bandwidth may be not feasible in most real applications. Several bandwidth broadening approaches have been adopted to circumvent this gain-bandwidth dilemma, which will be discussed in detail in Chapter 5.

Table 1.6

Output Performances of Recent Piezoelectric MEMS Energy Harvesters

References	Piezoelectric material	Input acceleration (g)	Frequency (Hz)	Effective Volume* (mm^3)	Power (μW)	Power Density ($\mu W/cm^3$)
Yen et al. [18]	Sputtered AlN, d_{31}	1.0	853	0.821b	0.17	207b
Elfrink et al. [97]	Sputtered AlN, d_{31}	2.0	572	10.225b	60.0	5,868b
Fang et al. [99]	Sol-gel PZT, d_{31}	1.0	608	0.196b	2.16	11,020b
Shen et al. [101]	Sol-gel PZT, d_{31}	2.0	462.5	0.652	2.15	3,272
Shen et al. [102]	Sol-gel PZT, d_{31}	0.75	183.8	0.769	0.32	416
Lee at al. [114]	Aerosol PZT, d_{31}	2.0	255.9	0.425b	2.10	4,961b
Lee at al. [114]	Aerosol PZT, d_{33}	2.0	214	0.612.b	1.29	2,108b
Lei at al. [119]	Screen printed PZT, d_{31}	1.0	235	12.888b	14.0	1,086b
Aktakka et al. [121]	Thinned bulk PZT, d_{31}	1.5	154	17.955b	205.0	11,417b
Tang et al. [124]	Thinned bulk PZT, d_{31}	1.0	514.1	0.401	11.56	18,857b
Tang et al. [125]	Thinned PMN-PT, d_{33}	1.5	406	0.418	7.18	17,182

a The effective volume is the sum of the volume of the cantilever beam and the volume of the proof mass.
b Estimated from data in reference.

References

[1] Erturk, A., and D. J. Inman, *Piezoelectric Energy Harvesting*, New York: John Wiley & Sons, 2011.

[2] J. F. Nye, *Physical Properties of Crystals*, New York: Oxford University Press, 1990.

[3] IEEE Standard on Piezoelectricity, ANSI/IEEE Std 176, 1987.

[4] Ghodssi, R., and P. Lin, *MEMS Materials and Process Handbook*, New York: Springer, 2011.

[5] Bowen, C. R., et al., "Piezoelectric and Ferroelectric Materials and Structures for Energy Harvesting Applications," *Energy Environ. Sci.*, Vol. 7, 2014, pp. 25–44.

[6] Erturk, A., and D. J. Inman, "Piezoelectric Energy Harvesting," *Energ. Convers. Manage.*, Vol. 50, No. 7, 2009, pp. 1847–1850.

[7] Roundy, S., and P. K. Wright, "A Piezoelectric Vibration Based Generator for Wireless Electronics," *Smart Mater. Struct.*, Vol. 13, No. 4, 2004, pp. 1131–1142.

[8] Anton, S. R., and H. A Sodano, "A Review of Power Harvesting Using Piezoelectric Materials (2003–2006)," *Smart Mater. Struct.*, Vol. 16, No. 3, 2007, pp. R1–R21.

[9] Wang, Q. -M., et al., "Theoretical Analysis of the Sensor Effect of Cantilever Piezoelectric Benders," *J. Appl. Phys.*, Vol. 85, No. 3, 1999, pp. 1702–1712.

[10] Lee, B. S., et al., "Piezoelectric MEMS Generators Fabricated with an Aerosol Deposition PZT Thin Film," *J. Micromech. Microeng.*, Vol. 19, No. 6, 2009, p. 065014.

[11] Kim, S. -B., et al., "Comparison of MEMS PZT Cantilevers Based on d_{31} and d_{33} Modes for Vibration Energy Harvesting," *J. Micorelectromech. S.*, Vol. 22, No. 1, 2013, pp. 26–33.

[12] Beckert, W., and W. S. Kreher, "Modelling Piezoelectric Modules with Interdigitated Electrode Structures," *Comput. Mater. Sci.*, Vol. 26, 2003, pp. 36–45.

[13] Bowen, C., et al., "Optimisation of Interdigitated Electrodes for Piezoelectric Actuators and Active Fiber Composites," *J. Electroceram.*, Vol. 16, No. 4, 2006, pp. 263–269.

[14] Blevins, R. D., *Formulas for Natural Frequencies and Mode Shape*, Malabar, FL: Robert E. Krieger Publishing, 1979.

[15] Mitcheson, P. D., et al., "Architectures for Vibration-Driven Micropower Generators," *J. Microelectromech. S.*, Vol. 13, No. 3, 2004, pp. 429–440.

[16] Roundy, S. J., "Energy Scavenging for Wireless Sensor Nodes with a Focus on Vibration to Electricity Conversion," Ph.D. Thesis, University of California, Berkeley, 2003.

[17] Kim, S. G., S. P. Priya, and I. Kanno, "Piezoelectric MEMS for Energy Harvesting," *MRS Bull.*, Vol. 37, 2012, pp. 1039–1050.

[18] Yen, T. -T., et al., "Corrugated Aluminum Nitride Energy Harvesters for High Energy Conversion Effectiveness," *J. Micromech. Microeng.*, Vol. 21, No. 8, 2011, p. 085037.

[19] Wang, X., et al., "Direct-Current Nanogenerator Driven by Ultrasonic Waves," *Science*, Vol. 316, No. 5821, 2007, pp. 102–105.

[20] Koka, A., Z. Zhou, and H. A.Sodano, "Vertically Aligned $BaTiO_3$ Nanowire Arrays for Energy Harvesting," *Energ. Environ. Sci.*, Vol. 7, 2014, pp. 288–296.

[21] Radhakrishan, G., "Properties of AlN Films Grown at 350 K by Gas-Phase Excimer Laser Photolysis," *J. Appl. Phys.*, Vol. 78, No. 6000, 1995.

[22] Uchida, H., et al., "Structural and Chemical Characteristics of (Ti,Al) N Films Prepared by Ion Mixing and Vapor Deposition," *Vaccum*, Vol. 80, 2006, pp. 1356–1361.

[23] Takahashi, N., Y. Matsumoto, and T. Nakamura, "Investigations of Structure and Morphology of the AlN Nano-Pillar Crystal Films Prepared by Halide Chemical Vapor Deposition Under Atmospheric Pressure," *J. Phys. Chem. Solids*, Vol. 67, No. 4, 2006, pp. 665–668.

[24] Sato, A., et al., "Characterization of AlN:Mn Thin Film Phosphors Prepared by Metalorganic Chemical Vapor Deposition," *J. Crys. Growth*, Vol. 298, 2007, pp. 379–382.

[25] Brown, P. D., et al., "Structural Characterisation of Al Grown on Group III-Nitride Layers and Sapphire by Molecular Beam Epitaxy," *J Crys Growth*, Vol. 234, 2002, pp. 384–390.

[26] Iwata, S., et al., "Growth and Optical Properties of AlN Homoepitaxial Layers Grown by Ammonia-Source Molecular Beam Epitaxy," *J Crys Growth*, Vol. 301-302, 2007, pp. 461–464.

[27] Lal, K., et al., "Frequency Dependent Conductivity of Aluminum Nitride Films Prepared by Ion Beam-Assisted Deposition," *Thin Solid Films*, Vol. 434, 2003, pp. 264–270.

[28] Matsumoto, T., and M. Kiuchi, "Zinc-Blende Aluminum Nitride Formation Using Low-Energy Ion Beam Assisted Deposition," *Nucl. Instrum. B.*, Vol. 242, 2006, pp. 424–426.

[29] You, Y. Z., and D. Kim, "Influence of Incidence Angle and Distance on the Structure of Aluminium Nitride Films Prepared by Reactive Magnetron Sputtering," *Thin Solid Films*, Vol. 515, 2007, pp. 2860–2863.

[30] Creighton, J. R., G. T. Wang, and M. E. Coltrin, "Fundamental Chemistry and Modeling of Group-III Nitride MOVPE," *J. Crystal Growth*, Vol. 298, 2007, pp. 2–7.

[31] Guo, Q. X., et al., "Growth Properties of AlN Films on Sapphire Substrates by Reactive Sputtering," *Vacuum*, Vol. 80, 2006, pp. 716–718.

[32] Hwang, B. H., et al., "Growth Mechanism of Reactively Sputtered Aluminum Nitride Thin Films," *Mat. Sci. and Eng. A*, Vol. 325, 2002, pp. 380–388.

[33] Medjani, F., et al., "Effect of Substrate Temperature and Bias Voltage on the Crystallite Orientation in RF Magnetron Sputtered AlN Thin Films," *Thin Solid Films*, Vol. 515, 2006, pp. 260–265.

[34] Chu, A. K., et al., "Influences of Bias Voltage on the Crystallographic Orientation of AlN Thin Films Prepared by Long-Distance Magnetron Sputtering," *Thin Solid Films*, Vol. 429, 2003, pp. 1–4.

[35] Dubois, M. A., and P. Muralt, "Stress and Piezoelectric Properties of Aluminum Nitride Thin Films Deposited onto Metal Electrodes by Pulsed Direct Current Reactive Sputtering," *J. Appl. Phys.*, Vol. 89, 2001, pp. 6389–6395.

[36] Abdallaha, B., et al., "Deposition of AlN Films by Reactive Sputtering: Effect of Radio Frequency Substrate Bias," *Thin Solid Films*, Vol. 515, 2007, pp. 7105–7108.

[37] Iborra, E., et al., "Piezoelectric Properties and Residual Stress of Sputtered AlN Thin Films for MEMS Applications," *Sensor. Actuat. A: Phys.*, Vol. 115, 2004, pp. 501–507.

[38] Ruffner, J. A., et al., "Effect of Substrate Composition on the Piezoelectric Response of Reactively Sputtered AlN Thin Films," *Thin Solid Films*, Vol. 354, 1999, pp. 256–261.

[39] Kamohara, T., et al., "Influence of Aluminum Nitride Interlayers on Crystal Orientation and Piezoelectric Property of Aluminum Nitride Thin Films Prepared on Titanium Electrodes," *Thin Solid Films,* Vol. 515, 2007, pp. 4565–4569.

[40] Kamohara, T., et al., "Growth of Highly C-Axis-Oriented Aluminum Nitride Thin Films on Molybdenum Electrodes Using Aluminum Nitride Interlayers," *J. Crystal Growth,* Vol. 275, 2005, pp. 383–388.

[41] Kamohara, T., et al., "Improvement in Crystal Orientation of AlN Thin Films Prepared on Mo Electrodes Using AlN Interlayers," *Ceram. Int.*, Vol. 34, No. 4, 2008, pp. 985–989.

[42] Kamohara, T., M. Akiyama, and N. Kuwano, "Influence of Molybdenum Bottom Electrodes on Crystal Growth of Aluminum Nitride Thin Films," *J. Crystal. Growth*, Vol. 310, 2008, pp. 345–350.

[43] Akiyama, M., et al., "Influence of Metal Electrodes on Crystal Orientation of Aluminum Nitride Thin Films," *Vacuum*, Vol. 74, 2004, pp. 699–703.

[44] Lee, J. -B., et al., "Effects of Bottom Electrodes on the Orientation of AlN Films and the Frequency Responses of Resonators in AlN-Based FBARs," *Thin Solid Films*, Vol. 447-448, 2004, pp. 610–614.

[45] Vartuli, C. B., et al., "Wet Chemical Etching of AlN and InAlN in KOH Solutions," *J. Electrochem. Soc.*, Vol. 143, 1996, pp. 3681–3684.

[46] Zhuang, D., and J. H. Edgar, "Wet Etching of GaN, AlN, and SiC: A Review," *Mater. Sci. Eng.*, Vol. 48, 2005, pp. 1–46.

[47] Chen, D., et al., "The Influence of the AlN Film Texture on the Wet Chemical Etching," *J. Microelectron*, Vol. 40, 2009, pp. 15–19.

[48] Saravanan, S., et al., "A Novel Surface Micromachining Process to Fabricate AlN Unimorph Suspensions and Its Application for RF Resonators," *Sensor. Actuat. A: Phys.*, Vol. 130–131, 2006, pp. 340–345.

[49] Sheng, T. Y., Z. Q. Yu and G. J. Collins, "Disk Hydrogen Plasma Assisted Chemical Vapor Deposition of Aluminum Nitride," *Appl. Phys. Lett.*, Vol. 52, 1988, pp. 576–578.

[50] Tanner, S. M., and V. V. Felmetsger, "Microstructure and Chemical Wet Etching Characteristics of AlN Films Deposited by AC Reactive Magnetron Sputtering," *J. Vac. Sci. Technol. A*, Vol. 28, 2010, pp. 69–76.

[51] Pearton, S. J., et al., "Ar^+-Ion Milling Characteristics of III-V Nitrides," *J. Appl. Phys*, Vol. 76, 1994, pp. 1210–1215.

[52] Pearton, S. J., et al., "Dry Etching of Thin-Film InN, AlN and GaN," *Semicond. Sci. Technol.*, Vol. 8, 1993, pp. 310–312.

[53] Pearton, S. J., C. R. Abernathy, and F. Ren, "Low Bias Electron Cyclotron Resonance Plasma Etching of GaN, AlN, and InN," *Appl. Phys. Lett.*, Vol. 64, 1994, pp. 2294–2296.

[54] Engelmark, F., G. F. Iriarte, I. V. Katardjiev, "Selective Etching of Al/AlN Structures for Metallization of Surface Acoustic Wave Devices," *J. Vac. Sci. Technol. B*, Vol. 20, 2002, pp. 843–848.

[55] Shul, R. J., et al., "High-Density Plasma Etch Selectivity for the III–V Nitrides," *Solid-State Electron.*, Vol. 42, No. 12, 1998, pp. 2269–2276.

[56] Chen, D., et al., "Dry Etching of AlN Films Using the Plasma Generated by Fluoride," *Vacuum*, Vol. 83, 2009, pp. 282–285.

[57] Khan, F. A., et al., "High Rate Etching of AlN Using BCl_3 / Cl_2 / Ar Inductively Coupled Plasma," *Mater. Sci. Eng. B*, Vol. 95, 2002, pp. 51–54.

[58] Saito, S., K. Sugita, and J. Tonotani, "Effect of CHF_3 Addition on Reactive Ion Etching of Aluminum Using Inductively Coupled Plasma," *Japan J. Appl. Phys.*, Vol. 44, No. 5A, 2005, p. 2971.

[59] Wang, J. S., Y. Y. Chend, and K. M. Lakin, "Reactive Ion Etching of Pizeoelectric Films for Acoustic Wave Device," *Ultrasonics Symposium Proceedings*, 1982.

[60] Klissurska, R. D., et al., "Effect of Nb Doping on the Microstructure of Sol-Gel Derived PZT Thin Films," *J. Am. Ceram. Soc.*, Vol. 78, 1995, pp. 1513–1520.

[61] Muralt, P., J. Baborowski, and N. Lederman, "Piezoelectric Micro-Electro-Mechanical Systems with $PbZr_xTi_{1-x}O_3$ Thin Films: Integration and Application Issues," *Piezoelectric Materials and Devices*, EPFL Swiss Federal Institute of Technology, Switzerland, 2002, pp. 231–260.

[62] Castellano, R. N., and L. G. Feinstein, "Ion-Beam Deposition of Thin Films of Ferroelectric PZT," *J. Appl. Phys.*, Vol. 50, 1979, pp. 4406–4411.

[63] S. B. Krupanidhi, et al., "RF planar magnetron sputtering and characterization of ferroelectric PZT films, J. Appl. Phys., Vol. 54, pp. 6601–6609, 1983

[64] Okuyama, M., et al., "Preparation of $PbTiO_3$ Ferroelectric Thin Film by RF Sputtering," *Japan. J. Appl. Phys.* Vol. 18, 1979, pp. 1633–1634.

[65] Sreenivas, J., M. Sayer, and P. Garrett, "Properties of DC Magnetron Sputtered PZT Thin Films," *Thin Solid Films,* Vol. 172, 1989, pp. 251–267.

[66] Kojima, M., et al., "Chemical Vapor Deposition of $PbTiO_3$ Films," *Japan. J. Appl. Phys.*, Vol. 22, 1983, pp. 14–17.

[67] Kwak, B. S., E. P. Boyd, and A. Erbil, "Metalorganic Chemical Vapor Deposition of $PbTiO_3$ Thin Films," *Appl. Phys. Lett.*, Vol. 53, 1988, pp. 1702–1704.

[68] Okada, M., et al., "Preparation of C-Axis Oriented $PbTiO_3$ Thin Films," *MOCVD, Ferroelectrics,* Vol. 91, 1989, pp. 181–192.

[69] Kuwabara, H., N. Menou, and H. Funakubo, "1 V Saturated $Pb(Zr,Ti)O_3$ Films with (111) Orientation Using Lattice-Matched (111) $SrRuO_3$/(111) Pt Bottom Electrode Prepared by Pulsed Metal Organic Chemical Vapor Deposition," *Appl. Phys. Lett.*, Vol. 93, 2008, p. 152901.

[70] Otani, Y., S. Okamura, and T. Shiosaki, "Recent Developments on MOCVD of Ferroelectric Thin Films," *J. Electroceram.*, Vol. 13, 2004, pp. 15–22.

[71] Budd, K. D., S. K. Dey, and D. A. Payne, "Sol-Gel Processing of PT, PZ, PZT, and PLZT Thin Films," *Br. Ceram. Proc.*, Vol. 36, 1985, p. 107.

[72] Dey, S. K., K. D. Budd, and D. A. Payne, "Thin Film Ferroelectrics of PZT by Sol-Gel Processing," *IEEE Trans. on UFFC,* Vol. 35, 1988, pp. 80–81.

[73] Schwartz, R. W., et al., "Sol-Gel Processing of PZT Thin Films: A Review of State-of-the-Art and Process Optimization Strategies," *Integr. Ferroelectr.*, Vol. 7, No. 1-4, 1995, pp. 259–277.

[74] Schwartz, R. W., "Chemical Solution Deposition of Perovskite Thin Films," *Chem. Mater.*, Vol. 209, 1997, pp. 2325–2340.

[75] Spierings, G. A. M. C., et al., "Preparation and Ferroelectric Properties of PZT Thin Films by Spin Coating and Metal Organic Decomposition," *J. Appl. Phys.*, Vol. 70, 1991, pp. 2290–2298.

[76] Tuttle, B. A., et al., "Microstructural Evolution of PZT Thin Films Prepared by Hybrid Metallo-Organic Decomposition," *J. Mater. Res.*, Vol. 7, 1992, pp. 1876–1882.

[77] Otsubo, S., et al., "Preparation of Pb(Zn $_{0.52}$Ti$_{0.48}$)O$_3$ Films by Laser Ablation," *Japan. J. Appl. Phys.*, Vol. 29, 1990, p. L133.

[78] Roy, D., S. B. Krupanidhi, and J. P. Dougherty, "Excimer Laser Ablated Lead Zirconate Titanate Thin Films," *J. Appl. Phys.*, Vol. 69, 1991, pp. 7930–7932.

[79] Florian, C., "PZT Thin Film Growth and Chemical Composition Control on Flat and Novel Three-Dimensional Micromachined Structures for MEMS Devices," Thesis, EPFL, 2007.

[80] Kumar, V., Y. Ohya, and Y. Takahashi, "Sol-Gel Processing of Ferroelectric $PbTiO_3$ and $Pb(Zr_{0.5}Ti_{0.5})O_3$ Thin Films," *Japan. J. Appl. Phys.*, Vol. 37, 1998, pp. 4477–4481.

[81] Gurkovich, S. R., and J. B. Blum, "Preparation of Monolithic Lead-Titanateby Sol–Gel Process," in *Ultrastructure Processing of Ceramics, Glasses and Composites*, New York: Wiley/Interscience, 1984.

[82] Ledermann, N., et al., "{100}-Textured, Piezoelectric Pb (Zr$_x$, Ti$_{1-x}$) O$_3$ Thin Films for MEMS: Integration, Deposition and Properties," *Sensor. Actuat. A: Phys.*, Vol. 105, 2003, pp. 162–170.

[83] Al-Shareef, H. N., et al., "Electrodes for Ferroelectric Thin Films," *Integr. Ferroelectr.*, Vol. 3, 1993, pp. 321–332.

[84] Vijay, D. P., and S. B. Desu, "Electrodes for PZT Ferroelectric Thin Films," *J. Electrochem. Soc.*, Vol. 40–4, 1993, p. 14026.

[85] Ramesh, R., et al., "Ferroelectric LaSrCoO/PbZrTiO/LaSrCoO Hetero Structures on Silicon Via Template Growth," *Appl. Phys. Lett.*, Vol. 63, 1993, pp. 3592–3594.

[86] Maeder, T., L. Sagalowicz and P. Muralt, "Stabilized Platinum Electrodes for Ferroelectric Film Deposition Using Ti, Ta and Zr Adhesion Layers," *Jpn. J. Appl. Phys.* Vol. 37, 1993, pp. 2007–2012.

[87] Willems, G. J., et al., "Nucleation and Orientation of Sol–Gel PZT Thin Films on Pt Electrodes," *Integr. Ferroelectr.*, Vol. 15, 1997, pp. 19–28.

[88] Mancha, S., "Chemical Etching of Thin Film PLZT," *Ferroelectrics*, Vol. 135, 1992, pp. 131–137.

[89] Miller, R., and J. Bernstein, "A Novel Wet Etch for Patterning PZT Thin Films," *Integr. Ferroelectr.*, Vol. 29, 2000, pp. 225–231.

[90] Zheng, K., J. Lu, and J. Chu, "A Novel Wet Etching Process of Pb(Zr,Ti)O3 Thin Films for Applications in Microelectromechanical System," *J. Appl. Phys.*, Vol. 43, 2004, pp. 3934–3937.

[91] Baborowski, J., et al., "Mechanisms of PZT Thin Film Etching with ECR/RF Reactor," *Int. Ferroelectrics*, Vol. 31, 2001, pp. 261–271.

[92] Chung, C. W., "Reactive Ion Etching of Pb (Zr$_x$Ti$_{1-x}$) O$_3$ Thin Films in an Inductively Coupled Plasma," *J. Vac. Sci. Technol. B*, Vol. 16, 1998, pp. 1894–1900.

[93] Koo, S., et al., "Etching Properties of Pb (ZrxTi1-x) O_3 Lead–Zirconate–Titanate Thin Films in an Inductively Coupled Plasma Cl_2/Ar and BCl_3/Ar Gas Chemistries," *J. Vac. Sci. Technol. B*, Vol. 1622, 2004, pp. 1519–1523.

[94] Polcawich, R. G., "Design, Fabrication, Test, and Evaluation of RF MEMS Series Switches Using Lead Zirconate Titanate (PZT) Thin Film Actuators," Ph.D. Thesis, Pennsylvania State University, 2007.

[95] Renaud, M., et al., "Fabrication, Modelling and Characterization of MEMS Piezoelectric Vibration Harvesters," *Sensor. Actuat. A: Phys.*, Vol. 380, 2008, pp. 145–146.

[96] Trollier-McKinsky, S., and P. Muralt, "Thin Film Piezoelectrics for MEMS," *J. Electroceramics*, Vol. 12, No. 1-2, 2004, pp. 7–17.

[97] Elfrink, R., et al., "Vibration Energy Harvesting with Aluminum Nitride-Based Piezoelectric Devices," *J. Micromech. Microeng.*, Vol. 19, No. 9, 2009, p. 094005.

[98] Andosca, R., et al., "Experimental and Theoretical Studies on MEMS Piezoelectric Vibrational Energy Harvesters with Mass Loading," *Sensor. Actuat. A: Phys.*, Vol. 178, 2012, pp. 76–87.

[99] Fang, H. -B., et al., "Fabrication and Performance of MEMS-Based Piezoelectric Power Generator for Vibration Energy Harvesting," *Microelectron. J.*, Vol. 37, No. 11, 2006, pp. 1280–1284.

[100] Liu, J. -Q., et al., "A MEMS-Based Piezoelectric Power Generator Array for Vibration Energy Harvesting," *Microelectron. J.*, Vol. 39, No. 5, 2008, pp. 802–806.

[101] Shen, D., et al., "The Design, Fabrication and Evaluation of a MEMS PZT Cantilever with an Integrated Si Proof Mass for Vibration Energy Harvesting," *J. Micromech. Microeng.*, Vol. 18, No. 5, 2008, p. 055017.

[102] Shen, D., et al., "Micromachined PZT Cantilever Based on SOI Structure for Low Frequency Vibration Energy Harvesting," *Sensor. Actuat. A: Phys.*, Vol. 154, No. 1, 2009, pp. 103–108.

[103] Liu, H., et al., "Piezoelectric MEMS Energy Harvester for Low-Frequency Vibrations with Wideband Operation Range and Steadily Increased Output Power," *J. Micorelectromech. S.*, Vol. 20, No. 5, 2011, pp. 1131–1142.

[104] H. Liu, et al., "A New S-Shaped MEMS PZT Cantilever for Energy Harvesting from Low Frequency Vibrations Below 30 Hz," *Microsyst. Technol.*, Vol. 18, No. 4, 2012, pp. 497–506.

[105] Kanno, I., et al., "Piezoelectric Properties of C-Axis Oriented Pb (Zr, Ti) O_3 Thin Films," *Appl. Phys. Lett.*, Vol. 70, 1997, pp. 1378–1380.

[106] Reilly, E. K., and P. K. Wright, "Modeling, Fabrication and Stress Compensation of an Epitaxial Thin Film Piezoelectricmicroscale Energy Scavenging Device," *J. Micromech. Microeng.*, Vol. 19, 2009, p. 095014.

[107] Morimoto, K., et al., "High-Efficiency Piezoelectric Energy Harvesters of C-Axis-Oriented Epitaxial PZT Films Transferred onto Stainless Steel Cantilevers," *Sensor. Actuat. A: Phys.*, Vol. 163, No. 1, 2010, pp. 428–432.

[108] Isarakorn, D., et al., "The Realization and Performance of Vibration Energy Harvesting MEMS Devices Based on an Epitaxial Piezoelectric Thin Film," *Smart Mater. Struc.*, Vol. 20, No. 2, 2011, p. 025015.

[109] Harigai, T., H. Adachi and E. Fujii, "Vibration Energy Harvesting Using Highly (001)-Oriented Pb (Zr,Ti) O_3 Thin Film," *J. Appl. Phys.*, Vol. 107, No. 9, 2010, p. 096101.

[110] Arndt, G., et al., "Dynamic Analytical Modelling and Piezoelectric Characterization of a Pb (Zr, Ti) O_3/SiN Cantilever with Losses," *J. Micromech. Microeng.*, Vol. 20, No. 11, 2010, p. 115019.

[111] Do, Y. H., et al., "Preparation on Transparent Flexible Piezoelectric Energy Harvester Based on PZT Films by Laser Lift-Off Process," *Sensor. Actuat. A: Phys.*, Vol. 200, 2013, pp. 51–55.

[112] Jeon, Y. B., et al., "MEMS Power Generator with Transverse Mode Thin Film PZT," *Sensor. Actuat. A: Phys.*, Vol. 122, No. 1, 2005, pp. 16–22.

[113] Park, J. C., J. Y. Park, and Y. -P. Lee, "Modeling and Characterization of Piezoelectric-Mode MEMS Energy Harvester," *J. Micorelectromech. S.*, Vol. 19, No. 5, 2010, pp. 1215–1222.

[114] Lee, B. S., et al., "Piezoelectric MEMS Generators Fabricated with an Aerosol Deposition PZT Thin Film," *J. Micromech. Microeng.*, Vol. 19, No. 6, 2009, p. 065014.

[115] Kim, S. B., et al., "Comparison of MEMS PZT Cantilevers Based on d_{31} and d_{33} Modes for Vibration Energy Harvesting," *J. Micorelectromech. S.*, Vol. 22, No. 1, 2013, pp. 26–33.

[116] Liao, Y., and H. A. Sodano, "Model of a Single Mode Energy Harvester and Properties for Optimal Power Generation," *Smart Mater. Struc.*, Vol. 17, No. 6, 2008, p. 065026.

[117] Erturk, A., and D. J. Inman, "An Experimentally Validated Bimorph Cantilever Model for Piezoelectric Energy Harvesting from Base Excitations," *Smart Mater. Struc.*, Vol. 18, No. 2, 2009, p. 025009.

[118] Kok, S. -L., N. M. White, and N. R. Harris, "Fabrication and Characterization of Free-Standing Thick-Film Piezoelectric Cantilevers for Energy Harvesting," *Meas. Sci. Technol.*, Vol. 20, No. 12, 2009, p. 124010.

[119] Lei, A., et al., "MEMS-Based Thick Film PZT Vibrational Energy Harvester," *MEMS 2011*, 2011, pp. 125–128.

[120] Tanaka, K., et al., "Wafer Bonding of Lead Zirconate Titanate to Si Using an Intermediate Gold Layer for Microdevice Application," *J. Micromech. Microeng.*, Vol. 16, 2006, pp. 815–820.

[121] Aktakka, E. E., R. L. Peterson, and K. Najafi, "Thinned-PZT on SOI Process and Design Optimization For Piezoelectric Inertial Energy Harvesting," *Transducers 2011*, 2011, pp. 1649–1652.

[122] Xu, X. H., and J. R. Chu, "Preparation of a High-Quality PZT Thick Film with Performance Comparable to Those of Bulk Materials for Applications in MEMS," *J. Micromech. Microeng.*, Vol. 18, 2008, p. 065001.

[123] Wang, Z. H., J. M. Miao, and C. W. Tan, "Acoustic Transducers with a Perforated Damping Backplate Based on PZT/Silicon Wafer Bonding Technique," *Sensor. Actuat. A: Phys.*, Vol. 149, 2009, pp. 277–283.

[124] Tang, G., et al., "Fabrication and Analysis of High-Performance Piezoelectric MEMS Generators," *J. Micromech. Microeng.*, Vol. 22, No. 6, 2012, p. 065017.

[125] Tang, G., et al., "Development of High Performance Piezoelectric d_{33} Mode MEMS Vibration Energy Harvester Based on PMN-PT Single Crystal Thick Film," *Sensor. Actuat. A: Phys.*, Vol. 205, 2014, pp. 150–155.

2

Electromagnetic MEMS Vibration Energy Harvesting

Electromagnetism has been utilized to generate electricity since the early 1930s, not long after Faraday's fundamental breakthrough in electromagnetic induction. Over the last decade, a variety of micro/macro electromagnetic energy harvesters, varying in scale from a few cubic millimeters to a few cubic centimeters, have been investigated. The power generation ranges from microwatts to watts have demonstrated a promising feasibility for future applications. Hence, this chapter focuses on the elaboration of recent electromagnetic transducers for the harvesting of kinetic (vibration) energy [1–3]. The fundamental principle and modeling of electromagnetism will be introduced. Additionally, characterization and fabrication of coils and magnets of different architectures will be presented. Finally, the primary content of this chapter is to highlight and summarize recent achievements in this field. Although related, other prototypes for large-scale electromagnetic energy harvesting are not covered.

2.1 Basic Principle and Modeling

2.1.1 Electromagnetic Subsystem

The basic principle for most of the electromagnetic energy harvesters is based on Faraday's law of electromagnetic induction as demonstrated in Figure 2.1. The voltage, or electromotive force (emf), induced in a conductive loop of wire is proportional to the negative of the rate of change of magnetic flux:

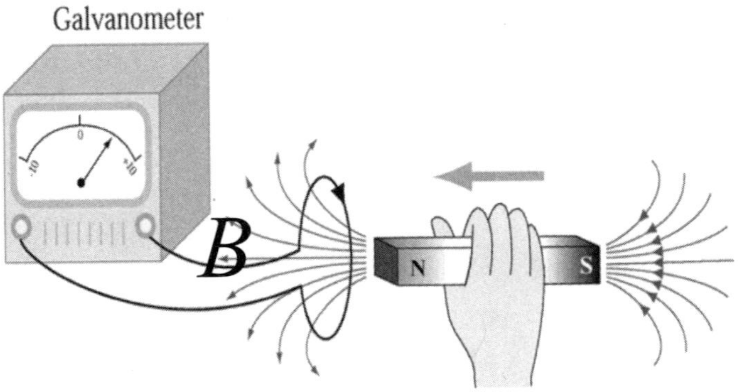

Figure 2.1 Faraday's law of electromagnetic induction.

$$\varepsilon = -\frac{d\phi}{dt} \tag{2.1}$$

where ε is the generated voltage or induced emf and ϕ is the magnetic flux through a wire loop. For most implementations of electromagnetic energy harvesters, the circuit consists of a coil of wire of multiple turns and the magnetic field is created by permanent magnets. Hence, the voltage induced in an N turn coil is given by:

$$\varepsilon = -\frac{d\Phi}{dt} = -\sum_{i=1}^{N} \int B dA_i \tag{2.2}$$

where Φ is the total magnetic flux of the N turn coil, which is generally defined as the sum of the magnetic flux for each individual turn, A_i indicates the area enclosed by the ith wire loop, and B is the magnetic flux density over the area of the ith turn. Equation (2.2) can be expanded to give:

$$\varepsilon = -\frac{d\Phi}{dz}\frac{dz}{dt} = k_t \dot{z} \tag{2.3}$$

where k_t is the transduction factor, which equals the magnetic flux gradient. Furthermore, the emf voltage can be expressed as:

$$\varepsilon = -\sum_{i=1}^{N} \frac{dB}{dt} A_i - \sum_{i=1}^{N} B \frac{dA_i}{dt} \tag{2.4}$$

From this equation, it is evident that the electromagnetic induction can be induced either by changing the magnetic field within a constant area or changing the area within a constant magnetic field. Thereby, it offers a wide range of possible implementations of the electromagnetic coupling. Figure 2.2(a, b) shows two typical energy harvesting configurations through the change of the magnetic flux: the moving magnet with a static coil (configuration I) and the moving coil with a static magnet (configuration II). Both of them consist of a spring-mass-damping system with spring stiffness *K*, damping factor *C*, and proof mass *M*. In Figure 2.2(a), a permanent magnet is used as the proof mass and it moves relative to a static coil. In contrast, Figure 2.2(b) shows a proof mass coated with winding coil and it moves relative to a static permanent magnet. In both cases, the electrical current will be generated due to the change of the magnetic flux across the coil. In the case that a winding coil with *N* turns and a rectangular cross section moves across a constant magnetic field, the rate of the change of overlapping area is:

$$N \cdot dA/dt = Nl \cdot dz/dt = Nl\dot{z} \tag{2.5}$$

Hence, the emf voltage becomes:

$$\varepsilon = -NBl\dot{z} = k_t \dot{z} \tag{2.6}$$

2.1.2 Mechanical Subsystem

The mechanical model in Figure 2.2 is considered to determine and predict the dynamic performance of the device. The variables $y(t)$ and $z(t)$ are the displacements of the housing and the effective mass, respectively. For a sinusoidal excitation $y = Y\sin\omega t$, where Y and ω are the amplitude and angular frequency of the vibration, respectively, the differential equation of motion is obtained as:

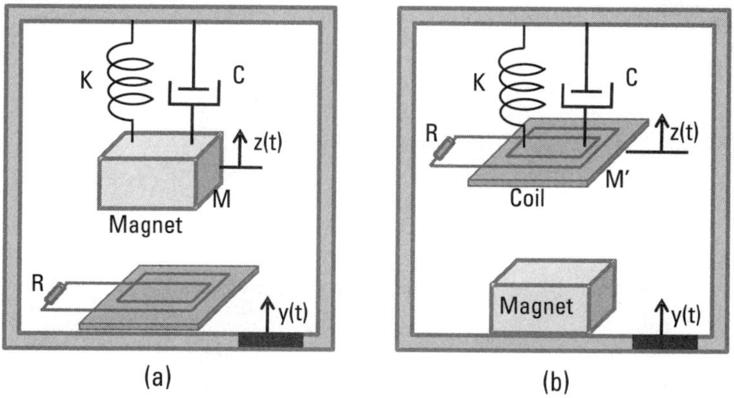

Figure 2.2 Typical energy harvesting configurations: (a) moving magnet with static coil and (b) moving coil with static magnet.

$$M\ddot{z} + C\dot{z} + Kz = -M\ddot{y} = MY\omega^2 \sin(\omega t) \tag{2.7}$$

where z is the relative displacement of the mass with respect to the housing. The solution to (2.7) can be easily found in the frequency domain using Laplace transformation:

$$\left(Ms^2 + Cs + K\right)Z(s) = -Ms^2Y(s) \tag{2.8}$$

The transfer function $G(s)$ can be obtained by rearranging (2.8):

$$G(s) = \frac{Z(s)}{Y(s)} = \frac{-Ms^2}{\left(Ms^2 + Cs + K\right)} \tag{2.9}$$

With the natural frequency of the undamped oscillation system $\omega_n = \sqrt{K/M}$ and the normalized damping factor $\varsigma = C/2m\omega_n$, the solution to (2.7) can be obtained by substituting $s = j\omega$ to (2.9) as:

$$z(t) = \frac{MY\omega^2}{K - M\omega^2 + j\omega C}\sin\omega t \tag{2.10}$$

where the amplitude Z and phase angle φ of the relative displacement of the mass is given as:

$$Z = \frac{MY\omega^2}{\sqrt{\left(K - M\omega^2\right)^2 + (C\omega)^2}} = \frac{Y\left(\omega/\omega_n\right)^2}{\sqrt{\left(1-\left(\omega/\omega_n\right)^2\right)^2 + \left(2C\left(\omega/\omega_n\right)\right)^2}} \tag{2.11}$$

$$\tan\varphi = \frac{C\omega}{K - M\omega^2} = \frac{2\xi\left(\omega/\omega_n\right)}{1-\left(\omega/\omega_n\right)^2} \tag{2.12}$$

The velocity of mass is obtained by the first derivative of (2.10):

$$\dot{z}(t) = \frac{\omega Y\left(\omega/\omega_n\right)^2}{\sqrt{\left(1-\left(\omega/\omega_n\right)^2\right)^2 + \left(2C\left(\omega/\omega_n\right)\right)^2}}\cos\omega t \tag{2.13}$$

2.1.3 Power Generation of the Overall System

The mechanical and electromagnetic subsystem of the vibration harvester can be combined to an overall system model. The electromagnetic domain (emf and induced current) and the mechanical domain (input force and relative velocity of the mass) are related to the transduction factor k_t. For a closed circuit with coil resistance R_C and load resistance R_L, the emf-induced current will cause an opposite magnetic field according to Lenz's law. The feedback electromechanical force is given by:

$$F = k_t i = \frac{k_t^2}{R_C + R_L}\dot{z} \tag{2.14}$$

By substituting (2.6) into (2.14), the feedback electromechanical force can be represented as:

$$F = \frac{(NBl)^2}{R_C + R_L}\dot{z} = C_e\dot{z} \tag{2.15}$$

Equation (2.15) demonstrates that the electromagnetic transduction mechanism can be represented by a dissipative velocity proportional to electromagnetic damping coefficient C_e. The maximum power generated is thereby related to the power dissipated in the damper:

$$P = \tfrac{d}{dt}\int_0^z F(t)dz = \frac{1}{2}C_e\dot{z}^2 \tag{2.16}$$

By substituting (2.13) into (2.16), the generated electrical power becomes:

$$P = \frac{m\zeta_e Y^2\omega^3(\omega/\omega_n)^3}{\left[1-(\omega/\omega_n)^2\right]^2 + \left[2(\zeta_e+\zeta_m)(\omega/\omega_n)\right]^2} \tag{2.17}$$

where the damping ratio (normalized damping factor) has been divided into the electromechanical part ς_e and parasitic part $\varsigma_m\left(\varsigma = \varsigma_e + \varsigma_m\right)$.

This basic equation was first published by Williams and Yates in [4]. When the vibration harvester operates at resonance, (2.17) can be further reduced to:

$$P = \frac{m\zeta_e Y^2\omega_n^3}{4(\zeta_e+\zeta_m)^2} \tag{2.18}$$

Equation (2.18) indicates that the output power depends on the moving mass and the damping coefficients. By setting $dP/d\varsigma_e = 0$, the optimal ratio is given by $\varsigma_e = \varsigma_m$ as illustrated in Figure 2.3 [5]. However, for an oscillating sytem with a high oscillation mass and an extremely low electromagnetic damping, the mass displacement could reach its maximum alllowable displacement limit. The maximum power could be reduced under such a constrained condition [6, 7]. The minimum allowable damping factor can by obtained by the rearrangement of the steady-state amplitude in (2.11) as:

$$\zeta_{e,\min} = \frac{1}{2\omega_c}\sqrt{\left(\frac{Y\omega_c^2}{Z_L}\right)^2 - \left(1-\omega_c^2\right)^2} - \zeta_m \tag{2.19}$$

where ω_c denotes the ratio of excitation frequency and resonant frequency ($\omega_c = \omega/\omega_n$), Z_L is the maximum allowable displacement. The minimum allowable damping factor in (2.19) can be substituted into (2.18) to obtain the dissipated power in the displacement constrained condition:

$$P_{con} = \frac{m\zeta_{e,\min}Y^2\omega_n^3}{4\left(\zeta_{e,\min} + \zeta_m\right)^2} \tag{2.20}$$

Consequently, the maximum dissipated power in constrained displacement condition can be given at resonance ($\omega = \omega_n$ or $\omega_c = 1$) as:

$$P_{con,\max} = \frac{1}{2}m\omega_n^3 Z_{\max}\left(Y - 2\zeta_m Z_{\max}\right) \tag{2.21}$$

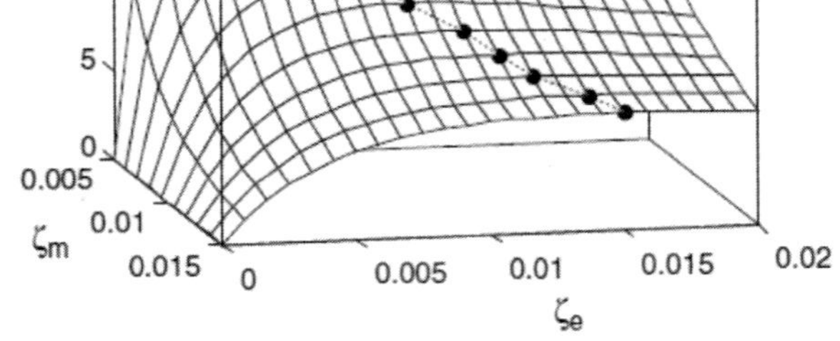

Figure 2.3 Output power as a function of the electromechanical and parasitic damping [5].

2.2 Characterization of Coils and Magnets

2.2.1 Wire-Wound and Microfabricated Coils

The induction coil is an essential part of an electromagnetic energy harvester. It is therefore of great importance to calculate coil parameters for device optimization. An air core coil is normally fabricated by filament winding method. A fundamental consideration of coil design is the number of turns N and resistance R_C, which determines the voltage and power developed by an electromagnetic energy harvester. The number of turns is governed by the geometry of the coil (i.e., the diameter of the wire and the density of the wound wire). The insulated wound wire will not fill the coil volume entirely and the percentage of conductive copper within a coil is given by its fill factor, f, (i.e., $A_{wire} = f \times A_{coil}/N$), where A_{wire} is the area of the wire and A_{coil} is the overall cross-sectional area of the coil. The copper fill factor depends on insulation thickness, tightness, and shape of winding, while typically a figure of 50%–60% can be assumed for a scramble wound coil [8]. For miniaturized electromagnetic energy harvesting, fine copper wire is desirable for minimizing the coil size while maximizing the number of turns. However, the finer the wire, the greater its resistance per unit length.

For a conventionally wire-wound coil as shown in Figure 2.4(a), the number of turns, length, and resistance of the coil can be predicted by:

$$V_T = \pi\left(r_o^2 - r_i^2\right)h \tag{2.22}$$

$$L_W = \frac{f \cdot V_T}{\pi\omega_d^2 / 4} \tag{2.23}$$

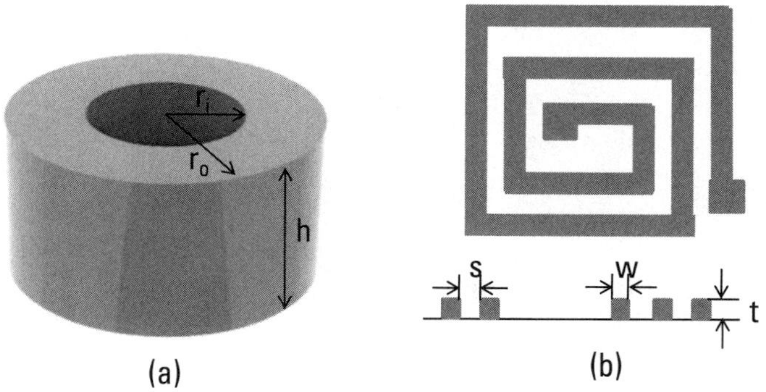

Figure 2.4 Illustration of (a) wire-wound coil and (b) microfabricated square spiral coil.

$$N = \frac{L_W}{r_i + (r_o - r_i)/2} \tag{2.24}$$

$$R_C = \rho \frac{L_W}{A_W} = \rho \frac{\pi (r_o - r_i) N^2}{f (r_o - r_i) h} \tag{2.25}$$

where r_i and r_o are the coil inner radius and outer radius, respectively, h is the coil thickness, w_d is the wire diameter, A_c is the area of coil, L_w is the length of wire, N is the number of turns, V_T is the coil volume, f is the coil fill factor, ρ is the resistivity of the coil material, and R_c is the coil resistance. For a general analysis, since the resistive impedance of the coil is significantly larger than the inductive impedance at frequencies less than 1 kHz, the inductance of the coil can be neglected. It can be seen from (2.25) that, for a constant number of turns, the coil resistance is proportional to the inverse of the coil dimension. Inversely, by applying (2.25), the length of the wire can be determined by measuring the coil resistance.

In addition to the aforementioned wire-wound coil in a cylindrical shape, microcoils consisting of multiple layers of planar coils are frequently employed in MEMS-based electromagnetic energy harvesters. As seen in Figure 2.4(b), they are generally fabricated by using photolithography techniques to define square or circular spiral patterns on various substrates, such as silicon, glass, or printed circuit board (PCB). Such coils are used for a range of applications including on-chip inductors, sensors with detection coils, or actuators producing with the magnetic field. The minimum dimension of an individual coil turn of the spiral coils, such as width, thickness, and spacing achievable, is actually dependent on the fabrication technology used. For example, a standard PCB technology may limit the wire spacing to be larger than 150 μm with a wire thickness of 35 μm, whereas a silicon-based microfabrication technique could achieve a minimum metal wire spacing of 1 to 2 μm with a typical thickness of 1 μm. In the practical silicon-based coil fabrication, the minimum spacing are usually taken to be the same as the wire thickness, although it is feasible to achieve an aspect ratio of up to 10 between the wire thickness and minimum spacing by more advanced techniques. The essential difference between the wire-wound coil and the microfabricated coil can be interpreted as the following. The wire-wounding is a 3-D approach, whereas the microfabrication is a planar (2-D) technology, which is restricted in the number of layers achievable in practice.

2.2.2 Fabrication of Electromagnetic Coils

As one of the essential components of the electromagnetic MEMS energy harvesters, both wire-wound and microfabricated coils have been widely employed. The wire-wound coil can be implemented by using a filament-winding approach based on precise x-y positioning and z-axis height controlling. Figure 2.5 shows a variety of wire-wound coils with different shapes, such as air core winding around a cylindrical or spherical bobbin [9–11]. Dozens or even hundreds of coil turns can be wound by such approach, which is therefore taken as a benefit to produce more electrical power through the electromagnetic coil.

However, microcoil fabrication processes were reported previously for numerous other devices such as magnetic sensors [12, 13], magnetic actuators [14], micromotors [15], transformers, and energy harvesters [16]. A typical fabrication process flow of a double-layer metal coil by using photolithography techniques is shown in Figure 2.6 [17, 18]. First, an ultrathin adhesion/seed layer is sputtered onto a silicon/glass substrate, followed by the spin-coating and patterning of a photoresist layer. Afterward, a metal layer such as copper (Cu), gold (Au), or aluminum (Al) is electroplated and patterned. Then the metal via is electroplated and patterned to be a channel connecting the two layers of the coil. In the next step, the photoresist layer is stripped and the seed layer underneath the wire spacing is etched. It is followed by the spin-coating of an insulation layer. By repeating the above steps, the second metal coil layer will be patterned on top. It should be noted that a thermal oxide layer (about 1 μm thick) is normally grown on the silicon substrate to provide isolation between the metal layer and silicon. In some cases, the patterning of the photoresist and metal coil can be conducted reversely [19, 20]. The first metal layer is deposited on a silicon substrate with thermal oxide insulation layer. Then the coil pattern is etched by using a patterned photoresist layer. Subsequently, an insulation layer and pad opening is etched and formed. The steps are repeated to form the second metal coil layer. In some works, SU-8 photosensitive epoxy has been

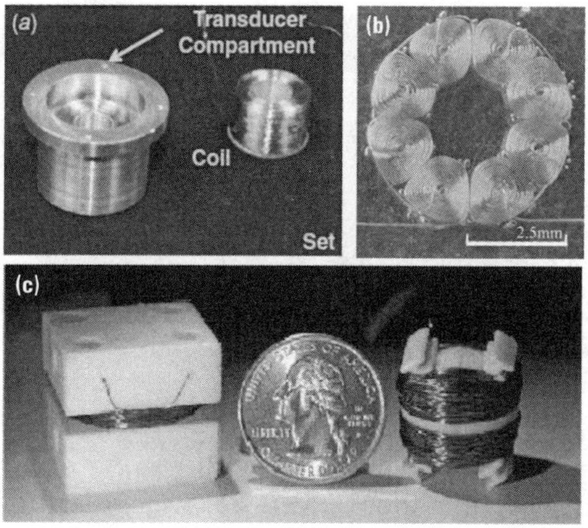

Figure 2.5 Wire-wound coils with different shapes. (Copyright: IOP Publishing, [9–11].)

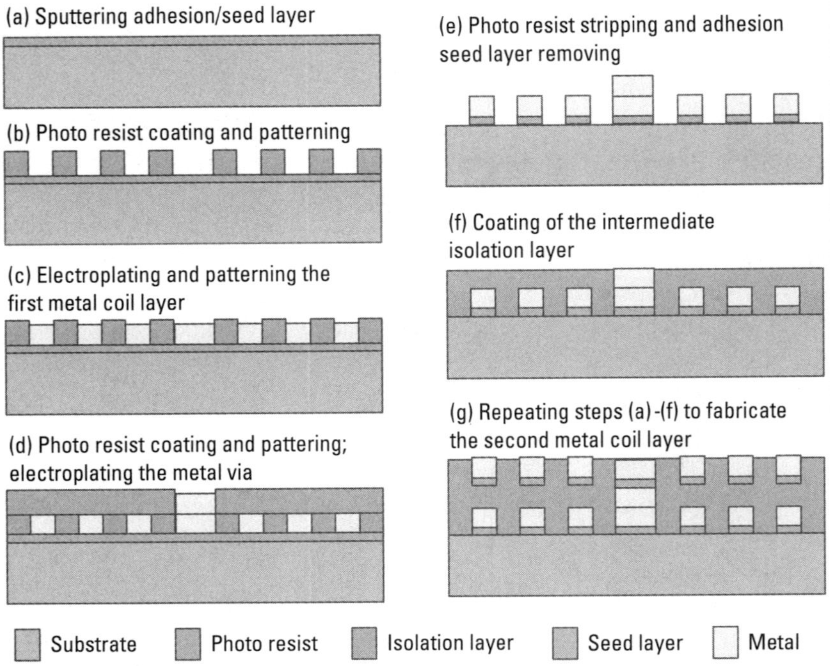

Figure 2.6 Process flow of a double-layer microcoil.

used as an insulation layer as well as a photoresist layer to pattern the opening via [21–23].

2.2.3 Magnetic Materials

The electromagnetic energy harvesting requires a magnetic field that is generated by the use of permanent magnets or electromagnets. However, an electromagnet requires a current flow to produce a magnetic field and thus consume power. Hence, it is not suitable for the small-scale, low-power, energy-harvesting devices. Permanent bulk magnets are made from hard ferromagnetic (or ferromagnetic) materials that are subjected to special processing in a powerful magnetic field during manufacture, to align their internal microcrystalline structures, making them very hard to demagnetize.

A ferromagnetic material is composed of a large number of magnetic domains. The magnetic domain is formed by large numbers of atom's moments grouped together within which their magnetic moments are aligned in a particular direction. The strong magnetic properties of a ferromagnetic material are gained due to the presence of magnetic domain. Taking the magnet as a whole, in a virgin state, each domain oriented in a different direction and the net magnetic field is zero. Once a magnetizing force is applied, the domains become aligned and thus produce a strong magnetic field within the magnet. Iron, nickel, and cobalt are examples of ferromagnetic materials, which have a large, positive susceptibility to an external magnetic field. They exhibit a strong attraction to magnetic fields and are able to retain their magnetic properties after the external field has been removed.

Figure 2.7 shows a typical hysteresis loop with graphic illustrations of the magnetic domain alignment during the magnetization and demagnetization process. The maximum magnetization after full saturation is called saturation magnetization (M_s). When the magnetizing field is removed, the magnetic material can remain a significant portion of magnetization, which is known as remanence or remanent magnetization (M_r). Intrinsic coercivity (H_c) represents the ability of a permanent magnet to resist demagnetization. Some ferromagnetic materials are classified as soft materials because they can be easily magnetized and also easily demagnetized (e.g., annealed iron). Hard magnetic materials require stronger magnetizing fields and are also very hard to demagnetize (e.g., alnico and ferrite). Hard materials have high coercivity, whereas soft materials have low coercivity.

The magnetic field produced by a permanent magnet is typically denoted by the magnetic flux density, *B*, which is a vector quantity representing the magnetic flux per unit area of a section normal to the direction of flux, which is measured in Tesla (T) in SI units. Magnetization, *M*, is a vector representing the magnetic moment per unit volume of the material. The SI unit for *M* is Ampere-turns (A/m). The magnetic field strength (or magnetizing force) is denoted as *H*, which is measured in A/m in SI units. The magnetic flux density

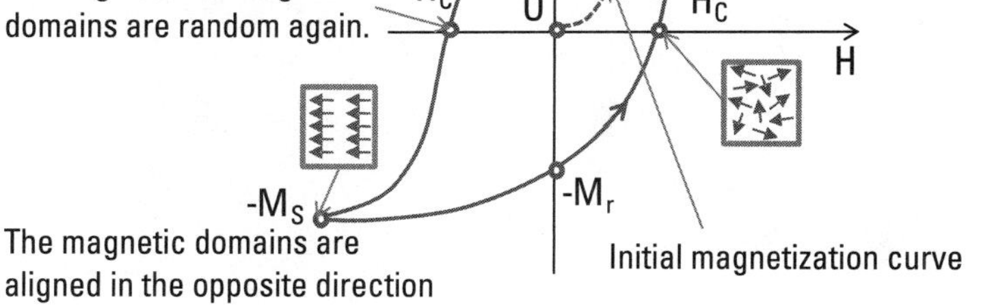

Figure 2.7 Hysteresis loop of the magnetic domain alignment during the magnetization and demagnetization process.

B, magnetization M, and applied magnetic field H are related based on the following equations:

$$B = \mu_m (H + M) \tag{2.26}$$

In free space, M is equal to zero. Hence, the terms B and H are linked by:

$$B = \mu_m H \tag{2.27}$$

where μ_m is the product of the permeability of free space ($\mu_0 = 4\pi \times 10^{-7}$Tesla*m/A) times the permeability of the material. A useful figure of merit for comparing magnetic materials is the maximum energy product, BH_{MAX}, calculated from a materials magnetic hysteresis loop. For a cylindrical magnet as seen in Figure 2.8(a), the longitudinal magnetic flux density along the central z axis of the cylinder can be expressed by [24]:

$$B(z) = -\frac{B_r}{2}\left(\left(\frac{z+l}{\sqrt{(z+l)^2 + R_2^2}} - \frac{z-l}{\sqrt{(z-l)^2 + R_2^2}}\right) - \left(\frac{z+l}{\sqrt{(z+l)^2 + R_1^2}} - \frac{z-l}{\sqrt{(z-l)^2 + R_1^2}}\right)\right) \tag{2.28}$$

where B_r is the residual magnetic flux density (1.1–1.4T for NdFeB magnets); R_1 and R_2 are the inner and outer radius of the cylindrical magnet; and the total length of the magnet is $2l$. For a hexahedral magnet as seen in Figure 2.8(b), the magnetic flux density along a distance d can be given by [25]:

$$B(d) = \frac{B_r}{\pi}\left(\tan^{-1}\left(\frac{ab}{2d\sqrt{a^2 + b^2 + 4d^2)}}\right) - \tan^{-1}\left(\frac{ab}{2(d+c)\sqrt{a^2 + b^2 + 4(d+c)^2)}}\right)\right) \tag{2.29}$$

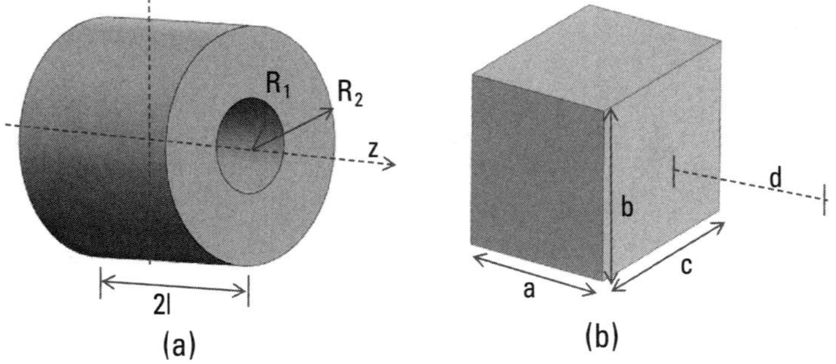

Figure 2.8 Permanent magnets with (a) cylindrical and (b) hexahedral shapes.

where *c* is the length of the magnet through magnetized direction, *a* and *b* are the height and width of magnet, and *d* is the distance from the magnet.

For a vibration-based energy-harvesting application, the particular properties of interest are the magnetic field strength, the flux density, and the coercive force. Typically, there are four types of magnets available: alnico, ceramic (hard ferrite), samarium cobalt (SmCo), and neodymium iron boron (NdFeB). Each type is subdivided into a range of grades and each has its own magnetic properties. Due to the much higher magnetic fields, SmCo and NdFeB become the most commercialized magnet materials [26, 27], which are known as rare earth magnets. Both types are fabricated using powder metallurgy processes and sintered to form a wide variety of magnet geometries. NdFeB has the highest maximum energy product ($BH_{max} \sim 400$ kJ/m^3) but suffers from low-working temperatures ($T_{curie} \sim 310$) and poor corrosion resistance. It is more suitable for a low-temperature vibration-based energy harvester. Whereas SmCo exhibits a combination of a high energy product ($BH_{max} \sim 240$ kJ/m^3) and good thermal stability, high working temperature, and corrosion resistance. It is selected for the integration with a combustion-driven generator, where operating temperatures may be several hundred degrees Celsius.

For small-dimensional energy harvesters, microfabrication techniques would be considered for the deposition and patterning of magnets. Sputtering and electroplating have been used for the deposition of microscale magnets. Electroplating as a more cost-effective technique is favorable for the deposition of thick layers, whereas Jiang et al. [28] successfully deposited 20-μm-thick NdFeB/Ta magnetic films by using a multilayer dc magnetron sputtering technique. In the multilayer film, the grain size is successfully controlled by the thickness of a single NdFeB layer. In Figure 2.9, Zhang et al. [29] used a batch process to integrate magnets onto a microelectromagnetic energy harvester. The etched deep trenches of the silicon wafer were covered by parylene layer. The NdFeB powders and wax powders were filled into the trenches and the residual powders were removed by a lift-off process. The wax-bonded NdFeB powders are only left in the 300-μm-deep trenches by the lift-off process. Later on, the micromagnets were magnetized along the in-plane direction.

As a matter of fact, the properties achievable for microfabricated magnets, such as coercive force (H_c), remanence (B_r), and energy density (BH_{max}), are considerably lower than those achievable for bulk-sintered rare earth magnets (i.e., SmCo and NdFeB). For instance, Ng et al. [29] deposited a 90-μm-thick magnet with the highest coercivity and remanence of 160 kA/m and 0.5T. These values are significantly low compared to a bulk magnet of around 800-kA/m coercivity and 1T remanence. In addition, thick (tens of micrometers) micromagnets with good properties are difficult to achieve, since the properties tend to degrade as the layer thickness increases. Hence, further works are

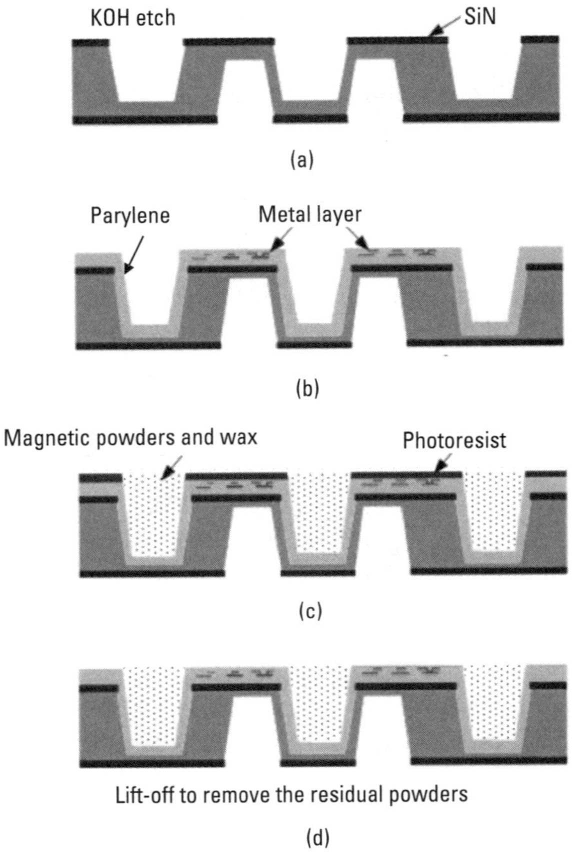

Figure 2.9 Microfabrication of the NdFeB magnets.

required to make the microfabricated permanent magnets an alternative to discrete sintered magnets.

2.3 Review of Existing Electromagnetic Energy Harvesters

The macroscale/microscale electromagnetic energy harvesters reported can be broadly classified into three categories: rotational, vibrational, and nonresonant devices, as shown in Figure 2.10 [31]. Rotational energy harvesters are designed for continuous rotational motion of magnetic rotor with respect to a coil-based stator under a steady driving torque. In contrast, vibrational energy harvesters usually operating at resonance mode are designed to harness power from environmental vibrations due to a relatively small displacement between magnet and coil. Third, a nonresonant device converts linear motion into nonlinear or nonresonant motion by using particular spring-mass configurations. Other rec-

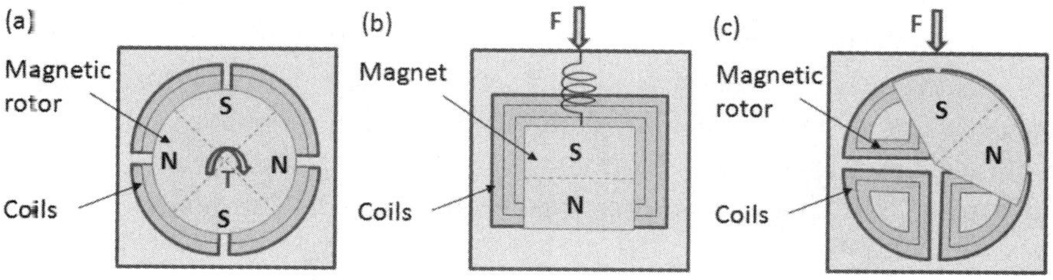

Figure 2.10 Schematic illustration of (a) rotational, (b) vibrational, and (c) nonresonant electromagnetic energy harvesters.

ommended reviews of the related literature are Beeby et al. [32] and Mitcheson [33].

2.3 1 Rotational Energy Harvesters

A rotational energy harvester relies on a continuous rotational energy source, such as from a fluid turbine or heat engine. The device with small size often operates at high rotational speed, which enables a relatively high-power density in contrast to its macroscale counterpart. In 1996, Wiegele introduced the design concept of a microturbo-generator device [34], which involves the coupling of an impulse turbine to a switched reluctance generator via a single rotor. The gas force causes the rotor to turn at high speeds. With proper current switching and control, this rotational motion can be converted into an electrical power output. However, the power generation portion was not implemented in the work.

Senesky and Sanders [35] from the University of California-Berkeley presented and tested a millimeter-scale power generator for the use with a MEMS internal combustion Wankel engine. The majority of the generator structure is built from discrete millimeter-scale components, with only the rotor being microfabricated. The rotation of a soft magnetic rotor beneath the stator poles varied the reluctance in the magnetic circuit, which induces a time-varying flux and hence voltage in a bobbin winding of the stator. The device was demonstrated to have an open-circuit voltage of 2.63V and a maximum output power of 375 μW at 13.3 krpm. Holmes et al. [21] from Imperial College reported an axial-flow micro turbine power generator. The device comprises a polymer rotor with embedded permanent magnets sandwiched between the upper and lower stators with electroplated planar coils. The fabrication process is a combination of silicon micromachining, electroplating, and laser etching. In operation, the magnetic rotor introduces a time-varying magnetic flux to the stator windings. The prototype with a diameter of 7.5 mm was demonstrated to deliver an output power of 1.1 mW at rotation speed of 30 krpm from an air-stream of 35 L/min. The device was integrated with a microfabricated axial-flow microturbine to produce a compact power conversion for both power generation and flow-sensing applications. Pan and Wu [10] from National Sun Yat-Sen University, Taiwan, reported a similar axial-flux generator structure. A four-layer stacked copper winding for the stator was fabricated by a unique filament winding method, and eight discrete, arc-shaped NdFeB magnets were used for the magnetic rotor. The single-phase prototype produced a maximum power of 0.41 mW at 2.2 krpm.

Other works based on three-phase, axial-flux, rotational generators are intended for integration with microscale turbine engines. In 2006, Raisigel et al. adapted a previously reported planar micromotor [36] for the use of a microgenerator [37]. By using photolithography, double-layer electroplated three-phase copper coils were micromachined into a silicon wafer. An 8-mm-diameter disc of SmCo or NdFeB was micromachined and selectively magnetized with up to 15 pairs of north and south poles [38]. Compared with the generator reported by Holmes et al. [21], this device possesses a similar operation mechanism, but with a mechanically continuous magnetic rotor, rather than discrete magnets. A self-stabilized microturbo generator driven by compressed air was demonstrated to deliver an electric power of 14.6 mW at 58 krpm. High speed electrical tests exhibit an output power of 5W per stator at 380 krpm.

A joint group from Georgia Tech and the Massachusetts Institute of Technology (MIT) developed a series of microrotational generators [22, 23, 39–42] using a combination of microfabrication and precision assembly. The generators are three-phase, axial-flux, synchronous machines, each consisting of an eight-pole surface-wound stator and permanent magnet rotor as seen in Figure 2.11. The devices integrated a magnetic back iron on the rotor so as to increase the magnetic field in the air gap. They incorporate an off-the-shelf gas-driven spindle to power the machines. At a rotational speed of 120 krpm, the first-generation device reported by Arnold et al. [22] delivered a 2.5-W power conversion, which was converted to 1.1-W dc power via a passive rectification circuit. A second-generation device was followed to produce 8-W dc power at 305 krpm [41]. This generator was also tested at elevated temperatures for compatibility within a hot microengine environment [42]. Herrault et al. [43] presented a different generator with a smallest volume of only 6.6 mm and a dc power output of 3.6 mW at 392 krpm was characterized.

Table 2.1 summarized the performance of the reported rotational energy harvesters. In terms of power density, the micromachined devices by Raisigel [37] indicate superior performance (i.e., up to 220 W/cm^3) as a result of the compact size and high rotational speeds. This indicates that without back irons, higher power densities are achievable and the fabrication complexity is greatly reduced. Hence, ironless designs may be preferable for highly integrated structures. Moreover, static magnetic forces between the rotor and stator are eliminated because of the bearing design.

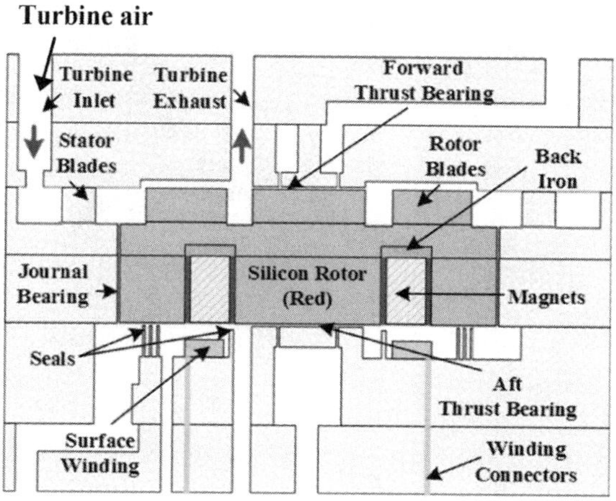

(a)

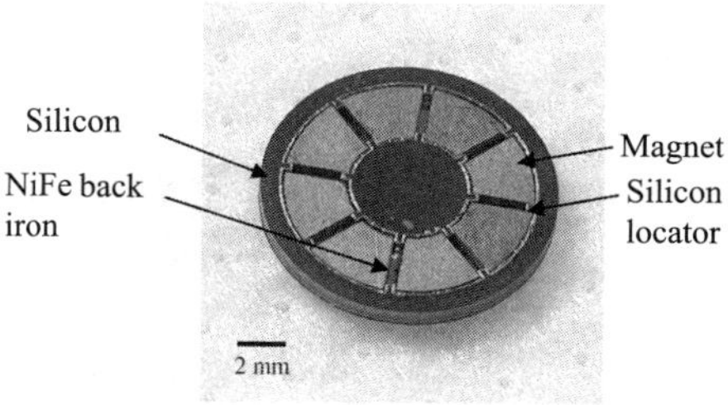

(b)

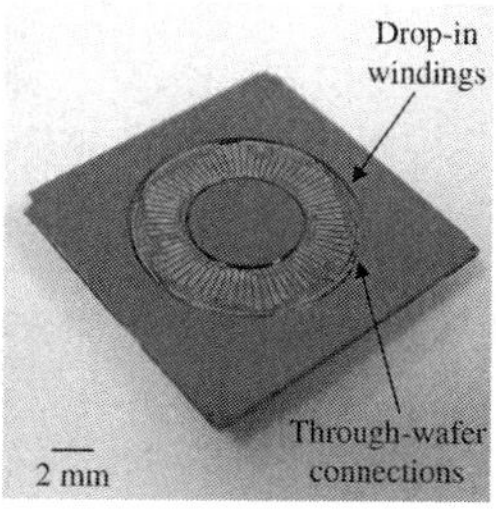

(c)

Figure 2.11 Microrotational generators reported by Herrault et al. [23].

2.3.2 Vibrational Energy Harvesters

Vibrational energy harvesters are attractive because they generate power without the need for prime mover source (i.e., turbine or engine). The basic configuration of a vibrational energy harvester utilizes a mass-spring-damper system such that a magnet and coil move with respect to each other under the excitation of an external vibration. The maximum power of the device is achieved at

Table 2.1
Performance of the Reported Rotational Electromagnetic Energy Harvesters

References	Volume (cm^3)	Rotational Speed (krpm)	Open Circuit Voltage (V_{rms})	Normalized Voltage (mV_{rms}/krpm)	Maximum Power (W)	Power Density (W/cm^3)
Pan et al. [10]	0.033	2.2	0.079	36	4.1×10^{-4}	1.2×10^{-2}
Homes et al. [21]	0.041	30	0.42	14	1.1×10^{-3}	2.7×10^{-2}
Arnold et al. [22]	0.146	120	0.50	4.2	1.3	8.9
Senesky et al. [35]	1.0	13.3	1.86	140	3.8×10^{-4}	3.8×10^{-4}
Raisigel et al. [37]	0.023	380	4.18	11	5.0	220
Arnold et al. [41]	0.136	305	1.60	5.4	8.0	59
Herranit et al. [43]	0.0066	392	0.051	0.13	3.6×10^{-3}	0.55

resonance (i.e., the input vibration frequency matches the mechanical resonant frequency). In comparison to rotational energy harvesters, vibrational devices operate at lower frequencies and thus have lower power densities. To incorporate the magnet and coil into a linear or nonlinear mass-spring-damper system, there are a wide range of configurations reported. Basically, the spring configurations of the vibrational energy harvesters can be grouped into four categories: membrane or planar springs, cantilever beam springs, multiple spring systems, and nonlinear or nonresonant springs.

2.3.2.1 Vibrational Energy Harvesters with Membrane or Planar Springs

The earliest microscale implementation of this type of magnetic energy harvesters was reported by Williams et al. from the University of Sheffield [4, 44–46]. As shown in Figure 2.12(a), the spring-mass configuration is formed by a millimeter-size bulk SmCo magnet attached to a polyimide circular membrane on the upper gallium arsenide (GaAs) wafer. The planar Au coil is integrated on the backside of the lower wafer using a lift off process. The two wafers were bonded together using silver epoxy to form the assembled energy harvester. The electromagnetic device with an overall size of around 5 mm × 5 mm × 1 mm generated a peak power of 0.3 μW for vibration amplitude of 0.5μm and frequency of 4.4 kHz [45].

Similar designs to these membrane-based harvesters have been demonstrated by other researchers. Huang et al. [47] reported a two-wafer device with electroplating copper coils and a magnet suspended by an electroplated nickel-iron spring. The device is targeted at generating power from human motion

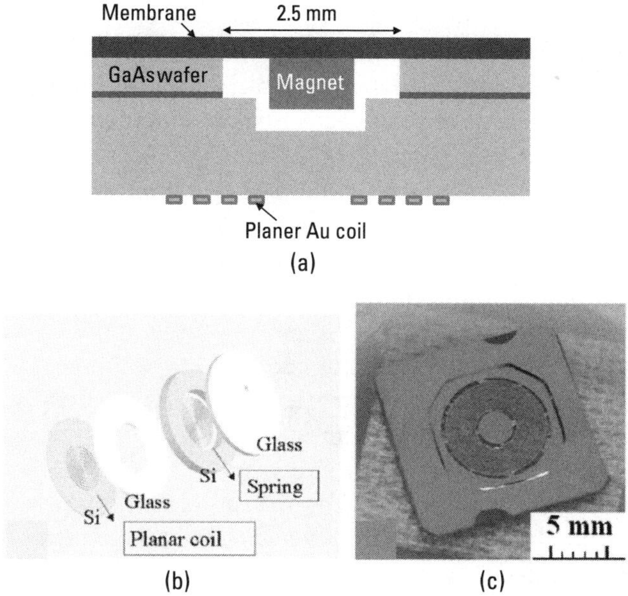

Figure 2.12 (a) Schematic drawing of the vibrational energy harvesters with membrane spring; and (b, c) fully microfabricated harvester with circular spiral spring by Pan et al. (Copyright: Elsevier, 2006, [55].)

and is reported to generate 0.16 μW from a finger tap. At the resonance of 100 Hz, the device produced 1.4-μW power at an excitation amplitude of 50 mm (acceleration of 19.7 m/s^2). A group from the University of Barcelona has described a similar construction by using a Kapton membrane, a PCB structure with Al coils, and a NdFeB magnet. Due to the poor assembly approach introducing large damping effects and poor electromagnetic coupling, the initial prototype generated only 0.2-W for 6.8-μm vibration amplitude at 360 Hz [48]. Later on, the improved version of the device generated about 55 μW at excitation frequency of 38 Hz and amplitude of 5 μm [49].

A research group from the Chinese University of Hong Kong developed a magnet-spring configuration composed of a NdFeB magnet suspended by a laser-micromachined copper spring structure [50–54]. The wire-wound coil is fixed on the housing of the structure. The device has an overall volume of around 1 cm^3 with a relatively low resonant frequency of 64 Hz. Several variations have been reported and a maximum power output of 830 μW was achieved by Ching et al. [53] at the vibration amplitude of 200 μm at its third mode (resonance frequency of 110 Hz). The device was also demonstrated to integrate with a power management circuit into an AA-battery-sized form factor [54].

Pan et al. [55] from National Sun Yat-Sen University reported a fully microfabricated structure using electroplated FePt permanent magnets and a spiral silicon spring structure. As shown in Figure 2.12(b), the device was similar to the spiral-spring design developed by Ching et al. [53], but possessed an advancement towards miniaturization. A peak voltage of 40 mV and an output power of 100 μW were obtained at 60 Hz (vibration amplitude not reported). Similar designs have been reported by other researchers in recent years, and their common spring-mass configuration can be summarized as the magnetic mass suspended by a spring structure or membrane moves relative to the stationary wound coil [17, 56–58]. The main problems for these devices are the vulnerable assembly process between the bulk magnet and the fragile spring.

Other than the structures with a moving magnet and a stationary coil, researchers also adopted the opposite constructing way for the electromagnetic energy harvesters. Based on the relative movement of a planar inductor with respect to a stationary permanent magnet, Sardini et al. [59] developed a variety of efficient electromagnetic energy harvesters. The devices used different polymeric materials as spiral vibrational springs, which have low resonant frequencies from about 100 Hz to about 30–40 Hz due to the low Young's modulus. Measurements showed that the maximum instantaneous voltage and power generated were 183 mV and 290 μW, respectively, for the linear system at 102 Hz, while the maximum output of 378-mV voltage and 153-μW power were produced by the nonlinear system at 41 Hz. Liu et al. [19, 20] developed multifrequency energy harvesters based on a similar configuration. The device consists of a movable circular mass patterned with double-layer Al coils and a permanent magnet attached on a supporting beam. Due to the circular-ring structure, the device exhibits three resonant modes along 3-D directions.

2.3.2.2 Vibrational Energy Harvesters with Cantilever Beam Springs

A research team from the University of Southampton demonstrated a cantilever spring-based harvester design [60, 61]. As seen in Figure 2.13(a), the cantilever beam is clamped at one end and a pair of NdFeB magnets is attached to a U-shaped core at the free end of the beam. The enameled copper coil of a 0.2-mm diameter and 27 turns is fixed in between the poles of the magnets. The device with a size of 240 mm^3 was found to generate power of 0.53 mW for a vibration amplitude of 13 μm at 320 Hz [60]. Glynne-Jones et al. [61] further developed a similar prototype with four magnets and slightly larger volume of 840 mm^3. It delivered a maximum power of 4 mW and an averaged power of 157 μW when attached to an automobile engine. This cantilever-based, four-magnet configuration was miniaturized by Beeby et al. [62], utilizing discrete magnets, conventionally wound coil, and machined components. As seen in Figure 2.13(b), the device with a small volume of 0.15 cm^3 arranged four magnets on an etched cantilever with the wound coil located within the moving magnetic field. The coil properties and magnet size were optimized and the final device with the coil of 2,300 turns produced the maximum power of 46 μW and volt-

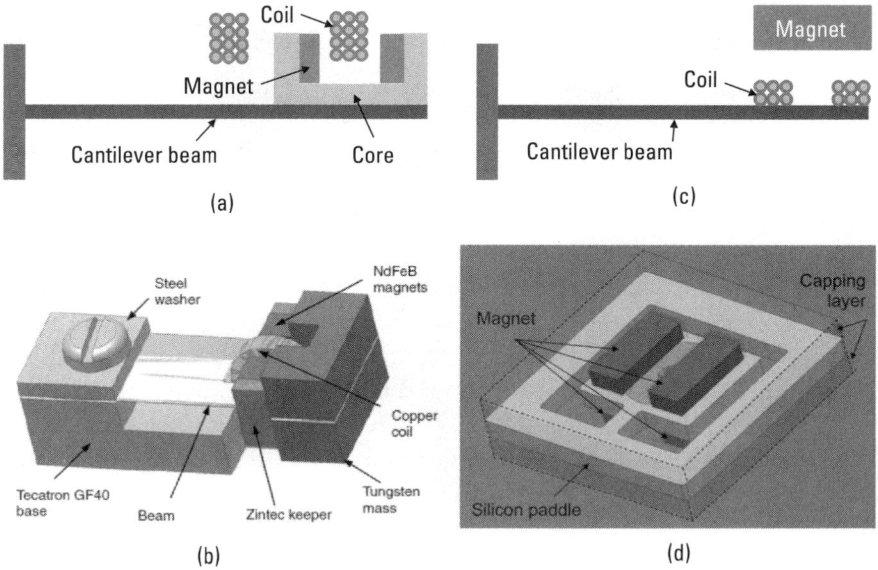

Figure 2.13 Schematic drawing and prototype of cantilever beam based vibrational energy harvesters with (a, b) moving magnets and (c, d) moving coils. (Copyright: IOP Publishing, [62], Copyright: Elsevier, 2008, [69].)

age of 428 mVrms at its resonant frequency of 52 Hz from just 0.59 m/s^2 acceleration. Further improvements by Torah et al. [63], including the increment of the magnet size and rearrangement of the center of gravity, increased the power output to 58 μW at the same excitation level. Beeby et al. [64] also developed a macroscale harvester with a similar configuration, called the Perpetuum PMG7. It was designed to resonate at 50 Hz and produced about 3-mW ac power at 0.5 m/s^2 RMS acceleration. The active mass of the generator is 85g and its volume is 41.3 cm^3. The cantilever spring with permanent magnet attached was also employed by Bouendeu et al. [65] and Peralta et al. [66]. The former harvester was equipped with four-wire wound microcoils together that were able to supply a 500 Ω load with 422 μW of electrical power. The latter one realized the fabrication of a microharvester with a 3-D pyramidal high density coil.

The research group from the University of Southampton used an alternative method to integrate microcoils with the vibrational cantilever beam, while leaving the magnet stationary as seen in Figure 2.13(c). Beeby et al. [67] and Koukharenko [68, 69] attempted to assemble a traditionally wound coil within a micromachined silicon cantilever. The microcantilever supporting a paddle contained a DRIE circular recess to accommodate a 600-turn coil wound with 25-μm-thick enameled copper wire. The silicon coil layer was sandwiched between two capping chips that support four NdFeB magnets as shown in Figure 2.13(d). The cantilever and coil structure was designed to vibrate laterally in the plane of the paddle with a resonant frequency of 9.5 kHz. An optimized power of 122 nW was delivered to 100 Ω load at the acceleration of 0.4g. The low power output is due to the high-damping levels of the microstructure. Kulkarni et al. [70] simulated the performance of the harvesters with various electroplated magnet configurations. Theoretically, a 250-turn coil vibrating at 7.4 kHz, moving in the magnetic field with amplitude of 240 μm could produce 950 mV and 85 μW. However, this analysis ignores parasitic damping and assumes high-quality electroplated magnetic films with a flux density of 1.6T.

Later on, Kulkarni et al. [71] reported and compared three different prototypes by partial microfabrication and assembly. Prototype A used a wire-wound copper coil and prototype B used an electroplated copper coil both on a DRIE silicon cantilever paddle. Prototype C used moving NdFeB magnets in between two microfabrication coils. A standard MEMS process has been conducted to fabricate the integrated coil and paddle. For prototype A, the maximum output power was 148 nW at resonant frequency of 8.08 kHz and acceleration of 3.9 m/s^2, while prototype B gave a maximum power of 23 nW for an acceleration of 9.8 m/s^2 at its resonant frequency of 9.83 kHz. With the lowest volume of 0.1 cm^3, the power generation is actually a substantial improvement in the microfabricated silicon-based harvesters reported in literature. However, the parasitic damping is still too large in case of silicon-based harvesters. Therefore, most of the input power is used in keeping the beam vibrating, leaving a small portion for electrical power generation. The output performance using an integrated coil was improved by prototype C. It generated a maximum power of 586 nW measured across a 110 Ω load resistance at a frequency of 60 Hz and acceleration of 8.829 m/s^2.

2.3.2.3 Vibrational Energy Harvesters with In-Plane Folded Springs

In addition to the aforementioned spring configurations adopted by the vibrational harvesters, there is another category of devices utilizing in-plane folded spring configurations as shown in Figure 2.14. Jiang et al. [28] described a MEMS-based electromagnetic energy harvester, which is fabricated by bonding an in-plane vibrator with embedded micromagnets and a substrate with integrated microcoils, as seen in Figure 2.14(b). The micromagnets are formed by using a sputtering deposition of NdFeB/Ta multilayered magnetic films with a thickness of 10 μm and silicon molding techniques. The high-aspect-ratio folded microsprings are patterned and formed by photolithography and DRIE processes, and the serially connected microcoils with multiple units are fabricated by electroplating process. The maximum voltage of the device at resonance of 115 Hz is approximately 2 mV, which corresponds to a power density of only 1.2 nW/cm^3. Zhu et al. [72] at the University of Southampton presented a similar planar electromagnetic energy harvester integrated with a Halbach array. Halbach array is a special arrangement of permanent magnets that doubles the

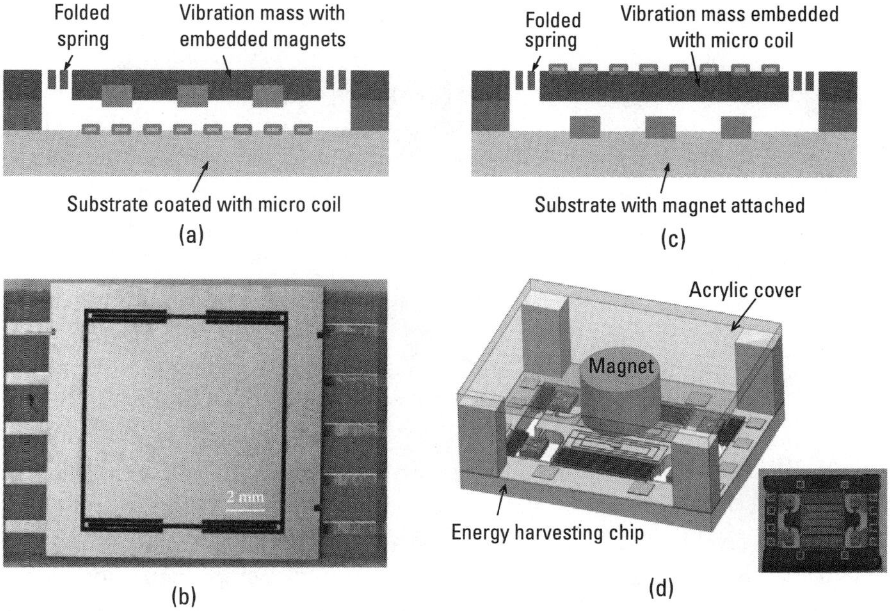

Figure 2.14 Schematic drawing and prototype of in-plane vibrational energy harvesters with (a, b) a moving magnet (Copyright: IOP Publishing, [28]), and (c, d) a moving coil.

magnetic field on one side of the array while cancelling the field of the other side to near zero. This arrangement can improve the electromagnetic coupling in a limited space. The energy harvester has a resonant frequency of 44.9 Hz and generated an average power of over 120 μW when excited at 0.3g (1g = 9.8m·s^{-2}). On this basis, Roundy et al. [73] reported a thinner planar energy harvester implemented with a multipole magnetic sheet. The single magnet with multipole reduced the assembly costs compared to an arrangement of multiple magnets. Furthermore, the planar coils are implemented in a standard low-cost printed circuit board (PCB) process.

The above reviewed in-plane vibrational energy harvesters utilize permanent magnet as the proof mass attached on the folded spring, while the devices are able to generate electricity due to the net displacement between the moving magnets and stationary coils. However, this architecture is kind of difficult to be achieved in the microscale due to the integration issue. A counterpart approach is to assemble the bulk magnets to housing while fabricate the MEMS spring structure with moving microcoils as shown in Figure 2.14(c). Figure 2.14(d) shows an in-plane electromagnetic harvester device proposed by Liu et al. [74]. It consisted of a MEMS suspension structure patterned with microcoils and a cylindrical magnet attached on an acrylic cover. By using folded springs and large proof mass, the EM-EH device is capable of achieving a relatively low operating frequency of 82 Hz.

Lastly, several commercial products have utilized vibrational magnetic energy harvester schemes. Perpetuum Ltd. (www.perpetuum.co.uk) is a United Kingdom-based spin-off from the University of Southampton. They developed the PMG17 harvesters for industrial condition monitoring applications, which are capable of delivering 3.5-mW power from acceleration of just 0.1g, or 40 mW at 1g, at frequencies of 100 or 120 Hz [75]. A competing product by Ferro Solution in the United States can produce 9.3-mW power for 0.1g vibrations at 21 Hz [76].

Table 2.2 summarizes the output performance of these vibrational devices. Power density (W/cm^3) is calculated from the power generation (W) divided by the active volume (cm^3). It is found that the two commercial electromagnet harvester devices provide the maximum power under common environmental vibration scenarios (100 Hz, 1.4g [75] and 21 Hz, 0.1g [76]). However, they are considerably large in size. Furthermore, the maximum power density of more than 2 mW/cm^3 is possible to be achieved [60], while at a relatively high acceleration level of 10g. Additionally, the vibrational designs with permanent magnet as the moving mass yield significantly high power than those without the magnetic mass, which indicates a design preference for electromagnetic

Table 2.2

Performance of the Reported Vibrational Electromagnetic Energy Harvesters

References	Volume (cm^3)	Vibration Frequency (Hz)	Vibration Acceleration (g)	Open Circuit Voltage (V_{rms})	Maximum Power	Power Density (W/cm^3)
Shearwood et al. [45]	0.025	4400	39	—	3×10^{-7}	1.2×10^{-5}
Williams et al. [53]	1	110	9.7	2.2 (pk)	8.3×10^{-4}	8.3×10^{-4}
Pan et al. [55]	0.45	60	—	0.04 (pk)	1.0×10^{-4}	2.2×10^{-4}
El-hami et al. [60]	0.24	322	10	0.013	5.3×10^{-4}	2.2×10^{-3}
Glynne-Jones et al. [61]	0.84	322	5.4	0.009	3.7×10^{-5}	4.4×10^{-5}
Beeby et al. [64]	0.06	357	0.43	0.03	2.85×10^{-6}	4.8×10^{-5}
Koukarenko et al. [68]	0.1	1600	0.4	—	1.0×10^{-7}	1.0×10^{-6}
Perpetuum [75]	130	100	1.4	15.6	4.0×10^{-2}	3.1×10^{-4}
Ferro Solutions [76]	75	21	0.1	—	9.3×10^{-3}	1.2×10^{-4}
Kulah et al. [87]	~2.3	1	—	0.006 (pk)	4.0×10^{-6}	1.7×10^{-6}

energy harvesters. Nevertheless, this is only applicable to those devices with relatively large volume. For those require small volumes, the design including microfabricated cantilevers patterned with microcoil and assembled stationary magnet would be a prior consideration.

2.3.3 Nonlinear/Nonresonant Energy Harvesters

The resonant-based harvesters achieve their maximum power generation at only one particular frequency and thus cannot easily track a time-varying vibration frequency. Therefore, nonlinear or nonresonant energy harvesters with particular spring-mass configurations have been developed, for harvesting vibration energy with wider frequency spectrum or higher power efficiency.

First, it can be accomplished using an eccentric rotor, which will rotate under the forced acceleration of the pivot point. As demonstrated in Figure 2.10(c), these devices respond to vibrational source, but generate power via rotational motion of the eccentric rotor. In 1988, Seiko in Japan introduced a well-established nonresonant harvester to power wristwatches, called the Automatic Generating System (AGS) [77–79]. The human motion induces the rotation of the eccentric rotor, and the rotation is speeded-up by a gear train to spin a small electromagnetic harvester. The device can produce 5–10-μW power on average when the watch is worn and 1 mW when the watch is forcibly shaken. The ETA company also supported an electromagnetic generator for wristwatch coined the Autoquartz [80, 81]. The device contains a proof mass and winds up a spring that pulses a microgenerator at its optimal rate of 15,000 rotations per minute for 50 ms, yielding 6-mA bursts at greater than 16V that are integrated on a storage capacitor. Spreemann et al. [82] reported a similar nonresonant harvester that is made use of the rotation of an eccentric rotor (Figure 2.15). When the pendulum rotates the embedded magnets, a change of magnetic flux in the stator coils will be induced and therefore produce an output voltage according to Faraday's law. A 1.5 cm^3 volume prototype is capable of generating 0.4–3 mW for vibration frequencies ranging from 30 to 80 Hz.

Second, researchers also attempted to directly utilize magnetic repulsive force for nonlinear springs [83–85]. Such a prototype generally consists of two permanent magnets fixed at the two ends of a tube. A magnetic mass is suspended in the middle of the tube by the magnetic repulsive forces from the top and bottom magnets. They are applicable to harvest energy from vibrations of low-frequency and large amplitude, such as hand-shaking, walking, or running [86].

Additionally, frequency upconversion (FUC)-based electromagnetic energy harvesters were initially presented by Kulah and Najafi [87] and later on were investigated and improved by Galchev et al. [88, 89]. The device normally contains at least two spring-mass systems with one working at low frequency

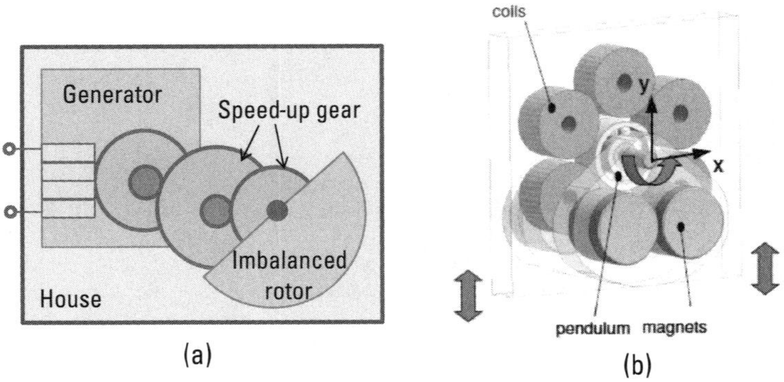

Figure 2.15 Schematic and prototype of the energy harvesters with eccentric rotor. (Copyright: IOP Publishing, [82].)

and the others at high frequency. Such a spring configuration was initially proposed for upconverting low-frequency environmental vibrations into a high-frequency self-oscillation of the system and hence providing more efficient energy generation at low frequencies. Besides the aforementioned particular spring-mass configurations, there are indeed a large number of related electromagnetic energy harvesters reported in the literature, including nonresonant, wideband, multifrequency, nonlinear, and FUC. Since the main topic of this chapter is the traditional microscale electromagnetic harvesters, those advanced architectures of nonlinear/nonresonant energy harvesters will be covered and discussed in Chapter 5.

References

[1] Priya, S., and D. J. Inman, *Energy Harvesting Technologies*, New York: Springer, 2009.

[2] Beeby, S., and N. White, *Energy Harvesting for Autonomous Systems*, Norwood, MA: Artech House, 2010.

[3] Spreemann, D., and Y. Manoli, *Electromagnetic Vibration Energy Harvesting Devices*, New York: Springer, 2012.

[4] Williams, C. B., and R. B. Yates, "Analysis of a Micro-Electric Generator for Microsystems," *Sensor. Actuat. A: Phys.*, Vol. 52, No. 1-3, 1996, pp. 8–11.

[5] Roundy, S. J., "Energy Scavenging for Wireless Sensor Nodes with a Focus on Vibration to Electricity Conversion," Ph.D. thesis, University of California, Berkley, 2003.

[6] Mitcheson, P. D., et al., "Architectures for Vibration-Driven Micropower Generators," *IEEE/ASME J. Microelectromech. Syst.*, Vol. 13, No. 3, 2004, pp. 429–440.

[7] Mitcheson, P. D., "Analysis and Optimisation of Energy-Harvesting Micro-Generator Systems," Ph.D. thesis, University of London, London, 2005.

[8] McLyman, W. T., *Transformer and Inductor Design Handbook*, 2nd ed., New York: Marcel Dekker, 1988.

[9] Galchev, T. V., et al., "Harvesting Traffic-Induced Vibrations for Structural Health Monitoring Of Bridges," *J. Micromech. Microeng.*, Vol. 21, No. 10, 2011, p. 104005.

[10] Pan, C. T., and T. T. Wu, "Development of a Rotary Electromagnetic Microgenerator," *J. Micromech. Microeng.*, Vol. 17, No. 1, 2007, pp. 120–128.

[11] Bowers, B. J., and D. P. Arnold, "Spherical, Rolling Magnet Generators for Passive Energy Harvesting from Human Motion," *J. Micromech. Microeng.*, Vol. 19, No. 9, 2009, p. 094008.

[12] Puers, R., G. Vandevoorde, and D. De Bruyker, "Electrodeposited Copper Inductors for Intraocular Pressure Telemetry," *J. Micromech. Microeng.*, Vol. 10, No. 2, 2000, pp. 124–129.

[13] Seidemann, V., and S. Buettgenbach, "Fabrication Technology for Closely Coupled Micro Coils with Integrated Flux Guidance and Their Application to Proximity and Magnetoelastic Force Sensors," *IEEE Sensors J.*, Vol. 3, No. 5, 2003, pp. 615–621.

[14] Lagorce, L. K., O. Brand, and M. G. Allen, "Magnetic Microactuators Based on Polymer Magnets," *J. Microelectromech. Syst.*, Vol. 8, No. 1, 1999, pp. 2–9.

[15] Watanabe, Y., et al., "A New Fabrication Process of a Planar Coil Using Photosensitive Polyimide and Electroplating," *Sensor. Actuat. A: Phys.*, Vol. 54, 1996, pp. 733–738.

[16] Brunet, M., et al., "Thick Photoresist Development for the Fabrication of High Aspect Ratio Magnetic Coils," *J. Micromech. Microeng.*, Vol. 12, No. 4, 2002, pp. 444–449.

[17] Wang, P. -H., et al., "Design, Fabrication and Performance of a New Vibration-Based Electromagnetic Micro Power Generator," *Microelectron. J.*, Vol. 38, No. 12, 2007, pp. 1175–1180.

[18] Sari, I., T. Balkan, and H. Kulah, "An Electromagnetic Micro Power Generator for Wideband Environmental Vibrations," *Sensor. Actuat. A: Phys.*, Vol. 145-146, 2008, pp. 405–413.

[19] Liu, H., et al., "Feasibility Study of a 3D Vibration-Driven Electromagnetic MEMS Energy Harvester with Multiple Vibration Modes," *J. Micromech. Microeng.*, Vol. 22, No. 12, 2012, p. 125020.

[20] Liu, H., Y. Qian, and C. Lee, "A Multi-Frequency Vibration-Based MEMS Electromagnetic Energy Harvesting Device," *Sensor. Actuat. A: Phys.*, Vol. 204, 2013, pp. 37–43.

[21] Holmes, A. S., G. D. Hong, and K. R. Pullen, "Axial-Flux Permanent Magnet Machines for Micropower Generation," *J. Micromech. Microeng.*, Vol. 14, No. 1, 2005, pp. 54–62.

[22] Arnold, D. P., et al., "Microfabricated High-Speed Axial-Flux Multiwatt Permanent-Magnet Generators- Part II: Design, Fabrication, and Testing," *J. Micorelectromech. S.*, Vol. 15, No. 5, 2006, pp. 1351–1363.

[23] Herrault, F., et al., "Fabrication and Performance of Silicon-Embedded Permanent-Magnet Microgenerators," *J. Micorelectromech. S.*, Vol. 19, No. 1, 2010, pp. 4–13.

[24] Zhang, T., et al., "Permanent-Magnet Longitudinal Fields for Magnetostrictive Devices," *J. Appl. Phys.*, Vol. 101, No. 3, 2007, p. 034511.

[25] Mizuno, M., and D. G. Chetwynd, "Investigation of a Resonance Microgenerator," *J. Micromech. Microeng.*, Vol. 13, 2003, pp. 209–216.

[26] Gilles, P. -A., et al., "Design of a Permanent Magnet Planar Synchronous Micromotor," *Proc. 35th Mtg. IEEE Ind. Appl. Soc.*, Vol. 1, 2000, pp. 223–227.

[27] Yang, C., "An Axial Flux Electromagnetic Micromotor," *J. Micromech. Microeng.*, Vol. 11, 2001, pp. 113–117.

[28] Jiang, Y., et al., "Fabrication of a Vibration-Driven Electromagnetic Energy Harvester with Integrated NdFeB/Ta Multilayered Micro-Magnets," *J. Micromech. Microeng.*, Vol. 21, No. 9, 2011, p. 095014.

[29] Zhang, Q., et al., "Microelectromagnetic Energy Harvester with Integrated Magnets," *Proceedings of the 16th International Solid-State Sensors, Actuators and Microsystems Conference (IEEE Transducers 2011)*, 2011, pp. 661–664.

[30] Ng, W. B., A. Takado, and K. Okada, "Electrodeposited CoNiReWP Thick Array of High Vertical Magnetic Anisotropy," *IEEE Trans. on Magn.*, Vol. 41, No. 10, 2005, pp. 3886–3888.

[31] Arnold, D. P., "Review of Microscale Magnetic Power Generation," *IEEE Trans. on Magn.*, Vol. 43, No. 11, 2007, pp. 3940–3951.

[32] Beeby, S. P., M. J. Tudor, and N. M. White, "Energy Harvesting Vibration Sources for Microsystems Applications," *Mease. Sci. Technol.*, Vol. 17, 2006, pp. R175–R195.

[33] Mitcheson, P. D., et al., "Energy Harvesting from Human and Machine Motion for Wireless Electronic Devices," *Proceedings of the IEEE*, Vol. 96, No. 9, 2008, pp. 1457–1486.

[34] Wiegele, T. G., "Micro-Turbo-Generator Design and Fabrication: A Preliminary Study," *Proc. 31st Intersociety Energy Conversion Engineering Conf. (IECEM'96)*, 1996, pp. 2308–2313.

[35] Senesky, M. K., and S. R. Sanders, "A Millimeter-Scale Electric Generator," *Proc. 2004 IEEE Annu. Industry Applications Soc. Meeting*, 2004, pp. 346–352.

[36] Achotte, N., et al., "Planar Brushless Magnetic Micromotors," *J. Microelectromech. Syst.*, Vol. 15, No. 4, 2006, pp. 1001–1014.

[37] Raisigel, H., O. Cugat, and J. Delamare, "Permanent Magnet Planar Micro-Generators," *Sensor. Actuat. A: Phys.*, Vol. 130–131, 2006, pp. 438–444.

[38] Gilles, P. -A., J. Delamare, and C. Cugat, "Rotor for a Brushless Micromotor," *J. Magn. Magn. Mater.*, Vol. 242–245, 2002, pp. 1186–1189.

[39] Das, S., et al., "Microfabricated High-Speed Axial-Flux Multiwatt Permanent-Magnet Generators—Part I: Modeling," *J. Microelectromech. Syst.*, Vol. 15, No. 5, October 2006, pp. 1330–1350.

[40] Zana, I., et al., "Magnetic Patterning of Permanent-Magnet Rotors for Microscale Motor/Generators," *Proc. 5th Int. Workshop on Micro and Nanotechnology for Power Generation and Energy Conversion Apps. (PowerMEMS 2005)*, November 2005, pp. 116–119.

[41] Arnold, D. P., et al., "Design Optimization of an 8-Watt, Microscale, Axial-Flux, Permanent-Magnet Generator," *J. Micromech. Microeng.*, Vol. 16, No. 9, September 2006, pp. S290–S296.

[42] Herrault, F., et al., "Multiwatt, Microscale, Permanent-Magnet Generators Operating at High Temperatures," *IEEE International Magnetics Conference (INTERMAG)*, 2006.

[43] Herrault, F., et al., "Ultraminiaturized Milliwatt-Scale Permanent Magnet Generators," *Proc. 14th Int. Conf. Solid-State Sensors, Actuators, and Microsystems (Transducers'07)*, 2007, pp. 899–902.

[44] Williams, C. B., and R. B. Yates, "Analysis of a Micro-Electric Generator for Microsystems," *Proc. Transducers'95/Eurosensors IX*, Vol. 1, June 1995, pp. 369–372.

[45] Shearwood, C., and R. B. Yates, "Development of an Electromagnetic Microgenerator," *Electron. Lett.*, Vol. 33, No. 22, October 1997, pp. 1883–1884.

[46] Williams, C. B., et al., "Development of an Electromagnetic Micro-Generator," *IEE Proc. Circuits, Devices Syst.*, Vol. 148, No. 6, December 2001, pp. 337–342.

[47] Huang, W. S., et al., "Design and Fabrication of a Vibrational Micro-Generator for Wearable MEMS," *Proceedings of Eurosensors XVII*, 2003, pp. 695–697.

[48] Serre, C., et al., "Vibrational Energy Scavenging with Si Technology Electromagnetic Inertial Microgenerators," *Microsyst. Technol.*, Vol. 13, No. 11-12, July 2007, pp. 1655–1661.

[49] Serre, C., et al., "Design and Implementation of Mechanical Resonators for Optimised Inertial Electromagnetic Generators," *Microsys. Technol.*, Vol. 14, 2008, pp. 653–658.

[50] Li, W. J., et al., "Infrared Signal Transmission by a Laser-Micromachined Vibration-Induced Power Generator," *Proc. 43rd IEEE Midwest Symp. Circuits and Systems*, 2000, pp. 236–239.

[51] Li, W. J., et al., "Dynamical Modeling and Simulation of a Laser-Micromachined Vibration-Based Micro Power Generator," *Int. J. Nonlin. Sci. Numer. Sim.*, Vol. 1, No. 5, 2000, pp. 345–353.

[52] Ching, N. N. H., et al., "PCB-Integrated Micro-Generator for Wireless Systems," *Proc. Int. Symp. Smart Structures and Microsystems*, Hong Kong, October 2000.

[53] Ching, N. N. H., et al., "A Laser-Micromachined Multi-Modal Resonating Power Transducer for Wireless Sensing Systems," *Sensor. Actuat. A: Phys.*, Vol. 97, April 2002, pp. 685–690.

[54] Lee, J. M. H., et al., "Development of an AA Size Energy Transducer with Micro Resonators," *Proc. IEEE Int. Symp. Circuits and Systems*, Vol. 4, 2003, pp. IV876–IV879.

[55] Pan, C. T., et al., "Fabrication and Analysis of a Magnetic Self-Power Microgenerator," *J. Magn. Magn. Mater.*, Vol. 304, No. 1, September 2006, pp. e394–e396.

[56] Khan, F., F. Sassani, and B. Stoeber, "Copper Foil-Type Vibration-Based Electromagnetic Energy Harvester," *J. Micromech. Microeng.*, Vol. 20, No. 12, 2010, p. 125006.

[57] Park, J. C., D. H. Bang, and J. Y. Park, "Microfabricated Electromagnetic Power Generator to Scavenge Low Ambient Vibration," *IEEE Trans. on Magn.*, Vol. 46, No. 6, 2010, pp. 1937–1942.

[58] Bouendeu, E., et al., "Design Synthesis of Electromagnetic Vibration-Driven Energy Generators Using a Variational Formulation," *J. Microelectromech. Syst.*, Vol. 20, No. 2, 2011, pp. 460–469.

[59] Sardini, E., and M. Serpelloni, "An Efficient Electromagnetic Power Harvesting Device for Low-Frequency Applications," *Sensor. Actuat. A: Phys.*, Vol. 172(2), 2011, pp. 475–482.

[60] El-Hami, M., et al., "Design and Fabrication of a New Vibration-Based Electromechanical Power Generator," *Sensor. Actuat. A: Phys.*, Vol. 92, No. 1–3, August 2001, pp. 335–342.

[61] Glynne-Jones, P., et al., "An Electromagnetic, Vibration-Powered Generator for Intelligent Sensor Systems," *Sensor. Actuat. A: Phys.*, Vol. 110, No. 1–3, February 2004, pp. 344–349.

[62] Beeby, S. P., et al., "A Micro Electromagnetic Generator for Vibration Energy Harvesting," *J. Micromech. Microeng.*, Vol. 17, No. 7, 2007, pp. 1257–1265.

[63] Torah, R. N., et al., "Energy Aware Wireless Microsystem Powered by Vibration Energy Harvesting," *Proceeding of Sensors*, 2007.

[64] Beeby, S. P., et al., "Experimental Comparison of Macro and Micro Scale Electromagnetic Vibration Powered Generators," *Microsyst. Technol.*, Vol. 13, No. 11-12, 2007, pp. 1647–1653.

[65] Bouendeu, E., et al., "A Low-Cost Electromagnetic Generator for Vibration Energy Harvesting," *IEEE Sensors J.*, Vol. 11, No. 1, 2011, pp. 107–113.

[66] Peralta, M., et al., "Analysis and Fabrication Steps for a 3D-Pyramidal High Density Coil Electromagnetic Micro-Generator for Energy Harvesting Applications," *Sensor. Actuat. A: Phys.*, Vol. 205, 2014, pp. 103–110.

[67] Beeby, S. P., et al., "Micromachined Silicon Generator for Harvesting Power from Vibrations," *Proceedings of Transducers 2005*, 2005, pp. 780–783.

[68] Koukharenko, E., et al., "Microelectromechanical Systems Vibration Powered Electromagnetic Generator for Wireless Sensor Applications," *Microsyst. Technol.*, Vol. 12, No. 10-11, 2006, pp. 1071–1077.

[69] Koukharenko, E., M. J. Tudor, and S. P. Beeby, "Performance Improvement of a Vibration-Powered Electromagnetic Generator by Reduced Silicon Surface Roughness," *Mater. Lett.*, Vol. 62, No. 4-5, 2008, pp. 651–654.

[70] Kulkarni, S., et al., "Vibration Based Electromagnetic Micropower Generator on Silicon," *J. Appl. Phys.*, Vol. 99, No. 8, 2006, p. 08P511.

[71] Kulkarni, S., et al., "Design, Fabrication and Test of Integrated Micro-Scale Vibration-Based Electromagnetic Generator," *Sensor. Actuat. A: Phys.*, Vol. 145-146, 2008, pp. 336–342.

[72] Zhu, D., et al., "A Planar Electromagnetic Vibration Energy Harvester with a Halbach Array," *Power MEMS 2011*, 2011.

[73] Roundy, S., and E. Takahashi, "A Planar Electromagnetic Energy Harvesting Transducer Using a Multi-Pole Magnetic Plate," *Sensor. Actuat. A: Phys.*, Vol. 195, 2013, pp. 98–104.

[74] Liu, H., et al., "An In-Plane Approximated Nonlinear MEMS Electromagnetic Energy Harvester," *J. Microelectromech. Syst.*, No. 99, 2013.

[75] Perpetuum PMG17-100 Data Sheet, http://www.perpetuum.co.uk/.

[76] Ferro Solutions Energy Harvester Data Sheet, http://www.ferrosi.com.

[77] Kayakawa, M., "Electronic Wristwatch with Generator," U.S. Patent 5001685, March 19, 1991.

[78] Kitahara, J., "Development of Small AGS (Automatic Power-Generator for Wrist-Watch)," (in Japanese) *Horolog. Inst. Jpn.*, No. 157, 1996, pp. 33–42.

[79] Matsuzawa, K., and M. Saka, "Seiko Human Powered Quartz Watch," *Proc. Prospector IX: Human Powered Systems Technologies*, Auburn, AL: Space Powered Institute, Auburn University, 1997, pp. 359–384.

[80] Paradiso, J. A., and T. Starner, "Energy Scavenging for mobile and Wireless Electronics," *Pervasive Comput.*, Vol. 4, No. 1, January-March 2005, pp. 18–27.

[81] Kinetron MGS Watch Data Sheet, http://www. kinetron.nl/.

[82] Spreemann, D., et al., "Non-Resonant Vibration Conversion," *J. Micromech. Microeng.*, Vol. 16, No. 9, 2006, pp. S169–S173.

[83] Mann, B. P., and N. D. Sims, "Energy Harvesting from the Nonlinear Oscillations Of Magnetic Levitation," *J. Sound Vib.,* Vol. 319, No. 1-2, 2009, pp. 515–530.

[84] Foisal, A. R. M., C. Hong, and G.-S. Chung, "Multi-Frequency Electromagnetic Energy Harvester Using A Magnetic Spring Cantilever," *Sensor. Actuat. A: Phys.*, Vol. 182, 2012, pp. 106–113.

[85] Dallago, E., M. Marchesi, and G. Venchi, "Analytical Model of a Vibrating Electromagnetic Harvester Considering Nonlinear Effects," *IEEE Trans. on Power Electr.*, Vol. 25, No. 8, 2010, pp. 1989–1997.

[86] Saha, C. R., et al., "Electromagnetic Generator for Harvesting Energy from Human Motion," *Sensors and Actuators A: Physical*, Vol. 147, 2008, pp. 248–253.

[87] Kulah, H., and K. Najafi, "Energy Scavenging from Low-Frequency Vibrations by Using Frequency Up-Conversion for Wireless Sensor Applications," *IEEE Sensors J.*, Vol. 8, March-April 2008, pp. 261–268.

[88] Galchev, T., H. Kim, and K. Najafi, "Micro Power Generator for Harvesting Low-Frequency and Nonperiodic Vibrations," *J. Microelectromech. Syst.*, Vol. 20, No. 4, August 2011, pp. 852–866.

[89] Galchev, T. V., et al., "Harvesting Traffic-Induced Vibrations for Structural Health Monitoring Of Bridges," *J. Micromech. Microeng.*, Vol. 21, No. 10, 2011, p. 104005.

3

Electrostatic MEMS Vibration Energy Harvesting

With the quick development of handheld electronics from mobile phones, PDAs, and cameras to health-care devices, harvesting energy from ambient vibrations to power such devices will be a potential solution due to the limitation of the battery lifetime. Typically, there are three kinds of energy conversion mechanisms to collect electricity from vibration, namely piezoelectric, electromagnetic, and electrostatic mechanisms. The choice of a specific mechanism depends on the application environment and the operating conditions, such as the vibration amplitude and resonant frequency of the excitation sources. The energy electromechanical coupling coefficient of electromagnetic mechanism decreases 10 times faster than that of the electrostatic mechanism when the volume of the MEMS energy harvester decreases to 1% of original size [1]. Therefore, the electrostatic mechanism will be a good choice for meeting the requirement of miniaturization. Meanwhile, with the nature of standard MEMS and packaging techniques, capacitive-based electrostatic mechanism harvesters can be fabricated at a large production and low cost due to high integration with electronics. This chapter introduces the basic working principle of electrostatic mechanism harvester under charge-constrained and voltage-constrained energy conversion cycles. Then three major working mechanism energy harvesters will be covered and compared. The new developed electret-free and electret-based electrostatic harvester has been demonstrated in detail, including the theoretical design, structural optimization, and new materials development.

3.1 Basic Principles

The basis of electrostatic energy conversion is based on variable capacitor. Typically, there are two types of energy conversion cycles, namely charge-constrained and voltage-constrained conversion methods. When the charge through the capacitor is constrained, the voltage will increase with the decreasing of the capacitance. When the voltage across the capacitor is constrained, charge will move from the capacitor with the decreasing of the capacitance [2].

3.1.1 Charge-Constrained Energy Conversion Cycle

The charge-constrained conversion cycle in the QV diagram is shown in Figure 3.1(a). When the initially uncharged capacitor of the capacitance C_{max} is brought to an effective potential difference of V_{start} (from point A to point B), the energy that is stored in the system after charging can be expressed as

$$E_B = \frac{1}{2} C_{\max} V_{start}^2 \tag{3.1}$$

With an aid of a switch in the energy-harvesting circuits [2], the electrodes are electrically isolated. As a result, the charges on the electrodes are forced to keep constant from point B to point C. Meanwhile, in this process because the capacitance decreases to a minimum value of C_{min}, the potential difference increases from V_{start} to V_{max}. When the capacitance arrives at C_{min}, the electrical energy stored in the system can be calculated as

$$E_C = \frac{1}{2} C_{\min} V_{\max}^2 \tag{3.2}$$

The last step in the energy conversion cycle is the discharging of the charge on the variable capacitor back into the charge reservoir, which is shown by the path from point C to point A in Figure 3.1(a). The total converted energy from mechanical to electrical energy in one conversion cycle is

$$E_{conv-charge} = E_C - E_B = \frac{1}{2} C_{\min} V_{\max}^2 - \frac{1}{2} C_{\max} V_{start}^2 \tag{3.3}$$

Because the capacitor is charge-constrained, the charges at points B and C are the same according to

$$Q_{\max} = C_{\min} V_{\max} = C_{\max} V_{start} \tag{3.4}$$

The converted energy in one energy cycle can be rewritten as

$$E_{conv} = \frac{1}{2}V_{start}^2 \frac{C_{\max}}{C_{\min}}(C_{\max} - C_{\min}) = \frac{1}{2}V_{start}V_{\max}\Delta C \tag{3.5}$$

This quantity is equal to the shaded area inside in the cycle in Figure 3.1(a). This converted energy will eventually be reduced due to any losses incurred in the power electronics.

3.1.2 Voltage-Constrained Energy Conversion Cycle

Figure 3.1(b) shows the voltage-constrained energy conversion cycle. From point A to point B, the capacitor is charged up to some high voltage V_{max} when the capacitance reaches C_{max}. The energy stored in the system can be expressed as

$$E_B = \frac{1}{2}V_{\max}Q_{\max} = \frac{1}{2}V_{\max}C_{\max}V_{\max} = \frac{1}{2}C_{\max}V_{\max}^2 \tag{3.6}$$

When the voltages on the electrodes are forced to keep constant from point B to point C, the capacitance decreases to a minimum value of C_{min} and the charge decreases from Q_{max} to Q_{min}. In this case, the capacitor will have to return the change in charge $Q_{max} - Q_{min}$ to the reservoir. To keep a constant voltage V_{max} between the reservoir and the capacitor, the capacitor will provide an amount of energy $(Q_{max} - Q_{min})V_{\max}$ for the reservoir, which results from this capacitance change forced by the mechanical source. When the capacitor is discharged into the reservoir, as shown from point C to point A in Figure 3.1(b), the electrical energy stored in the system can be calculated as

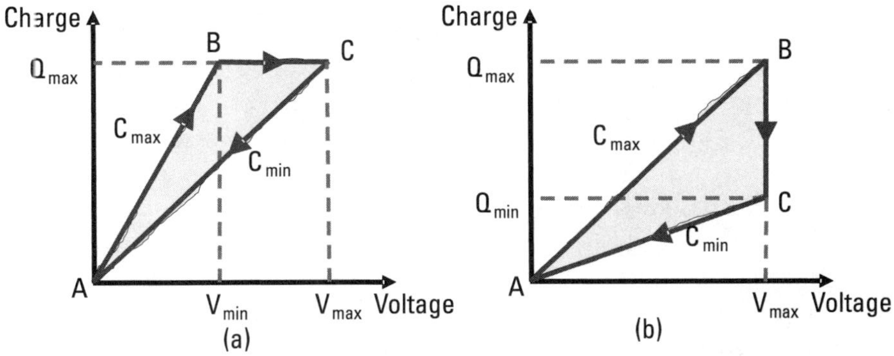

Figure 3.1 (a) The charge-constrained energy conversion cycle; (b) the voltage-constrained energy conversion cycle.

$$E_C = \frac{1}{2} V_{\max} Q_{\min} = \frac{1}{2} C_{\min} V_{\max}^2 \tag{3.7}$$

At the end of the cycle, the total converted energy from mechanical to electrical energy in one conversion cycle is

$$\begin{aligned} E_{conv-voltage} &= E_C + \left(Q_{\max} - Q_{\min}\right) V_{\max} - E_B \\ &= \frac{1}{2} C_{\min} V_{\max}^2 + \left(C_{\max} V_{\max} - C_{\min} V_{\max}\right) V_{\max} - \frac{1}{2} C_{\max} V_{\max}^2 \\ &= \frac{1}{2} \Delta C V_{\max}^2 \end{aligned} \tag{3.8}$$

This value is equal to the shaded area inside in the cycle in Figure 3.1(b). This converted energy will eventually be reduced due to any losses incurred in the power electronics. For the same values of C_{max} and C_{min}, a voltage-constrained cycle will be the one to convert the most amount of energy [3]. However, by incorporating a capacitor in parallel with the energy harvesting capacitor, the energy from the charge constrained system can approach that of the voltage-constrained system as the parallel capacitance approaches infinity [4]. Meanwhile, the voltage-constrained cycle requires a reservoir at voltage V_{max}, whereas the charge-constrained cycle only needs to be charged to V_{min}. This maximum voltage V_{max} is realized by virtue of charge conservation. The condition will be limited in very-low-power applications due to the low voltage levels typically [5].

3.1.3 Comparisons Among Three Major Electrostatic Mechanisms

Electrostatic energy harvesters employ either parallel-plate or comb finger electrodes as variable capacitors, which are biased with external voltage sources or charged by an electrets. The capacitor electrodes are separated or closed due to external vibration such that the variable capacitor will be a function of ambient vibrations. Normally there are three different types of electrostatic energy harvesters (i.e., in-plane overlap varying type, in-plane gap closing type, and out-of-plane gap closing gap), as illustrated in Figure 3.2. To optimize the effect of device configurations on energy output, the output energy density that is obtained by normalizing the output energy with the device volume is analyzed and compared in terms of three mechanisms [6]. The charge-constrained method is more popular over the voltage-constrained cycle as it requires just one external charge reservoir instead of two [7]. For this reason, the performance comparisons are performed and assumed in the case of the charge-constrained energy conversion cycle.

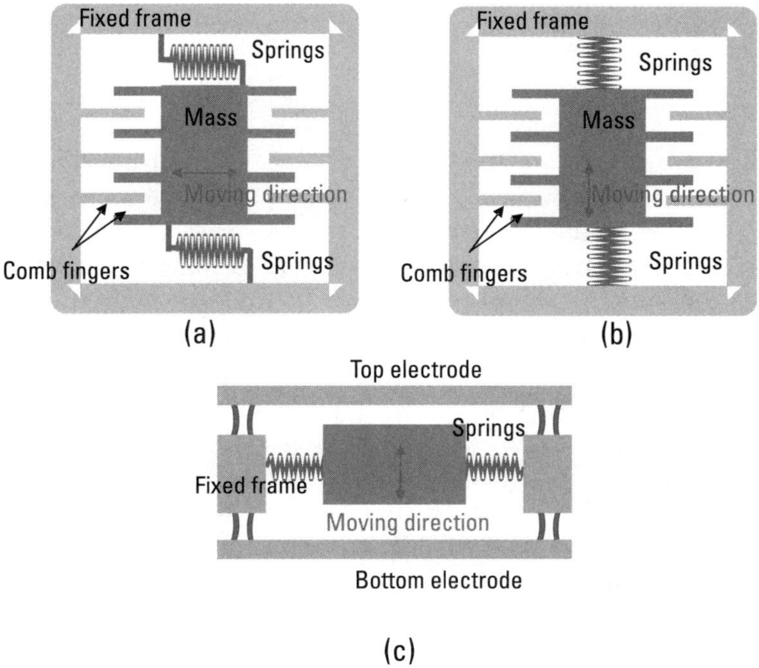

Figure 3.2 The three different electrostatic mechanisms: (a) in-plane overlap harvesters; (b) in-plane gap-closing harvesters; and (c) out-of-plane gap-closing harvesters.

The maximum and minimum capacitances under three electrostatic mechanisms are expressed as

$$C_{\max,inplaneoverlap} = 2\left(\frac{g_i}{\varepsilon_0 A} + \frac{2t}{\varepsilon_{Si_3N_4}\varepsilon_o A_i}\right) \tag{3.9}$$

$$C_{\min,inplaneoverlap} = 0 \tag{3.10}$$

$$C_{\max,inplanegapclose} = \frac{\varepsilon_o \varepsilon_{Si_3N_4} A_i}{2t} + \left(\frac{2g_i}{\varepsilon_o A_i} + \frac{2t}{\varepsilon_{Si_3N_4}\varepsilon_o A_i}\right)^{-1} \tag{3.11}$$

$$C_{\min,inplaneclose} = 2\left(\frac{g_i}{\varepsilon_0 A_i} + \frac{2t}{\varepsilon_{Si_3N_4}\varepsilon_o A_i}\right)^{-1} \tag{3.12}$$

$$C_{\max,outofplanegapclose} = \frac{\varepsilon_o \varepsilon_{Si_3N_4} A_o}{2t} \tag{3.13}$$

$$C_{\min,outofplanegapclose} = \left(\frac{2t}{\varepsilon_o \varepsilon_{Si_3N_4} A_o} + \frac{2g_o}{\varepsilon_0 A_o} \right)^{-1} \tag{3.14}$$

where ε_0 is the permittivity of free space, $\varepsilon_{Si_3N_4}$ is the dielectric constant of silicon nitride, g_i is the distance between adjacent comb electrodes when the electrodes are at their equilibrium positions, t is the thickness of the silicon nitride coating and A_i is the area of overlap between adjacent fingers in one set of comb electrodes, g_o is the distance between adjacent electrodes when the electrodes are at their equilibrium positions, and A_o is the area of the electrodes.

If the in-plane overlap and in-plane gap-closing converters were to be compared, it can be observed that the C_{max} of the in-plane overlap converter is very similar to the C_{min} of the in-plane gap-closing structure. When these two harvesters have similar geometries for similar structural dimensions, it can be concluded from this comparison that the energy output density for the in-plane overlap harvester is very much less than that of the in-plane gap closing mechanism [6]. Hence, the energy output density between the in-plane and out-of-plane gap-closing mechanisms is compared and shown in Figure 3.3. It is observed that the best energy output density for the in-plane gap-closing mechanism is always higher than that of the out-of-plane gap-closing mechanism for all load volumes between 5 mm^3 and 50 mm^3. Additionally, the ratio of the best energy output for the in-plane gap-closing mechanism to the out-of-plane gap-closing mechanism is approximately consistent at 1.8 for load volumes between 5 mm^3 and 50 mm^3.

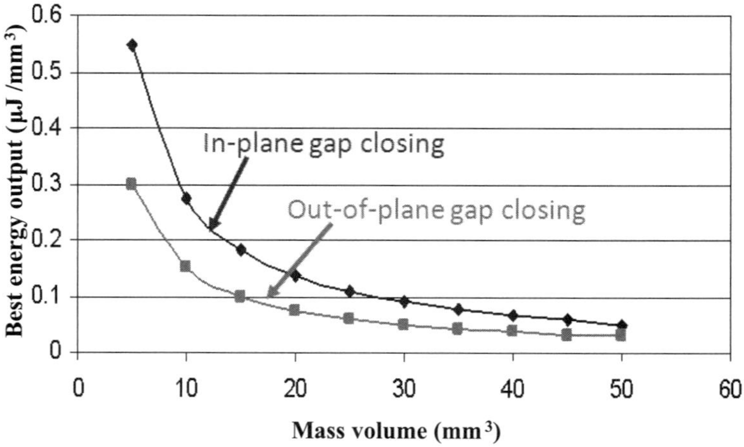

Figure 3.3 Curves of the optimum energy output density against the volume of movable mass [6].

3.2 Electret-Free Electrostatic Microharvesters

3.2.1 Three Major Mechanisms Harvesters

3.2.1.1 In-Plane Overlap Type Harvesters

The first electrostatic energy harvester based on the in-plane overlap was reported by Meninger et al. [7] at MIT in 2001. The output power of this harvester is simulated as 8 μW from a 2.5-kHz input motion. The obtained maximum energy can be easily be increased by moving to processes designed for higher voltage operation. In this structure the voltage is limited to 8V. Arakawa et al. at the University of Tokyo fabricated the prototype based on in-plane overlap, which is combined with a microthermal engine to improve output power. The harvester of volume of 800 mm^3 can deliver 6 μW when operating at 10 Hz [8]. This micromachining is not CMOS-compatible process due to the usage of CYTOP polymer. A 3,000 μm × 3,000 μm × 500 μm in-plane overlap harvester in charge-constrained cycle is fabricated based on the CMOS process [9]. The structural design and SEM photo are shown in Figure 3.4(a, b), respectively. The 2,000 μm × 2,200 μm × 450 μm (4.9 mg) vibrating mass is made of the CMOS handle layer so as to achieve a low-frequency operation. This device delivers 0.0924 μW under the excitation input of 10 μm at 105 Hz. To improve the integration for miniaturization and optimal operation, Ma et al. proposed a single-wafer floating-electrode energy harvester integrated with a micromechanical resonator and electronic device [10]. An insulated floating-electrode made of heavily doped polycrystalline silicon is deployed to replace permanently charge electrets typically for one the electrodes based on a standard CMOS floating-gate process. The floating electrode is charged by electron tunneling such as a conventional nonvolatile memory device. Figure 3.4(c, d) shows the view of the resonator over the fixed floating electrode and the SEM of a fabricated array harvester, respectively. An array of 20 devices produced a maximum power of 65 nW at 4.2 kHz at a displacement of 2.2 μm.

3.2.1.2 In-Plane Gap-Closing Type Harvesters

Despesse et al. reported an electrostatic structure of in-plane gap closing, which is fabricated by bonding and DRIE processes [11]. This structure with high electrical damping can scavenge the energy over a large spectrum below 100 Hz. A high electrical damping is realized by choosing the electrical stiffness close to the mechanical stiffness. A silicon device of a volume of 81 mm^2 × 0.4 mm with a 2 ×10^{-3} kg inertial mass excited is needed to obtain the output power of 70 μW under a vibration amplitude of 95 μm at 50 Hz. The device-sized reduction can obtained a higher capacity range and a higher-quality factor with a smaller system size. Chiu proposed a single silicon-on-insulator wafer process, as shown in Figure 3.5 [12, 13]. A 200-μm-thick device layer was employed for large capacitance. The initial and minimum finger gaps were 35 μm and 0.1

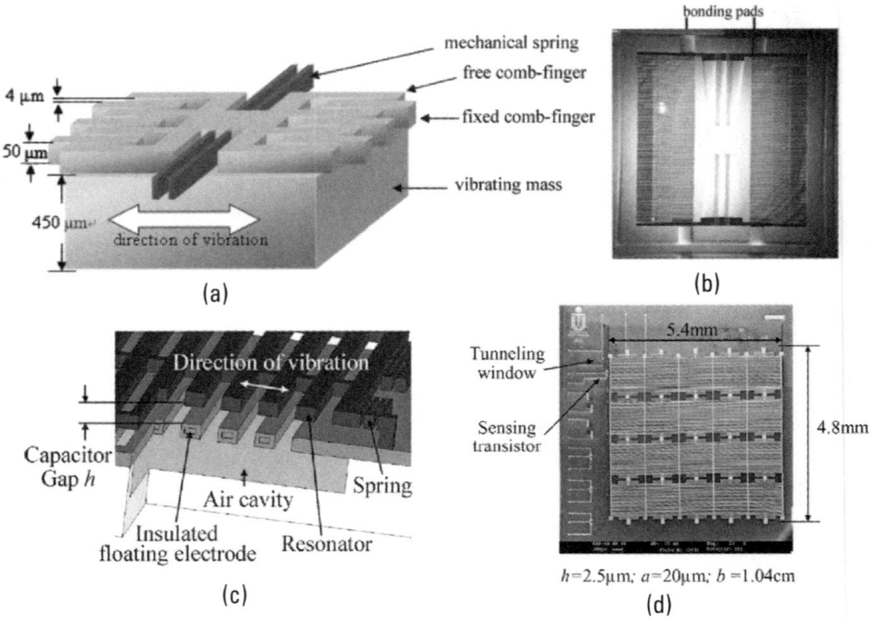

Figure 3.4 (a) The electrostatic comb-drive energy harvester; (b) SEM photo of the top view with bonding pad; (c) the integrated floating-electrode harvester; (d) SEM photo of a fabricated 5 × 4 array harvester [10].

μm, respectively. A steel ball was attached on the central plate for low resonant frequency to match the vibration source and improve the conversion efficiency. The device with area size of 1 cm^2 can generate the power output of 200 μW and output voltage of 40V at the input voltage of 3.3V and the optimal load of 8 MΩ.

3.2.1.3 Out-of-Plane Gap Closing Type Harvesters

Kuehne describes the design and analysis of a parallel-plate capacitive energy harvester [14]. This device is composed of a three-wafer sandwich structure. The upper glass wafer forms the capacitor gap and the counter electrode. The intermediate silicon wafer consists of the mechanical springs, the inertial mass, and the electrode. The lower glass wafer provides hermetic encapsulation and allows an operation mode under low-pressure conditions. By simulation, this harvester with a plate area of 36 mm^2 and a gap of 4 μm generates the maximum output power of 4.28 μW at the frequency of 1 kHz and the acceleration of 0.2g. An appropriate electronic circuit is established to significantly improve the low efficiency of only 23%. The additional DC/DC converter can maximize the transfer of energy between the capacitive converter and the energy storage circuit [15, 16]. An out-of-plane cantilever-based harvester is reported, which is with a base capacitance of 1 nF and a variable capacitance of between

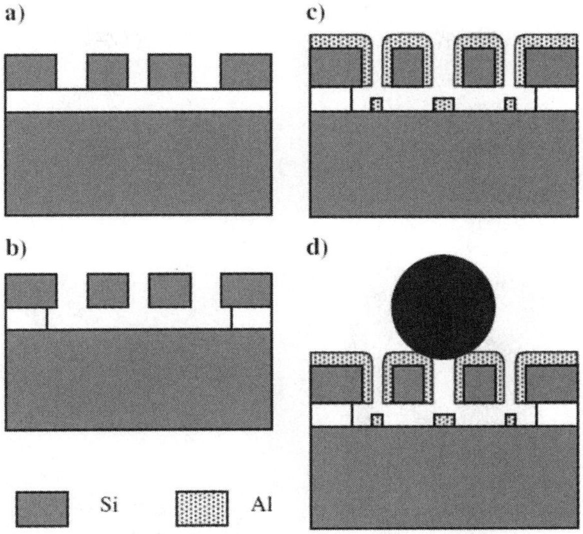

Figure 3.5 Fabrication process: (a) define structure by DRIE; (b) etch oxide by HF solution; (c) apply Al by thermal evaporation; (d) attach external mass [13].

30 pF and 350 pF, the resonant frequency of 45 Hz with a Q-factor of 30 [17]. The device was tested on a wall with a 1-μm displacement up to 100 Hz. The 120 nW is generated for the wall acceleration of 0.08 m s^{-2}.

3.2.2 New Developed Electrostatic Microharvester

Because the output power from electrostatic mechanism is very low, some new approaches are recently proposed by researchers to improve it, including multidimensional scavenging, integrated-circuit nonlinear structure and nonresonant structures.

3.2.2.1 Two-Dimensional and Three-Dimensional Electrostatic Energy Harvesters

Currently, most of the electrostatic mechanisms are designed to harvest kinetic energy from a single direction because the reported electrostatic energy harvesters comprise a set of springs that are softer regarding the moving direction of a mass along the first axis (i.e., a linear motion) and are stiffer in the two axes perpendicular to the first moving axis. However, the vibrations from the nature resources may present unpredictable or unexpected shifts in its dominant direction. For an isotropic external vibration source, the energy along the direction orthogonal to the dominant vibration direction will be lost. In most cases, the available energy is equally distributed over three freedom degrees. Thus, in average, the output energy from 2-D harvester is twice more efficient than that from the 1-D converter.

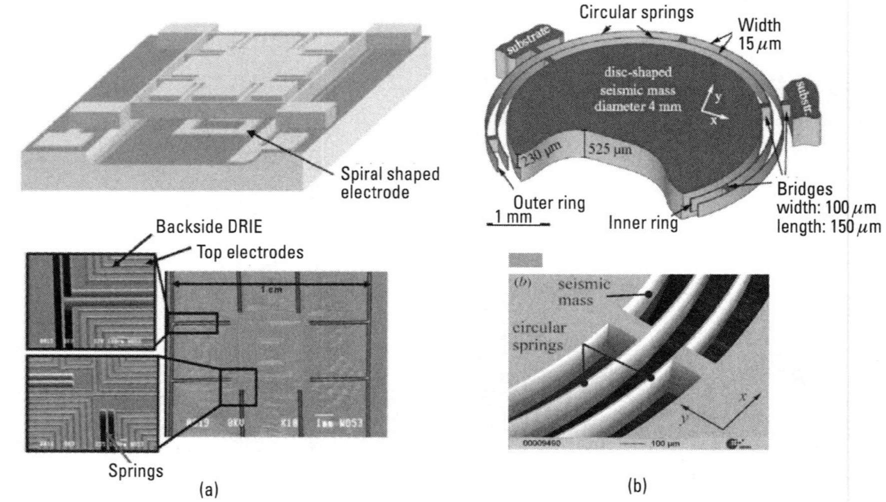

Figure 3.6 2-D resonator for energy harvesting: (a) 2-D linear structure [18]; (b) threefold rotational symmetry structure [19].

A 2-D linear electrostatic silicon-based harvester fabricated in a CMOS-compatible technology is proposed [18]. Figure 3.6(a) shows the design structure of an SEM image of this device. The area of this device is 1 cm^2 and resonant frequency is measured at 246 Hz. The 2.6 μW can be simulated at the optimized conditions when the input voltage is 6V. A concept of a resonant rotating mass fixed on a spiral spring has been proposed as an inertial mechanism for harvesting kinetic energy from a rotational motion (i.e., a kind of planar motion) [19, 20]. The design of the 2-D resonator is shown schematically in Figure 3.6(b) [19], which consists of a seismic mass suspended by a circular spring system. The in-plane resonant frequency of this device is calculated by finite-element analysis to be about 1,230 Hz when the height of the springs and bridges and the spaced gap between the various concentric components are 230 μm and 150 μm, respectively. The resonant frequency scales with the square root of the spring height. The device shows two perpendicular linear eigenmodes with resonant frequency values close to 370 Hz. The Q-factor of the devices is about 1,800 at atmospheric pressure. It increases to about 3,300 at 0.2 mbar. The eigenmode directions depend critically on material symmetry and geometrical imperfections due to fabrication. Compared to a 1-D harvester, the increasing of excitation directions and broadband excitation has improved the output power.

Rotary comb actuators (i.e., using a rotational in-plane overlap varying type of electrodes) with a pivot or a virtual anchor at the center of circle formed by the rotary comb have been investigated as actuators for driving a vertical micromirror to move along a trace formed by radial line scanning against the pivot [21, 22]. The rotary comb structure is typically formed by etching and releasing the vertical comb finger structure from the silicon device layer of a silicon-on-insulator (SOI) substrate. With the aid of integrated micromirrors, rotary comb actuators have been successfully applied to tunable external cavity diode lasers, variable optical attenuators, and optical switches [23–26]. Yang et al. proposed a MEMS energy harvester using rotary comb electrodes, which convert the kinetic energy of planar vibration from ambient sources and store it as electrical energy [27]. A schematic drawing of the one-sixth portion of the rotary comb-based electrostatic energy harvester is shown in Figure 3.7(a). Figure 3.7(b) shows the lumped model for one set of movable and fixed combs. The suspended silicon structure consists of the 12 sets of fixed combs, grid-like proof mass integrated with movable combs, and six ladder springs and stoppers for avoiding pull-in phenomenon between movable and fixed combs. The total capacitance can be calculated by Gauss's law:

$$C_{total} = C_{overlap} + C_{gap} \tag{3.15}$$

where $C_{overlap}$ is the overlap capacitance and C_{gap} is the gap capacitance.

The overlap capacitance for comb fingers is given by

$$C_{overlap} = \varepsilon t\left(\theta + \theta_0\right) \times A(n) \tag{3.16}$$

where

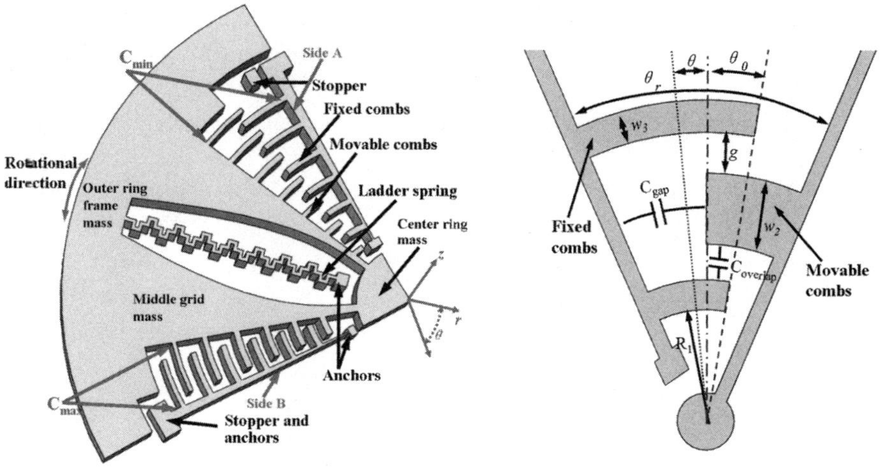

(a) (b)

Figure 3.7 The structural design of a rotary energy harvester: (a) schematic view of a 2-D rotary harvester; (b) the lumped models for one set of fixed and movable combs [27].

$$A(n) = \sum_{i=1}^{n} \left\{ \begin{aligned} &\left[\ln\left(\frac{R_1 + 2(i-1)\left(W_f + g\right)}{R_1 + 2(i-1)\left(W_f + g\right) - g} \right) \right]^{-1} \\ &+ \left[\ln\left(\frac{R_1 + (2i-1)\left(W_f + g\right)}{R_1 + (2i-1)\left(W_f + g\right) - g} \right) \right]^{-1} \end{aligned} \right\} - \left[\ln\left(\frac{R_1}{R_1 - g} \right) \right]^{-1}$$

$$W_f = \left(w_2 + w_3\right)/2$$

The gap capacitance can be calculated by

$$C_{gap} = \varepsilon t W_f B(n) \frac{1}{\dfrac{\theta_r - \theta_0}{2} - \theta} \tag{3.17}$$

where

$$B(n) = \sum_{i=1}^{n} \left\{ \frac{1}{R_1 + 2(i-1)\left(W_f + g\right)} + \frac{1}{R_1 + (2i-1)\left(W_f + g\right)} \right\}$$

where ε is the free-space permittivity with value of 8.854×10^{-6} pF/μm, t is the thickness of combs, R_1 is the radius of the concave surface of the first movable combs, g is the gap between fixed and movable combs, w_2 is the width of movable combs, and w_3 is the width of fixed combs. θ_r is the initial angle between fixed and movable arms.

Figure 3.8 shows the top view of the fabricated devices and enlarged view of ladder spring and combs. Figure 3.9(a, b) shows the prototype after wire bonding to a dual-in-line package (DIP) with 40 footprints and a vacuum-packaged prototype, respectively. The measured output powers under different accelerations in air and at the pressure level of 3 Torr are also shown in Figure 3.9(a, b). The maximum measured output power for the vibrations of 0.5g, 1g, 1.5g, 2g, and 2.5g is 0.11 μW, 0.17 μW, 0.24 μW, 0.3 μW, and 0.35μW, respectively, when the loading resistance matches the parasitic resistance of 80 MΩ at the resonant frequency. Compared to the one in air, the maximum electrical output power of 0.25-g acceleration at the pressure level of 3 Torr has obviously improved and arrives at 0.39 μW.

To broaden the scavenging directions of the harvester, a 3-D electrostatic harvester was designed and simulated by Janicek et al. [28]. This device employs a comb structure of electrodes and can work in all three axes. The very low resonant frequency of the structure is obtained using the modified long cantilever design. This structure is produced by SOI HARM 60-μm Tronics technology [29], as shown in Figure 3.10.

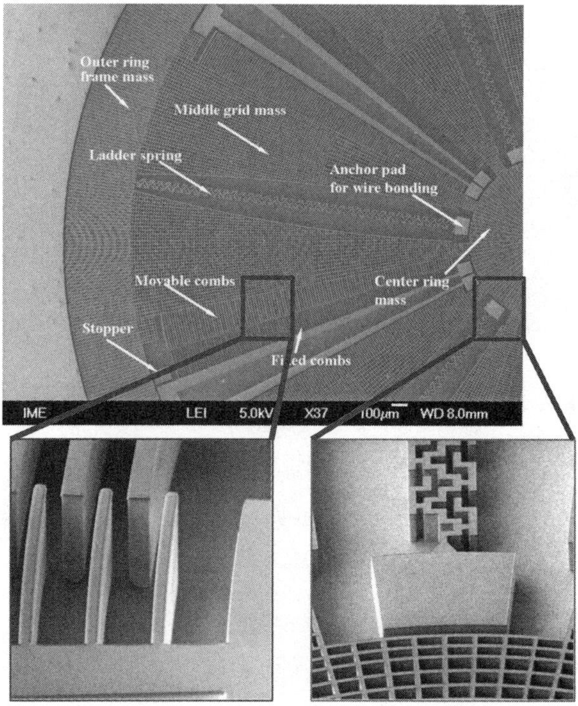

Figure 3.8 The fabricated energy harvester based on a rotary comb [27].

3.2.2.2 Circuit-Integrated Energy Harvesters

The output voltage of the harvester should be rectified for powering the device or charging the storing capacitor. Therefore, the highly efficient energy transfer depends on the electric interface between the harvester and the consuming load. Large losses must be avoided, especially in the rectifier, because the output power of a miniaturized energy harvester is typically in the microwatt range [30]. Normally, the discrete diodes are deployed as a simple half-wave or a full-wave rectifier. A prototype circuitry was proposed in [31] in which the timing switches were replaced by diodes. However, this model results in an excessive output power reduction due to the far-from-ideal characteristics of the diodes. More significantly, an even larger disadvantage is the serial voltage drop over the diodes about 0.6V to 1.2V in a full-wave rectifier circuit. To reduce this influence, an additional PMOS diode in parallel as a half-wave rectifier is proposed [32]. This causes a voltage drop of about 200–300 mV for standard CMOS. The threshold voltage of the diode-connected transistors is reduced by precharging the floating gates [33]. The overall voltage drop decreases to less than 200 mV. Due to the fact that the control circuitry consumes electri-

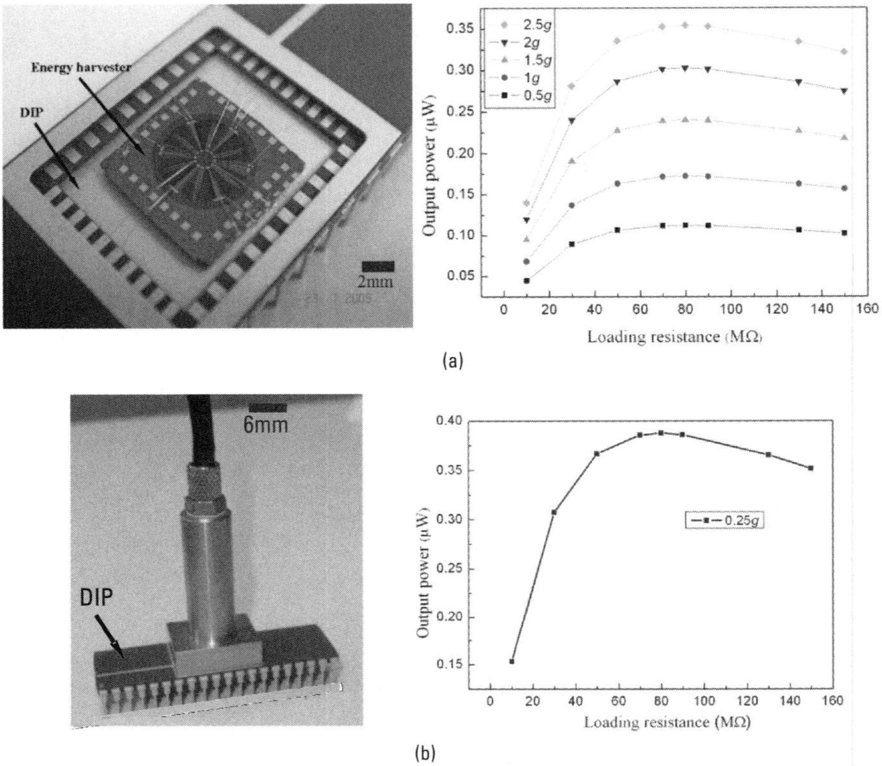

Figure 3.9 (a) The testing output power under air conditions; and (b) the testing power at a pressure level of 3 Torr [27].

cal power and thus reduces the overall efficiency, a full-wave rectifier with only a single comparator is a significant and useful improvement compared to the state of the art [34]. A two-stage active rectifier is proposed for the reduction of the voltage drop over the rectifier, the increase of the available output power, and a high conversion efficiency. The first stage is passive and used to convert the negative half-waves of the input sinusoidal wave into positive ones. The second stage is to control the current direction by blocking reverse current flow realized by a MOS diode. The active rectifier and the active diode have been implemented using a standard 0.35-μm CMOS process [35]. Figure 3.11(a) shows a chip micrograph with an overlaid true-to-scale layout and the dimensions of the active rectifier [34]. When the input voltage amplitude is between 1.25V and 3.75V at the operating frequency range from 1 Hz to several hundred kilohertz, the series voltage drop over the rectifier is less than 20 mV. The simple Schottky rectifier and the active diode deliver a maximum of about 112 μW and 134 μW, respectively. Compared to the half-wave Schottky recti-

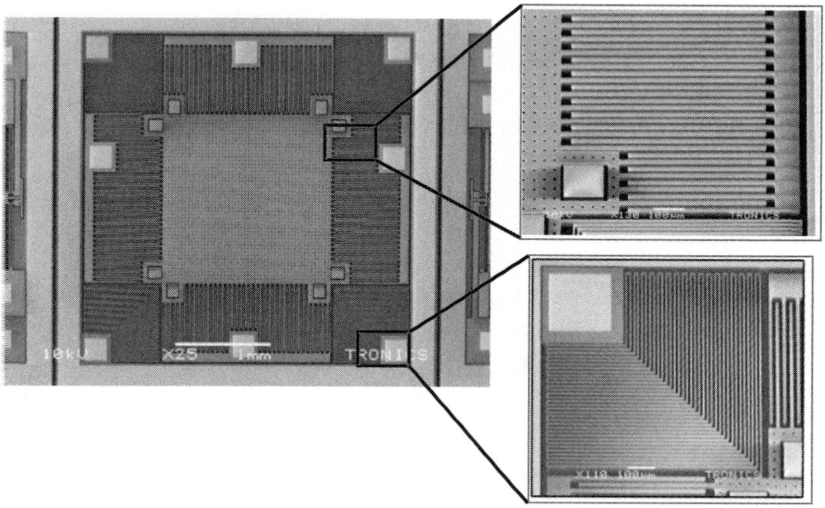

Figure 3.10 3-D prototype SEM of energy harvester [29].

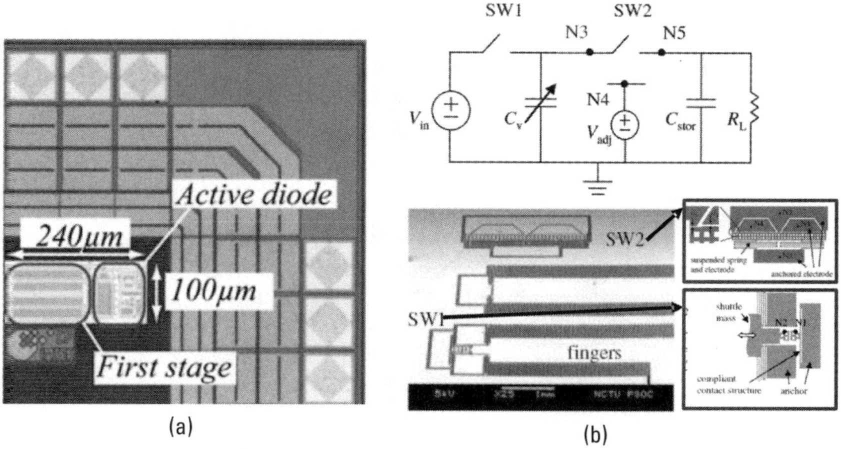

Figure 3.11 (a) Micrograph of the CMOS chip [34]; and (b) the harvester with integrated mechanical switch [35].

fier, the output power of 154 μW from the full-wave active rectifier is 38% more. The active diode has an efficiency of over 95% down to 15 kΩ.

Because the parasitic of the electronics can degrade the performance of the harvesters, a novel integrated mechanical switch is proposed for a resonant electrostatic energy harvester [36]. Figure 3.11(b) shows the integrated mechanical switches SW1 and SW2 of the harvester. SW1 should ideally be closed when the movable electrode of the variable capacitor is near the maximum displacement and opened immediately after charging is finished. SW2 should ideally be closed when the variable capacitor moves to the center position. At this position, the terminal voltage of C_v arrives at a maximum value V_{max}. Compared to the traditional design with diodes or active electric circuit components, the mechanical switches have the advantages of zero leakage current, low power consumption, and synchronous operation.

3.2.2.3 Wideband Electrostatic Energy Harvesters

For the linear resonant electrostatic energy harvesters, their narrow operating bandwidth limits their application in our environments because the vibration frequency of the excitation sources varies from one case to another. Many methods are proposed and developed to overcome this limitation by increasing their operation frequency range, including stopper effect [1, 36], nonlinear spring [37–39], and triangular electrode structure [40].

For practical application, the motion displacement of the proof mass should be limited due to large excited acceleration under the restricted motion space. Meanwhile, to avoid beam fracture at the large vibration amplitude, the proof mass motion must be limited. The designed end stopper may cause the device to behave as nonlinearity phenomena. To predict the output performance, the internal impacts on transducing structures of electrostatic harvester has been analyzed [41]. This impact model is represented by a parallel-connected spring and damper, and the stiffness value is obtained analytically from Herz's contact problem [42, 43]. Because the impact force model is incorporated into the mechanical domain of the electrostatic harvester model, it can handle a large amplitude displacement of the proof mass [36]. Figure 3.12(a) shows the structural design and SEM of bumps as end stoppers. The relative displacement δ between the proof mass and the end stops is simplified as being uniform in the entire contact area and can be calculated as [44, 45]

$$\delta = \left[1 + \ln \frac{t^3}{2\lambda R F_{im}}\right] \frac{2\lambda}{t} F_{im} \tag{3.18}$$

where t is the device layer thickness, F_{im} is the impact force, R is the bump radius, $\lambda = \frac{1-\upsilon^2}{\pi E}$, and E and υ are the Young's modulus and Poisson's ratio of the same contacting elastic material, respectively.

Figure 3.12(b) shows the up-sweep frequency response of output voltage. The output voltage increases with the increasing frequency up to 1,135 Hz. At this point, the proof mass contacts the end stops, causing a kink in the output voltage and a moderate further increase. The output voltage jumps down at the frequency of 1,425 Hz. This phenomenon is due to the abrupt increase in stiffness at the impacts. For sufficiently high acceleration power spectral density

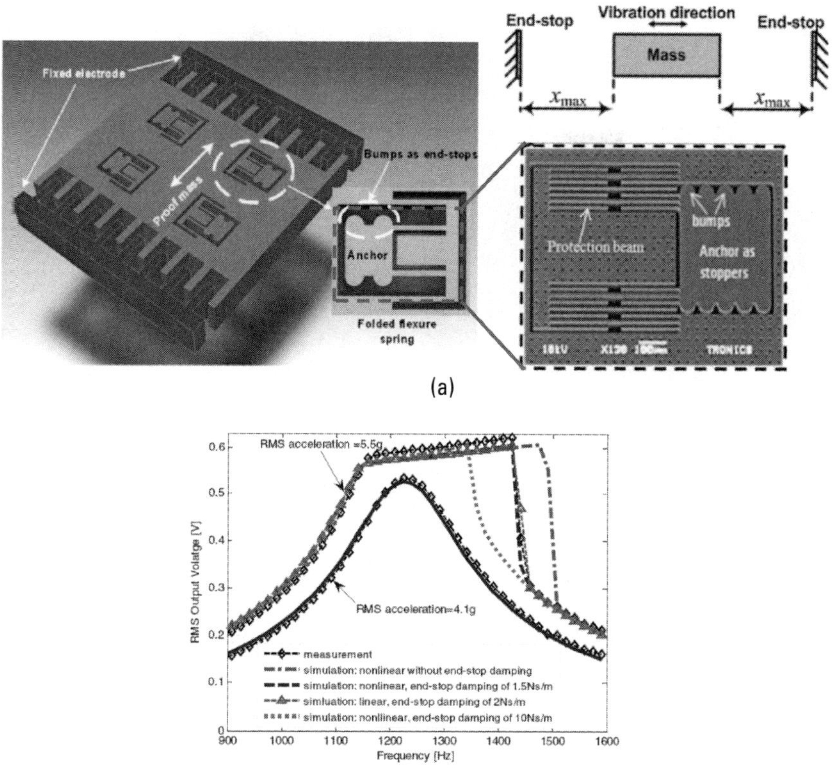

Figure 3.12 (a) Design and SEM of bumps as end-stops in an electrostatic energy harvester; and (b) up-sweep frequency response of RMS output voltage [36].

(PSD), the output power will be a soft limit on the output power and a bandwidth increases, for example, a factor 3.7 of a 3-dB bandwidth when increasing the acceleration PSD from 0.34×10^{-3} to 0.55×10^{-3} g^2 Hz^{-1}. The end-stop effects reduce the device efficiency down to 35.4% at 1.69×10^{-3} g^2 Hz^{-1}. To understand the dynamic behavior of the impact with the end stopper, a signal-flow model including a mechanical and an electrical domain is created, which are coupled by electrostatic forces [1].

Because the mechanical stopper may cause fatigue problems, the hardening of suspensions is exploited to broaden the bandwidth of harvesters [46–49]. However, the hardening suspensions cannot increase the amount of power harvester relative to that of a tuned linear device [50]. Both hardening and softening springs can widen the bandwidth of the harvesters and increase the output power. The softening spring is more promising because it gives a larger displacement, while a hardening spring restricts displacement. Nonlinear springs are deployed for a bandwidth-tolerant harvester [37]. Figure 3.13 depicts the

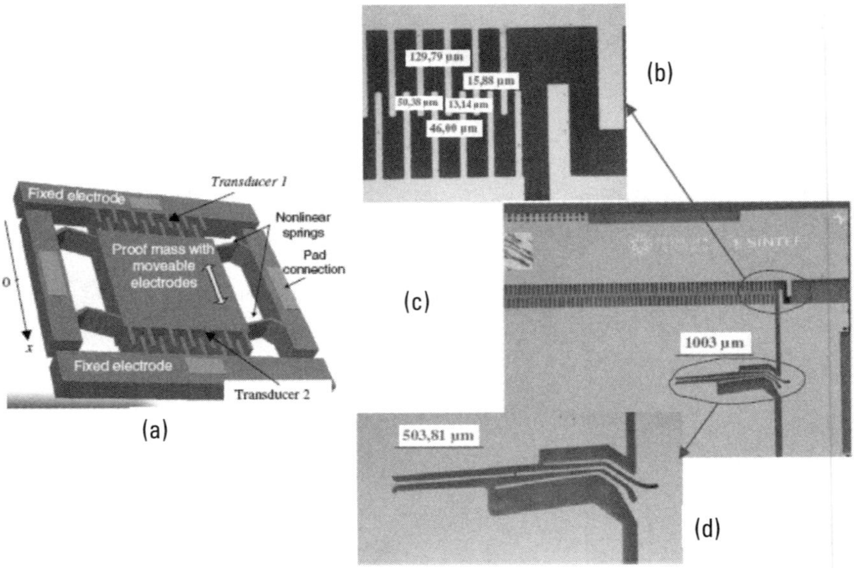

Figure 3.13 (a) Schematic of a MEMS electrostatic energy harvester with nonlinear springs; (b) more details of electrode fingers; (c) microscopic close-up showing a corner of the fabricated energy harvester; and (d) the nonlinear spring with its etch shield [37].

schematic and structural details of an energy harvester with a nonlinear spring. The nonlinearity is obtained by making an angle in the spring. At a broadband random vibration of 7.0×10^{-4} g^2 H^{-1}, the bandwidth increases by more than 13 times and the average harvesting output power increases by 68% compared to that of a linear vibration energy harvester. Apart from the nonlinear springs in the structural design for a large operation wideband, a triangular electrode structures for electrostatic harvester is proposed to increase its bandwidth [40]. Figure 3.14(a) demonstrates comparison of different electrode geometries with respect to the capacitance characteristic. A triangular electrode design (EG3) provides a nonlinear characteristic with a total capacitance change larger than a gap-closing design (EG2) and a change of capacitance per unit displacement larger than an area-overlap design (EG1). This nonlinear capacitance characteristic allows tuning the resonance frequency of the device. Figure 3.14(b) shows the interdigitated triangular electrode structures. When this fabricated device is connected with rectifier and energy storage [Figure 3.14(c)], a half-bandwidth of 25 Hz and 50 Hz is obtained under the excitation amplitude of 20 m s^{-2} at bias voltages of 10V and 20V, respectively.

Resonant vibration energy harvesters (VEHs) employ high-quality resonators that can convert vibration energy in a narrow frequency band. To understand the dynamic behavior, some coupled electromechanical analysis is

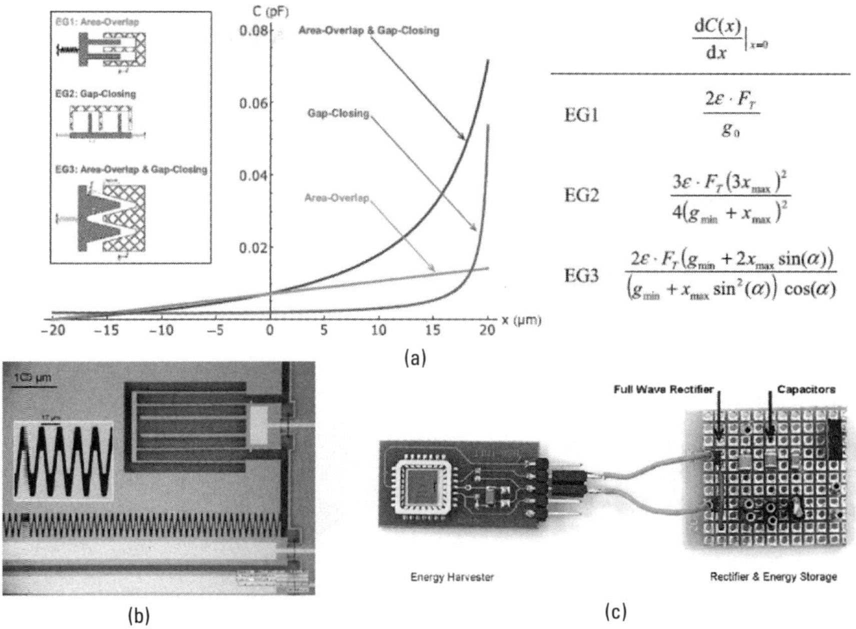

Figure 3.14 (a) Comparison of different electrode geometries with respect to the capicitance characteristic; (b) close-up view showing the interdigitated triangular electrode structures; and (c) the packaged energy harvester connected with rectifier and energy storage [40].

performed [51–53]. Blokhina et al. discussed and analyzed the limitation on converted power in resonant electrostatic vibration energy harvesters [54]. To solve the power limitation problem, nonresonant energy harvesters are developed [55, 56]. Bu et al. designed and fabricated a nonresonant energy harvester, which consists of three metal plates [55]. The inner plate is freestanding and is movable due to external vibrations. Both the upper and bottom electrodes are attached with electrets. When the inner plate moves upward, the amount of induced charges on the upper electrode decreases, and correspondingly the generated current flows through the first resistance into ground. Meanwhile, the amount of induced charges on the bottom plate increases, which results in current flowing from ground through the second resistance, and vice versa. Over the frequency range of 5 to 120 Hz, 3.9–7.0V and 0.2–0.7 μW are obtained.

3.3 Electret-Based Electrostatic Microharvesters

To get rid of the requirement of external bias, an electret layer is introduced to form an electrical potential and then electricity is harvested from alternatively changed relative positions of electrodes due to vibrations. An electret is an insulating material that exhibits a net electrical charge or dipole moment. The net charge or dipole moment in the electret can be used to provide a biasing electric field such as for a MEMS electret microphone [57, 58] and electrostatic energy harvester [59–62].

3.3.1 Electret Materials

Electret materials are grouped into SiO_2-based inorganic and polymer-based organic materials. SiO_2/Si_3N_4 electrets [63, 64] are used for capacitive energy harvesting because of their long charge storage stability, their ability to work in harsh environment [65], and their compatibility to CMOS processing. Naruse et al. developed a large-amplitude electret harvester with SiO_2 electrets [66]. SiO_2 has an extremely high surface charge density up to 10 mC m^{-2} [67]. To improve the long-term stability of the trapped charges, a double layer of SiO_2 and Si_3N_4 is proposed for investigating the charge stability [68]. However, it is difficult to obtain very high surface potential in SiO_2 electrets, since the thickness of low-impurity SiO_2 is limited by the fabrication process. The relative low-volume resistivity of 10^{16} Ω cm would result in a fast decay of the stored charges. Because ion implantation provides an exactly controlled charge density and a well-defined energy distribution of the charge density well below the surface, Meschdder et al. reported the properties of SiO_2 electret films charged by ion implantation for MEMS-based energy harvesting systems [69]. Although the oxide/nitride electrets could have higher charge densities [70], high-temperature processes often render them inferior to polymer counterparts in certain applications.

Boland et al. developed a rotational MEMS energy harvester with Teflon AF® as the electrets and obtained the power output of 25 μW [71–73]. Tsutsumino et al. proposed CYTOP-based energy harvesters [74, 75]. Among various electret materials, amorphous perfluorinated polymers such as Teflon and CYTOP® are research interests because of their ease of processing. Kashiwagi et al. successfully formed nanoclusters containing the organic siloxanes in the CYTOP film, as shown in Figure 3.15(a) [76]. The nanoclusters acted as the charge trap site enhance the surface charge density and the thermal stability of the trapped charges. The extremely high surface charge potential of −1.6 kV with the 15-μm thick film is obtained. Parylene HT as a new electrets material has been reported for electrostatic harvester, as shown in Figure 3.15(b), which includes the PEEK rotor block, the electrode stator and the device assembly [77]. The 7.32 μm of parylene HT® is deposited onto the PEEK rotor and then charged via corona charging. The harvester with parylene HT-coated PEEK rotors can collect 7.7 μW at 10 Hz and 8.23 μW at 20 Hz. Table 3.1 lists the properties of some common used electret materials.

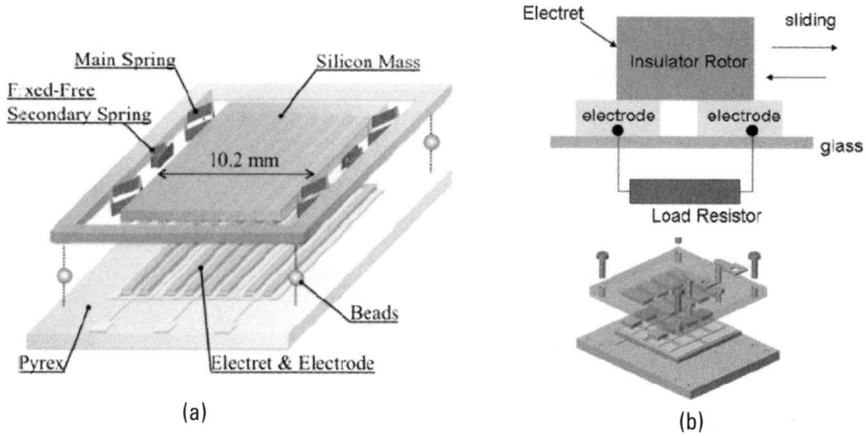

Figure 3.15 (a) Nano-cluster-based CYTOP electric energy harvester [76], and Parylene HT film electric energy harvester [77].

Table 3.1
Properties of Electret Materials

Electret Materials	Dielectric Constant	Volume Resistivity (Ω cm)	Dielectric Strength (kV mm^{-1})	Maximum Charge Density (mC m^{-2})
Thermal SiO_2	3.9	10^{16} [78]	500 [78]	16 [69]
LPCVD Si_3N_4	6–7	10^{14}	1000 [69]	—
Composite layer of thermal SiO_2 and APCVD Si_3N_4	—	—	—	11.51 [5]
CYTOP	2.1	10^{17} [79]	110 [79]	0.68 [75]
Teflon AF	1.9	$>10^{19}$	21	28 [71]
PTFE	2.1	$>10^{18}$	18 [69]	—

3.3.2 Theoretical Design of Electret Microharvesters

To design and optimize an electrets-based microharvester, a nonlinear dynamical model is created [80]. A series of constraints on the design of the device are adopted for reducing the number of free parameters. In this model, the nonlinear behavior of the harvester should be taken into account to get correct results. Generally, the main problem in this model is that they neglect the fringe effects that appear and become predominant when working in small dimensions. Nevertheless, Le et al. [81] developed an analytical model for computing the capacitance variation for an in-plane overlap plate transducer. Based on an analytic expression, it partially considers fringe effects and leads to better modelling of the structure. Boisseau et al. [82] deployed finite element method (FEM) software (Comsol Multiphysics) to compute the capacitance of the system considering all fringe effects.

The power output of the electret generators with no parasitic capacitance is theoretically defined as

$$P = \frac{\left(2nfA_{\max}\sigma d\right)^2 R_L}{\left[d + \varepsilon_E\left(g + 2nfA_{\max}R_L\varepsilon_0\right)\right]^2} \tag{3.19}$$

where σ is the charge density, n is the number of electrodes, A is the overlapping area, f is the frequency, ε_E is the relative permittivity of electret, ε_0 is the vacuum permittivity, d is the electret thickness, g is the separation gap, and R_L is the load resistor [71].

Based on this equation, the output power is proportional to the square of the surface charge density, and the square of the surface potential of electret. Therefore, the development of electret material with high surface charge density and high surface potential is crucial for high-performance electrets harvesters.

When the parasitic capacitance is considered, the effective power of the harvester can be calculated as the average of the time-dependent output power [83]

$$\begin{aligned} P &= \frac{2}{T}\int_0^{T/2} V(t)^2 / R_L \\ &= \frac{R_L V_{\max}^2}{\left(R_i + R_L\right)^3} \times \left(R_i + R_L - 4C_p nfR_iR_L \tanh\left(\frac{R_i + R_L}{4C_p nfR_iR_L}\right)\right) \end{aligned} \tag{3.20}$$

where C_p is the parasitic capacitance, R_i is internal resistor, and V_{max} is the maximum output voltage amplitude.

3.3.3 Fabrication of Electret Microharvesters

A cantilever-based electret energy harvester is proposed to validate the accurate analytical model, as shown in Figure 3.16(a) [84]. This beam is made of silicon and attached to the frame at one end. The mobile mass made of tungsten is attached to the other end. The electrets made from Teflon FEP (fluorinated ethylene propylene) [85] is charged to 1,400V. Figure 3.16(b) shows the deformation of the cantilever with design parameters. This device is theoretically possible to harvest up to 30 μW per gram of mobile mass at the vibration of 0.1g. Because the surface of the polymer must be exposed to the charging apparatus during the polarization process, there is a requirement of the absence of fabricated structures above the film. Furthermore, once charged, significant discharge can

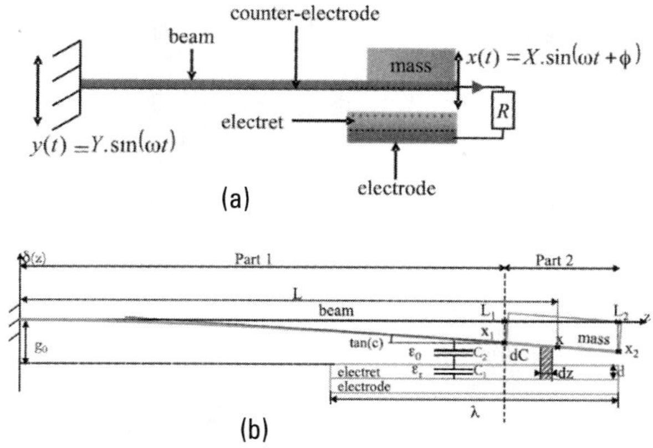

Figure 3.16 (a) Cantilever-based electric energy harvester; and (b) deformation of the cantilever for an imposed deflection [84].

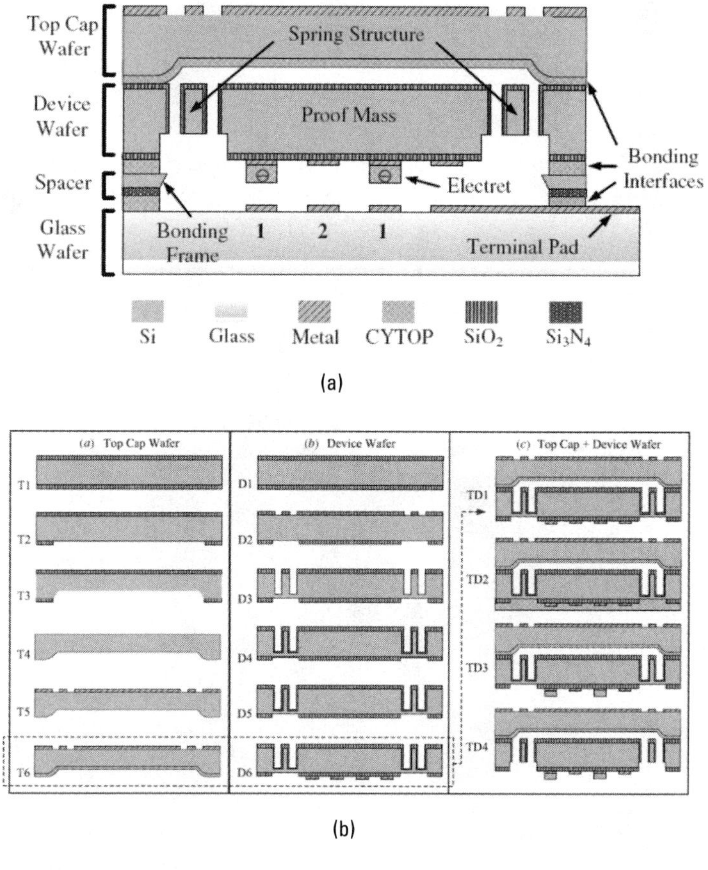

Figure 3.17 (a) Schematic view of the four-wafer stack structure for the energy harvester device; and (b) process flow of the energy harvester [87].

occur if exposed to temperatures near the glass transition of the polymer, thereby limiting subsequent device fabrication to low-temperature processes. In situ wafer-level polarization of electret films is developed to enable the integration of uncharged electret films into MEMS and subsequent postfabrication charging [86]. A wafer-level fabrication process is proposed to fabricate the MEMS electret energy harvester [87]. Figure 3.17(a) demonstrates the schematic view of a four-wafer stack structure. This device is composed of four wafers, namely, top cap, device, spacer, and glass. The process flow of the top-cap wafer, device wafer, and their bonding is shown in Figure 3.17(b). The bonding process is performed at a low temperature of 120°C for 1 hour to avoid any loss of the surface charge. The maximum output power of 32.5 nW was achieved with an external load of 17 MΩ, under a harmonic source motion with an acceleration rms amplitude of 0.03g (~0.3 m s^{-2}) at a frequency of 179.5 Hz.

In our environment, the natural frequency of most from vibration sources is very low. Low-frequency vibration is only sufficient to excite an extremely low output power. To overcome this difficult challenge, a new electrostatic micropower generator that can enable both separation gap control and long-range movement at low frequency and can increase the number of times power generation (electrostatic induction) is proposed [66]. Figure 3.18 shows an electrostatic micro-energy harvester supported on microball bearings. This harvester generates 40 μW of power output at a very low-frequency vibration (2 Hz). However, energy loss occurred because the mover collided with the edges of the device, which can be improved by refining this design. Bartsch et al. reported a 2-D electret-based resonant micro-energy harvester to extract energy from ambient vibrations with arbitrary planar motion directions. It shows that the

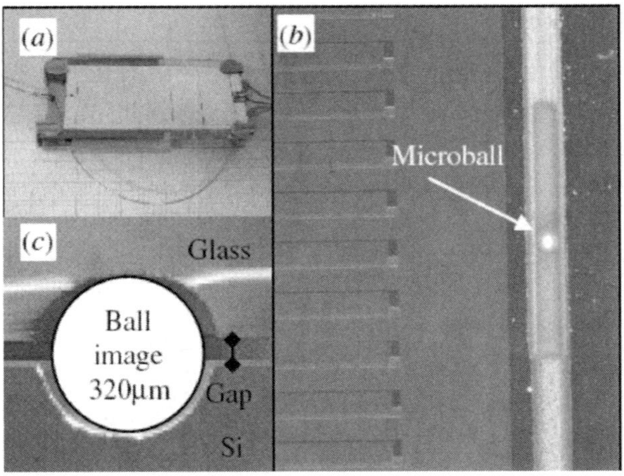

Figure 3.18 (a) Microphotograph of the prototype; (b) expanded view; and (c) cross-section image [66].

resonant frequencies and extracted harvester powers depending on the design of the circular spring system and capacitor dimensions within the 0.3–2-kHz and 1–100-pW ranges [80].

Compared with the solid spring-mass configuration, liquids are inherently more flexible and compliant suggesting their applications in nonresonant generators [88]. A liquid-rotor electret power generator employing mercury droplets and air-filled gaps in a capacitor was reported by Boland et al. [89]. The electret was made from a 25-μm-thick Teflon fluorinated ethylene propylene (FEP) glued to an electrode. Krupenkin and Taylor [90] further developed such ideas in investigating various designs of droplet arrays with an external bias. The ionic liquids can freely vary its position within a chamber, which results in changing the capacitance. By employing the similar ideas, a conductive liquid-based [91] and a dielectric liquid encapsulated energy harvester [92] are proposed. When the conductive droplets are rolling on thin-charge PTFE electrets, the film is poled without any extra charging process and the harvester can be charged by self-biased [88]. With a 1.2-mm mercury droplet rolling across the electret film of the prototype, a maximum output power was obtained at 0.18 μW and the peak value of the output voltage was 1.5V.

3.3.4 Performance Comparisons of Electret Microharvesters

To compare the output performance, a good way to normalize the output powers is to build a figure of merit [82, 84]. The output power is related to the active area and the available mechanical power from the ambient. The mechanical power is obtained by the inertial force ($Y\omega^2$) and the speed of the moving part of the energy harvester ($Y\omega$). The figure of merit can be calculated as

$$\lambda = \frac{P}{SY^2\omega^3} \tag{3.21}$$

where P is the output power, S is the active area of the energy harvester, and ω is the resonant angular frequency.

Table 3.2 compares the working conditions, device active area, and output power of published electret-based energy harvesters. The output powers of most energy harvesters are in the range of several tens of microwatts. To improve the output performance, the electret material is curial factor for electrets-based energy harvesters apart from optimized structural design. Currently, most electret material is CYTOP. Teflon FEP (fluorinated ethylene propylene) is a good electret material and charged to 1,400V, and the maximum normalized power of 38.75 is obtained [84]. In the future, for the application of electret-based energy harvester, long-term stability of electret materials and the device structure also should be developed and studied.

Table 3.2
Performance Comparisons of Electret-Based MEMS Energy Harvesters

Author	Ref.	Vibrations	Electret Materials	Electret Potential (V)	Active Surface (cm^2)	Output Power (P) (μW)	$I = P/SY^2\omega^3$
Sakane	61	1.2mm_{pp} @20 Hz	CYTOP	640	4	700	2.45
Yang	68	5μm_{pp} @560 Hz	SiO_2/Si_3N_4	400	0.3	46.14pW	5.65×10^{-6}
Tsutsmino	75	2mm_{pp} @20 Hz	CYTOP	1100	4	38	4.79×10^{-2}
Boisseau	84	20μm_{pp} @50 Hz	FEP	1400	4.16	50	38.75
Boland	93	1mm_{pp} @60 Hz	Teflon	850	0.12	6	3.73×10^{-2}
Lo	94	2mm_{pp} @60 Hz	CYTOP	300	4.84	2.26	8.72×10^{-5}
Lo	95	1mm_{pp} @50 Hz	Parylene-HT	1500	6	17.98	3.87×10^{-3}
Suzuki	96	2mm_{pp} @37 Hz	CYTOP	450	2.33	0.28	9.56×10^{-5}
Halvorsen	97	5.6μm_{pp} @596Hz	SiO_2/Si_3N_4	—	0.48	1	5.06×10^{-2}
Honzumi	98	1.2mm_{pp} @20 Hz	Parylene-C	52	0.01	90pW	3.32×10^{-5}
Miki	99	0.16μm_{pp} @1740 Hz	CYTOP	180	3	1	5.37×10^{-3}
Omron	100	2μm_{pp} @50 Hz	CYTOP	700	4	10	3.50×10^{2}
Kloub	101	1.2mm_{pp} @20 Hz	CYTOP	25	0.42	5	1.42
Sterken	102	1.2mm_{pp} @20 Hz	SiO_2/Si_3N_4	10	0.09	2nW	7.17×10^{-3}

References

[1] Hoffmann, D., B. Folkmer, and Y. Manoli, "Fabrication, Characterization and Modeling of Electrostatic Micro-Generators," *J. Micromech. Microeng.*, Vol. 19, 2009, p. 094001.

[2] Roundy, S., P. K. Wright, and J. Rabaey, "A Study of Low Level Vibrations as a Power Source for Wireless Sensor Nodes," *Comput. Commun.*, Vol. 26, 2003, pp. 1131–1144.

[3] Miranda J. O. M., "Electrostatic Vibration-to-Electric Energy Conversion," MIT dissertation, 2004.

[4] Beeby, S. P., M. J. Tudor, and N. M. White, "Energy Harvesting Vibration Sources for Microsystems Applications," *Meas. Sci. Technol.*, Vol. 17, 2006, pp. R175–R195.

[5] Anantha, P. C., et al., "Trends in Low Power Digital Signal Processing," *Proceedings of the 1998 IEEE International Symposium on Circuits and Systems*, 1998, Vol. 4, pp. 604–607.

[6] Lee, C. K., et al., "Theoretical Comparison of the Energy Harvesting Capability Among Various Electrostatic Mechanisms from Structure Aspect," *Sens. Actuators A*, Vol. 156, 2009, pp. 208–216.

[7] Meninger, S., et al., "Vibration-to-Electric Energy Conversion," *IEEE Trans. on Very Large Scale Integr. (VLSI) Syst.*, Vol. 9, 2001, pp. 64–76.

[8] Arakawa, Y., Y. Suzuki, and N. Kasagi, "Micro Seismic Power Generator Using Electret Polymer Film," *Power MEMS Conference 2004*, Japan, November 28–30, 2004, pp. 187–190.

[9] Sheu, G. J., S. M. Yang, and T. Lee, "Development of a Low Frequency Electrostatic Comb-Drive Energy Harvester Compatible to SoC Design by CMOS Process," *Sens. Actuators A*, Vol. 167, 2011, pp. 70–76.

[10] Ma, W., et al., "An Integrated Floating-Electrode Electric Microgenerator," *J. Microelectromech. Syst.*, Vol. 16, 2007, pp. 29–37.

[11] Despesse, G., et al., "Fabrication and Characterisation of High Damping Electrostatic Micro Devices for Vibration Energy Scavenging," *Proc. Design, Test, Integration and Packaging of MEMS and MOEMS*, 2005, pp. 386–390.

[12] Chu, Y. S., C. T. Kuo, and Y. Chiu, "A MEMS Electrostatic Vibration-to-Electricity Energy Converter," *Proceedings of PowerMEMS*, 2005, pp. 49–52.

[13] Chiu, Y., C. T. Kuo, and Y. S. Chu, "MEMS Design and Fabrication of an Electrostatic Vibration-to-Electricity Energy Converter," *Microsyst. Technol.*, Vol. 13, 2007, pp. 1663–1669.

[14] Kuehne I., et al., "Power MEMS: A Capacitive Vibration-to-Electrical Energy Converter with Built-In Voltage," *Sens. Actuators A*, Vol. 142, 2008, pp. 263–269.

[15] Ammar, Y., et al., "Wireless Sensor Network Node with Asynchronous Architecture and Vibration Harvesting Micro Power Generator," *Proceedings of the Oc-EUSAI Conference*, Grenoble, France, October 12–14, 2005.

[16] Ottmann, G., et al., "Adaptive Piezoelectric Energy Harvesting Circuit for Wireless Remote Power Supply," *IEEE Trans. on Power Electron.*, Vol. 17, 2002, pp. 669–676.

[17] Miyazaki, M., et al., "Electric-Energy Generation Using Variable-Capacitive Resonator for Power-Free LSI: Efficiency Analysis and Fundamental Experiment," *ISLPED'03*, 2003, pp. 193–198.

[18] Paracha, A. M., et al., "MEMS DC/DC Converter for 1D and 2D Vibration-to-Electricity Power Conversion," *Transducers 2009*, Denver, CO, June 21–25, 2009, pp. 2098–2101.

[19] Bartsch, U., J. Gaspar, and O. Paul, "Low-Frequency Two-Dimensional Resonators for Vibrational Micro Energy Harvesting," *J. Micromech. Microeng.*, Vol. 20, 2010, p. 035016.

[20] Yeatman, E. M., "Energy Harvesting from Motion Using Rotating and Gyroscopic Proof Masses," *Proc. of the Institution of Mech. Eng., Part C: J. Mech. Eng. Sci.*, Vol. 222, 2008, pp. 27–36.

[21] Jill, J. D., et al., "Widely Tunable External Cavity Diode Laser Based on a MEMS Electrostatic Rotary Actuator," *IEEE Proc. Optical Fiber Commun. Conf. and Exhibit (OFC2001)*, Vol. 2, 2001, pp. TuJ2-1–TuJ2-3.

[22] Yeh, J. A., C. N. Chen, and Y. S. Lui, "Large Rotation Actuated by In-Plane Rotary Comb-Drive with Serpentine Spring Suspension," *J. Micromech. Microeng.*, Vol. 15, 2005, pp. 201–206.

[23] Lim, T. S., et al., "Electrostatic MEMS Variable Optical Attenuator with Rotating Folded Micromirror," *IEEE J. Sel. Topic Quantum Electron.*, Vol. 10, 2004, pp. 558–562.

[24] Yeh, J. A., S. S. Jiang, and C. Lee, "MOEMS Variable Optical Attenuators Using Rotary Comb Drive Actuators," *IEEE Photon. Technol. Lett.*, Vol. 18, 2006, pp. 1170–1172.

[25] Tu, C. C., K. Fanchiang, and C. H. Liu, "Rotary Electrostatic Micromirror Switches Using Wafer-Scale Processing and Assembly," *Microsyst. Technol.*, Vol. 12, 2006, pp. 1099–1108.

[26] Hou Max, T. K., et al., "In-Plane Rotary Comb-Drive Actuator for a Variable Optical Attenuator," *J. Micro/Nanolith. MEMS MOEMS*, Vol. 7, 2008, p. 043015.

[27] Yang, B., et al., "A MEMS Rotary Comb Mechanism for Harvesting the Kinetic Energy of Planar Vibrations," *J. Micromech. Microeng.*, Vol. 20, 2010, p. 065017.

[28] Janicek, V., and M. Husak, "Designing the 3D Electrostatic Microgenerator," *Journal of Electrostatics*, Vol. 71, 2013, pp. 214–219.

[29] Janicek, V., et al., "Design and Fabrication of 3D Electrostatic Energy Harvester," *Radioengineering*, Vol. 21, 2012, pp. 231–238.

[30] Mitcheson, P. D., et al., "Performance Limits of the Three MEMS Inertial Energy Generator Transduction Types," *J. Micromech. Microeng.*, Vol. 17, 2007, pp. 211–216.

[31] Roundy, S., P. K. Wright, and J. M. Rabaey, *Energy Scavenging for Wireless Sensor Networks with Special Focus on Vibrations*, Dordrecht: Kluwer, 2003.

[32] Le, T. T., et al., "Piezoelectric Micro-Power Generation Interface Circuits," *IEEE J. Solid-State Circuits*, Vol. 41, 2006, pp. 1411–1420.

[33] Peters, C., et al., "High-Bandwidth Floating Gate CMOS Rectifiers with Reduced Voltage Drop," *Proc. IEEE Int. Symp. on Circuits and Systems ISCAS*, Seattle, WA, May 2008, pp. 2598–2601.

[34] Peters C., et al., "A CMOS Integrated Voltage and Power Efficient AC/DC Converter for Energy Harvesting Applications," *J. Micromech. Microeng.*, Vol. 18, 2008, p. 104005.

[35] austriamicrosystems, "0.35 μm CMOS process," AMS C35. http://www.ams.com/eng/support/foundry-design-support/MPW-foundry-services.

[36] Le, C. P., and E. Halvorsen, "MEMS Electrostatic Energy Harvesters with End-Stop Effects," *J. Micromech. Microeng.*, Vol. 22, 2012, p. 074013.

[37] Nguyen, D. S., et al., "Fabrication and Characterization of a Wideband MEMS Energy Harvester Utilizing Nonlinear Springs," *J. Micromech. Microeng.*, Vol. 20, 2010, p. 125009.

[38] Tvedt, L. G. W., D. S. Nguyen, and E. Halvorsen, "Nonlinear Behavior of an Electrostatic Energy Harvester Under Wide- and Narrowband Excitation," *J. Microelectromech. Syst.*, Vol. 19, 2010, pp. 305–316.

[39] Nguyes, D. S., and E. Halvorsen, "Nonlinear Springs for Bandwidth-Tolerant Vibration Energy Harvesting," *J. Microelectromech. Syst.*, Vol. 20, 2011, pp. 1225–1227.

[40] Hoffmann, D., B. Folkmer, and Y. Manoli, "Analysis and Characterization of Triangular Electrode Structures For Electrostatic Energy Harvesting," *J. Micromech. Microeng.*, Vol. 21, 2011, p. 104002.

[41] Le, C. P., et al., "Microscale Electrostatic Energy Harvester Using Internal Impacts," *J. Intell. Mater. Syst. Struct.*, Vol. 23, 2012, pp. 1–13.

[42] Johnson, K. L., *Contact Mechanics*, 9th ed., Cambridge: Cambridge University Press, 2003.

[43] Hearn, E. J., *Mechanics of Materials II: The Mechanics of Elastic and Plastic Deformation of Solids and Structural Materials*, 3rd ed., Portsmouth, NH: Heinemann, 1997.

[44] Puttock, M. J., and E. G. Thwaite, "Elastic Compression of Spheres and Cylinders at Point and Line Contact," *National Standards Laboratory Technical Paper*, Melbourne, Australia: Commonwealth Scientific and Industrial Research Organization, 1969.

[45] Norden, B. N., "On the Compression of a Cylinder in Contact with a Plane Surface," Institute for Basic Standards, Washington, D.C.: National Bureau of Standards, 1973.

[46] Ramlan, R., et al., "Potential Benefits of a Non-Linear Stiffness in an Energy Harvesting Device," *Nonlinear Dyn.*, Vol. 59, 2009, pp. 545–558.

[47] Burrow, S. G., et al., "Vibration Energy Harvesters with Non-Linear Compliance," *Proc. SPIE*, Vol. 6928, 2008, pp. 692807–692810.

[48] Marinkovic, B., and H. Koser, "Smart Sand—A Wide Bandwidth Vibration Energy Harvesting Platform," *Appl. Phys. Lett.*, Vol. 94, 2009, p. 103505.

[49] Marzencki, M., M. Defosseux, and S. Basrour, "MEMS Vibration Energy Harvesting Devices with Passive Resonance Frequency Adaptation Capability," *J. Microelectromech. Syst.*, Vol. 18, 2010, pp. 1444–1453.

[50] Ramlan, R., "Effects of Non-Linear Stiffness on Performance of an Energy Harvester Device," PhD Thesis, University of Southampton, 2009.

[51] Galayko, D., et al., *Proceedings of International Conference on Solid-State Sensors, Actuators and Microsystems IEEE Transducers*, 2011, p. 2382.

[52] Blokhina, E., et al., "Steady-State Oscillations in Resonant Electrostatic Vibration Energy Harvesters," *IEEE Trans. Circuits Syst., I: Regul. Pap.*, Vol. 60(4), pp. 875–884, 2013.

[53] Blokhina, E., et al., *Proceedings of IEEE International Symposium on Circuits and Systems*, 2012, p. 397.

[54] Blokhina, E., et al., "Limit on Converted Power in Resonant Electrostatic Vibration Energy Harvesters," *Applied Physics Letters*, Vol. 101, 2012, p. 173904.

[55] Bu, L., et al., "Non-Resonant Electrostatic Energy Harvester for Wideband Applications," *Micro & Nano Letters*, Vol. 8, 2013, pp. 135–137.

[56] Kiziroglou, M. E., C. He, and E. M. Yeatman, "Non-Resonant Electrostatic Energy Harvesting from a Rolling Mass," *Proceedings of the 5th International Workshop on Wearable and Implantable Body Sensor Networks*, Hong Kong, June 1–3, 2008, pp. 318–321.

[57] Hsieh, W., T. Hsu, and Y. Tai, "A Micromachined Thin-Film Teflon Electret Microphone," *Transducers '97, Solid-State Sensors and Actuators*, Chicago, IL: IEEE, 1997, p. 2B.02.

[58] Hsieh, W. H., T. -J. Yao, and Y. -C. Tai, "A High Performance MEMS Thin-Film Teflon Electret Microphone," *Transducers '99 10th Int. Conf. Solid-State Sensors and Actuators*, Sendai: Institute of Electrical Engineers of Japan, 1999, pp. 1064–1067.

[59] Sterken, T., et al., 2003 "An Electrets Based Electrostatic μ-Generator," *Proc. Transducers'03*, June 8–12, 2003, pp. 1291–1294.

[60] Arakawa, Y., Y. Suzuki, and N. Kasagi, "Micro Seismic Power Generator Using Electret Polymer Film," *Power MEMS Conference*, Kyoto, Japan, 2004, pp. 187–190.

[61] Sakane, Y., Y. Suzuki, and N. Kasagi, "The Development of a High-Performance Perfluorinated Polymer Electret and Its Application to Micro Power Generation," *J. Micromech. Microeng.*, Vol. 18, 2008, p. 104011.

[62] Naruse, Y., et al., "Electrostatic Micro Power Generation from Low-Frequency Vibration Such as Human Motion," *J Micromech and Microeng.*, Vol. 19, 2009, p. 094002.

[63] Sterken, T., et al., "Harvesting Energy from Vibrations by a Micromachined Electret Generator," *Int. Conf. Solid-State Sensors, Actuators and Microsystems Conf.*, 2007, pp. 129–132.

[64] Sterken, T., et al., "An Electret-Based Electrostatic Micro-Generator," *Transducers 12th Int. Conf. Solid-State Sensors, Actuators and Microsystems*, 2003, pp. 1291–1294.

[65] Zhang, S. J., et al., "AU: PROVIDE ARTICLE TITLE," *Int. Conf. on Power MEMS*, Freiburg, Germany, November 2007, pp. 105–108.

[66] Naruse, Y., et al., "Electrostatic Micro Power Generator from Low Frequency Vibration Such as Human Motion," *J. Micromech. Microeng.*, Vol. 19, 2009, p. 094002.

[67] Minami, T., et al., "SiO2 Electret Films Prepared by Various Deposition Methods," *Thin Solid Films*, Vol. 426, 2003, pp. 47–52.

[68] Chen, Z., Z. Lv, and J. Zhang, "PECVD SiO2/Si3N4 double layers electrets on glass substrate," *IEEE Trans. on Dielectr. Electr. Insul.*, Vol. 15, 2008, pp. 915–919.

[69] Mescheder, U., et al., "Properties of SiO2 Electret Films Charged by Ion Implantation for MEMS-Based Energy Harvesting Systems," *J. Micromech. Microeng.*, Vol. 19, 2009, p. 094003

[70] Amjadi, H., "Charge Storage in Double Layers of Thermally Grown Silicon Dioxide and APCVD Silicon Nitride," *IEEE Trans. Dielectr. Electr. Insul.*, Vol. 6, 1999, pp. 852–857.

[71] Boland, J., et al., "Micro Electret Power Generator," *IEEE 16th Ann. Int. Conf. Micro Electro Mechanical Systems, 2003 (MEMS-03)* (Kyoto), 2003, pp. 538–541.

[72] Boland, J. S., et al., "Arrayed Liquid Rotor Electret Power Generator Systems," *18th IEEE Int. Conf. Micro Electro Mechanical Systems 2005 (MEMS 2005),* 2005, pp. 618–621.

[73] Boland, J. S., and Y. C. Tai, "Liquid-Rotor Electret Micropower Generator," *Solid-State Sensor, Actuator, and Microsystems Workshop*, 2004.

[74] Tsutsumino, T., Y. Suzuki, and N. Kasagi, "Electromechanical Modeling of Micro Electret Generator for Energy Harvesting," *Transducers 2007, Int. Conf. Solid-State Sensors, Actuators and Microsystems,* 2007, pp. 863–866.

[75] Tsutsumino, T., et al., "Seismic Power Generator Using High-Performance Polymer Electret," *19th IEEE Int. Conf. Micro Electro Mechanical Systems*, 2006 (MEMS 2006), (Istanbul), 2006, pp. 98–101.

[76] Kashiwagi, K., et al., "Nano-Cluster-Enhanced High-Performance Perfluoro-Polymer Electrets for Energy Harvesting," *J. Micromech. Microeng.*, Vol. 21, 2011, p. 125016.

[77] Lo, H. W., and Y. C. Tai, "Parylene-Based Electrets Power Generators," *J. Micromech. Microeng.*, Vol. 18, 2008, p. 115025.

[78] Hohn, D., "Kapazitive Siliziumsensoren fur Horschallanwendungen," *Fortschrittsbericht VDI No. 60*, Dusseldorf: VDI-Verlag, 1986.

[79] Kamiyanagi, H., and S. Shibata, "Charging Phenomena of the Mediaum Close Implantation by a Carbonization of the Surface Layer of the Photo-Resist," *5th Int. Workshop on Junction Technology*, 2005, pp. 65–68.

[80] Peano, F., and T. Tambosso, "Design and Optimization of a MEMS Electrets-Based Capacitive Energy Scavenger," *IEEE J. Microelectromech. Syst.*, Vol. 14, 2005, pp. 429–435.

[81] Le, C. P., and E. Halvorsen, "Evaluation of Parameters for Electrostatic Energy Harvesters," *Proc. DTIP of MEMS/MOEMS*, 2009, pp. 286–291.

[82] Boisseau, S., G. Despesse, and A. Sylvestre, "Optimization of an Electrets-Based Energy Harvester," *Smart Mater. Struct.*, Vol. 19, 2010, p. 075015.

[83] Bartsch, U., et al., "Influence of Parasitic Capacitances on the Power Output of Electrets-Based Energy Harvesting Generators," *Power MEMS 2009*, Washington, D.C., December 1–4, 2009, pp. 332–335.

[84] Boisseau, S., et al., "Cantilever-Based Electrets Energy Harvesters," *Smart Mater. Struct.,* Vol. 20, 2011, p. 105013.

[85] Kotrappa, P., "Long Term Stability of Electrets Used in Electret Ion Chambers," *J. Electrost.,* Vol. 66, 2008, pp. 407–409.

[86] Kranz, M., M. G. Allen, and T. Hudson, "In Situ Wafer-Level Polarization of Electrets Films in MEMS Acoustic Sensor Arrays," *Sensors and Actuators* A, Vol. 188, 2012, pp. 181–189.

[87] Crovetto, A., F. Wang, and O. Hansen, "An Electrets-Based Energy Harvesting Device with a Wafer-Level Fabrication Process," *J. Micromech. Microeng.*, Vol. 23, 2013, p. 114010.

[88] Yang, Z. C., E. Halvorsen, and T. Dong, "Electrostatic Energy Harvester Employing Conductive Droplet and Thin-Film Electrets," *IEEE J. Microelectromech. Syst.*, 2014.

[89] Boland, J. S., et al., "Arrayed Liquid Rotor Electret Power Generator Systems," *Proc. 18th IEEE Int. Conf. Micro Electro Mech. Syst.*, February 2005, pp. 618–621.

[90] Krupenkin, T., and J. A. Taylor, "Reverse Electrowetting as a New Approach to High-Power Energy Harvesting," *Nat. Commun.*, Vol. 2, December 2011, pp. 448–450.

[91] Choi, D. -H., et al., "Liquid-Based Electrostatic Energy Harvester with High Sensitivity to Human Physical Motion," *Smart Mater. Struct.*, Vol. 20, No. 12, 2011, p. 125012.

[92] Bu, L., et al., "Liquid Encapsulated Electrostatic Energy Harvester for Low-Frequency Vibrations," *J. Intell. Mater. Syst. Struct.*, Vol. 24, No. 1, 2012, pp. 61–69.

[93] Boland, J. S., et al., "Arrayed Liquid Rotor Electret Power Generator Systems," *Proc. MEMS'05*, 2005, pp. 618–621.

[94] Lo, H. -W., R. Whang, and Y. -C. Tai, "A Simple Micro Electrets Power Generator," *Proc. MEMS'07*, 2007, pp. 859–862.

[95] Lo, H. W., and Y. C. Tai, "Parylene-HT-Based Electret Rotor Generator," *Proc. MEMS'08*, 2008, pp. 984–987.

[96] Suzuki, Y., et al., "Micro Electret Energy Harvesting Device with Analogue Impedance Conversion Circuit," *Proc. PowerMEMS'08*, 2008, pp. 7–10.

[97] Halvorsen, E., et al., "An Electrostatic Energy Harvester with Electret Bias," *Proc. Transducers'09, 2009*, pp. 1381–1384.

[98] Honzumi, M., et al., "Soft-X-Ray-Charged Vertical Electrets and Its Application to Electrostatic Transducers," *Proc. MEMS 2010*, 2010, pp. 635–638.

[99] Miki, D., et al., "Large-Amplitude MEMS Electret Generator with Nonlinear Spring," *Proc. MEMS 2010*, 2010, pp. 176–179.

[100] Shimizu, N., Omron Prototypes Compact, Simple Vibration-Powered Generator, Nikkei Electronics, November 18, 2008, http://techon.nikkeibp.co.jp/english/NEWS EN/20081117/, p. 161303.

[101] Kloub, H., et al., "A Micro Capacitive Vibration Energy Harvester for Low Power Electronics," *Proc. PowerMEMS'09*, 2009, pp. 165–168.

[102] Sterken, T., et al., "Harvesting Energy from Vibrations by a Micromachined Electret Generator," *Int. Conf. Solid-State Sensors, Actuators and Microsystems*, 2007, pp. 129–132.

4

Triboelectric Energy Harvesting

Triboelectric energy harvesting devices work on the principles of contact electrification and electrostatic induction. Contact electrification is the phenomenon of charging two surfaces due to friction when the surfaces are put in contact with each other. Triboelectricity has been known as a phenomenon for many decades but usually has negative connotations attached to it because of its negative effects in electronic devices. Recently, Professor Z. L. Wang's group first demonstrated contact electrification as a viable method to generate power [1–3]. Triboelectric energy harvesters have a lot of potential to be a feasible solution for power needs, not just at the microscale or nanoscale but also at a larger scale (e.g., harvesting energy from ocean waves, wind energy, and energy from motion of vehicles and pedestrians).

4.1 Working Principle

The principle of triboelectricity is based on charging of two surfaces when they are put in contact with each other. The surfaces involved in the contact electrification process will be referred to as triboelectric layers hereafter in the text. There have been many theories proposed in literature to explain this phenomenon [4]. One of the well-established models to explain the concept of electron transfer is based on difference in work functions of materials [5]. According to this model, the charge transfer occurs due to difference in work function of two metals, which can be given by:

$$V_{1/2} = -\frac{(\phi_1 - \phi_2)}{e} \quad (4.1)$$

where $V_{1/2}$ is the contact potential difference of metal 1 against metal 2, ϕ_1 is the work function of metal 1, ϕ_2 is the work function of metal 2 and e is the elementary charge. Figure 4.1 pictorially depicts this model given by Harper [5]. This model can further be extended to the metal-insulator and insulator-insulator contact by using the concept of effective work function for insulators [6, 7].

Based on the varying tendencies of different materials during contact electrification process, various materials can be arranged in an order known as the triboelectric series [8, 9]. The materials up in the series have a tendency to donate electrons relative to the materials that are placed down in the series as shown in Figure 4.2. The relative position of materials in triboelectric series can be used to find the relative polarity of charges generated when two materials are put in contact with each other.

On the basis of operating mechanism, triboelectric mechanism for energy harvesting can broadly be divided into two categories: out-of-plane contact-separation mechanism and in-plane sliding mechanism. The theory and working principle for both mechanisms are explained in the following sections.

Figure 4.1 Charge transfer due to the difference in work function between two metals for metal-metal contact [4].

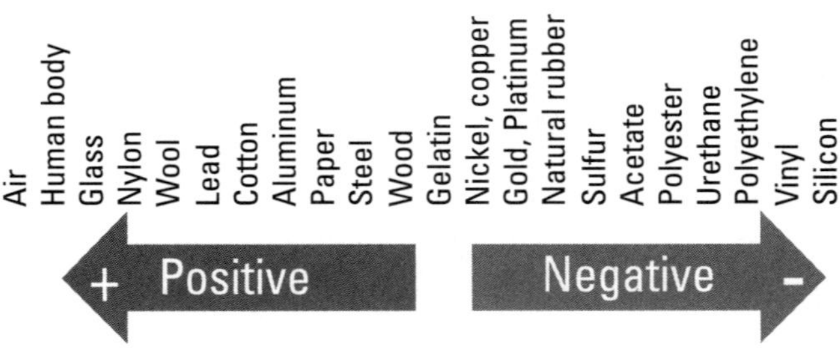

Figure 4.2 Triboelectric series for various materials arranged in order of their tendency to attract or donate electrons.

4.1.1 Out-of-Plane Contact-Separation Mechanism

In the out-of-plane mechanism, the triboelectric layers undergo contact and separation motion when the mechanism force or pressure is applied and released. During the contact, the triboelectric layers get charged with equal and opposite surface charge density due to contact electrification process. Due to the charges generated on triboelectric layers, there are charges induced on the metal electrodes in the energy harvesting device as shown in Figure 4.3 [10]. As the two layers move relative to each other, the potential difference between the two electrodes changes and there is current in the external circuit. This system can be treated as a system of parallel plate capacitors to develop a theoretical model. A general theoretical model was proposed by Niu et al. [10] for the out-of-plane contact-separation mechanism-based triboelectric energy harvesters. The open-circuit voltage and short-circuit current can be given by (4.2) and (4.3), respectively.

$$V_{oc} = \frac{\sigma x(t)}{\varepsilon_0} \tag{4.2}$$

$$I_{sc} = \frac{S\sigma d_0 v(t)}{\left(d_0 + x(t)\right)^2} \tag{4.3}$$

where V_{oc} is the open-circuit voltage, σ is the surface charge density, $x(t)$ is the relative distance between two triboelectric layers, ε_0 is the dielectric constant of air. I_{sc} is the short-circuit current, S is the area of triboelectric layers, $v(t)$ is the relative velocity of triboelectric layers, d_0 is the effective thickness constant given by $d_1/\varepsilon_{r1} + d_2/\varepsilon_{r12}$, d_1 and d_2 are thickness of triboelectric layers 1 and 2, and ε_{r1} and ε_{r12} are the relative dielectric constants of triboelectric layers 1 and 2.

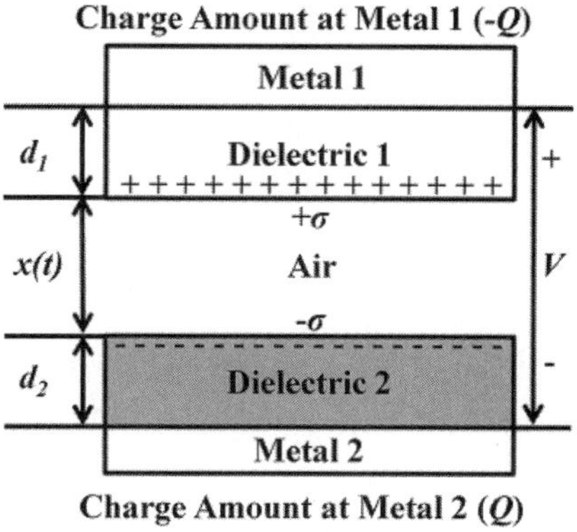

Figure 4.3 Schematic of an out-of-plane contact-separation-based triboelectric energy harvester [10].

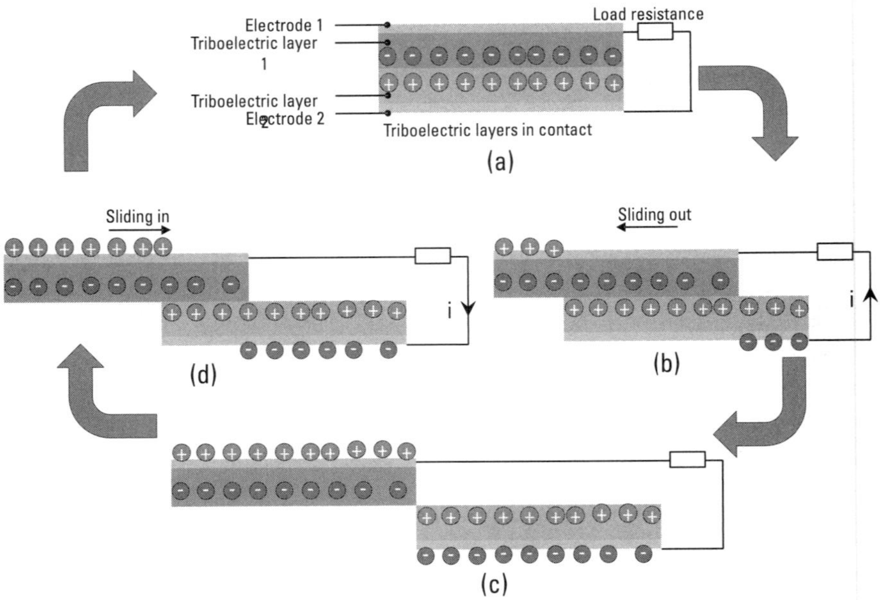

Figure 4.4 Schematic for in-plane sliding mechanism-based triboelectric energy harvester: both triboelectric layers (a) in contact; (b) sliding apart from each other; (c) at maximum separation; and (d) approaching each other.

4.1.2 In-Plane Sliding Mechanism

A schematic for in-plane sliding mechanism based triboelectric energy harvester is show in Figure 4.4. This mechanism involves relative sliding motion between triboelectric layers. As the triboelectric layers are in contact, there is equal and opposite surface charge generated on both the layers [Figure 4.4(a)]. As the triboelectric layer 1 moves out relative to triboelectric layer 2 as shown in Figure 4.4(b), it leads to a decrease in the contact area between two layers. Due to this relative movement and the charge configuration on both electrodes and triboelectric layers, there is an electric field that results in a higher potential at electrode 2 relative to electrode 1. There is a flow of electrons from the top electrode to the bottom electrode due to a potential difference resulting into a current from the bottom electrode to the top electrode. This flow of current continues till the two triboelectric layers are completely separated, as shown in Figure 4.4(c). Afterwards, the triboelectric layer 1 moves inwards relative to triboelectric layer 2, which results in current in the external circuit in the opposite direction of sliding out process, as shown in Figure 4.4(d). A theoretical model was proposed by Niu et al. [11] for the in-plane sliding mechanism-based triboelectric energy harvester. According to the model, the open-circuit voltage and short-circuit current can be expressed by (4.4) and (4.5), respectively.

$$V_{oc} = \frac{\sigma x}{\varepsilon_0 (l - x)} \left(\frac{d_1}{\varepsilon_{r1}} + \frac{d_2}{\varepsilon_{r2}} \right) \tag{4.4}$$

$$I_{sc} = \sigma w v(t) \tag{4.5}$$

where V_{oc} is the open-circuit voltage, σ is the surface charge density, x is the relative separation distance between triboelectric layers in lateral direction, l is the length of triboelectric layers in the direction of motion, ε_0 is the dielectric constant of air, d_1 and d_2 are thickness of triboelectric layers 1 and 2, ε_{r1} and ε_{r2} are the relative dielectric constants of triboelectric layers 1 and 2, I_{sc} is the short-circuit current, $v(t)$ is the relative velocity of triboelectric layers, and w is the width in the lateral direction perpendicular to motion direction of the triboelectric layer.

4.2 Materials and Fabrication

Performance of triboelectric energy-harvesting mechanism primarily depends on two things: choice of triboelectric materials and topography of the triboelectric layers. Choice of materials for triboelectric layers is important as it largely affects the amount of surface charge generated due to contact electrification phenomenon. Materials can be appropriately chosen according to their relative position in the triboelectric series (i.e., their relative tendency to attract or donate electrons). Surface topography is also a crucial factor in triboelectric mechanism as it is predominantly responsible for determining effective contact area between the triboelectric layers when they are in contact with each other. The performance of triboelectric mechanism is closely affected by the effective contact area [12], which makes surface topography an important factor to consider in the design and fabrication of triboelectric energy harvesters.

In literature, there have been various types of materials used in the triboelectric mechanism based energy harvesters. Moreover, different fabrication methods have been deployed to achieve various topographies and patterns at microscale and nanoscale on the triboelectric layers. A summary is presented in Table 4.1 about these materials and topographies used to improve the performance of triboelectric energy harvesters.

4.3 Development of Triboelectric Energy Harvesters

In past couple of years, many prototypes for triboelectric energy harvesters have been demonstrated by researchers, especially from Professor Z. L. Wang's group

Table 4.1
Materials and Surface Topographies for Triboelectric Energy Harvesters

Material	Surface Topography/Processing	Reference
TiO_2	Nanowires and nanosheet	[13]
Al	Nanopores	[14]
PTFE	Nanowires	[14]
SiO_2	Nanoparticles	[15]
PDMS	Pyramid-shaped structures using molding	[16]
Al	Nanostructures of average size ~200 nm	[16]
PDMS	Nanowire	[17]
Si	Pyramid-shaped texture	[17]
PTFE	Etched nanoparticle topography	[18]
Ag	Nanowire/nanoparticle composite	[19]
Fluorinated ethylene-propylene	Dry-etched nanowire structures	[20]
Kapton	Dispersed with PTFE nanoparticles	[21]
PDMS	Micropillars	[3]
Al	Nanopores	[22]
Poly-vinyl chloride	Nanowires	[23]
Cu	Micro roughness by coating Cu on etched Si structures	[24]
Kapton	Dry-etched nanowires	[25]
PDMS	Convex dome-shaped structures	[26]
Au	Nanoparticles	[27]

at Georgia Tech. One of the early prototypes was reported in 2012 by Fan et al. [1] based on triboelectric mechanism as shown in Figure 4.5(a). A transparent and flexible energy harvester was demonstrated that used patterned PDMS films as triboelectric layer. There were three differernt patterns used for topographies on PDMS films apart from the plain film. The three patterns used were line, cube, and pyramid-shaped structures. It was demonstrated that plain film, line, cube, and pyramid pattern-based devices performed in an increasing order as depicted in Figure 4.5(b). The device was also shown to be used as a pressure sensor that could be powered by itself. The device generated a maximum output voltage and current of 18V and 0.7 μA, respectively.

Yang et al. [3] demonstrated a triboelectric mechanism based energy harvester for detection of variation of time varying magnetic field as shown in

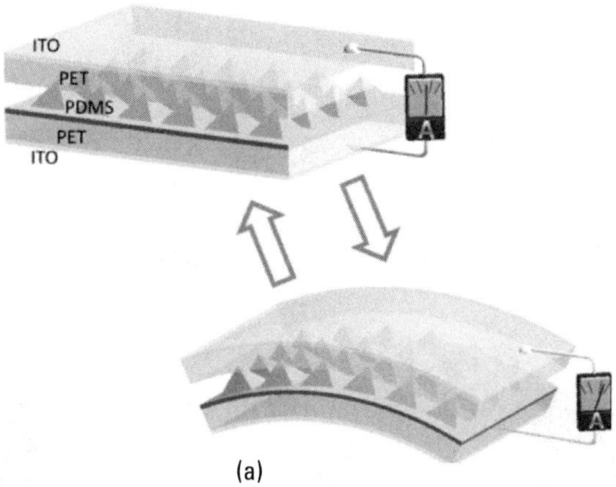

(a)

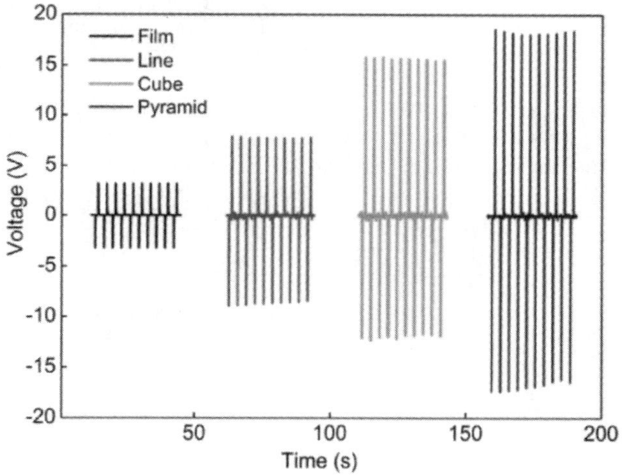

(b)

Figure 4.5 (a) Schematic drawing of one of the early prototypes of triboelectric mechanism based energy harvester and (b) voltage generated by the energy harvester using different topographies for patterned PDMS films. (Adapted with permission from [1]. Copyright: American Chemical Society, 2012.)

Figure 4.6. The device used polyethylene terephthalate film with indium tin oxide and PDMS with micropillar structures [Figure 4.6(d)] as the triboelectric layers. The detection sensitivity for change in the magnetic field was observed to be 0.363 ± 0.0004 ln(mV)/G. The device could detect a magnetic field of up to 3G for frequencies lower than 0.4 Hz.

A low-cost triboelectric energy harvester was reported by Bai et al. [22] that utilized a multilayered structure in a zigzag manner to increase the effective

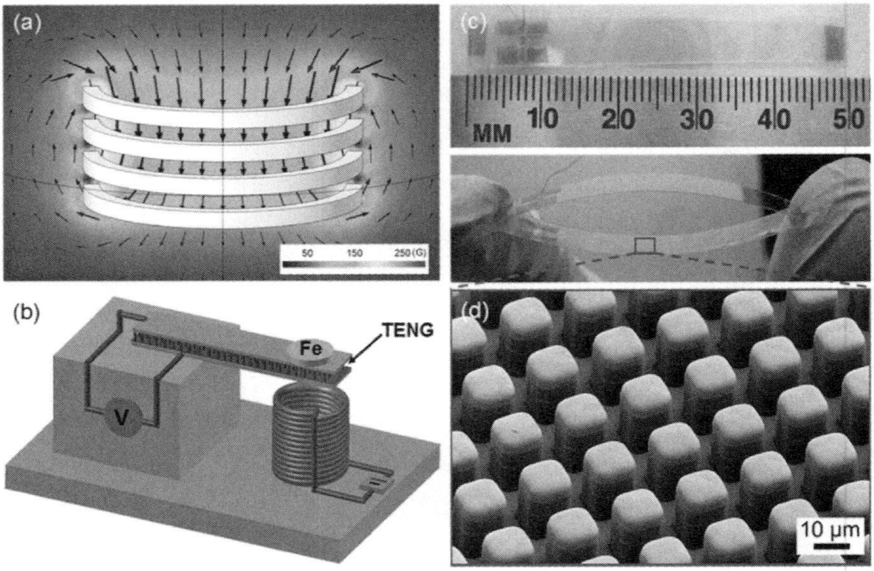

Figure 4.6 (a) Magnetic field variation in the coils for which sensor measurements were conducted, (b) schematic of self-powered magnetic sensor, (c) prototype of the triboelectric energy harvester, and (d) PDMS micropatterns used as triboelectric layer (Adapted with permission from [3]. Copyright: American Chemical Society, 2012.)

contact area without significantly changing the overall form factor of device (Figure 4.7). The reported device was assembled at the bottom of a shoe and was observed to generate enough power to light up a few LEDs using human motion. The energy-harvesting device generated a maximum open-circuit voltage of 215V and a short-circuit current of 0.66 mA. The maximum power density was reported to be 9.76 mW/cm^2. Zhu et al. [28] reported a shoe-based triboelectric energy technology to power consumer electronic devices. The triboelectric energy harvester was assembled with the insole, which could harvest pressure energy from the walking movement of a human. The prototype was demonstrated to light LED bulbs and also charge a cellphone. The energy harvester generated a maximum open-circuit voltage of 220V and a maximum short-circuit current of 600 μA.

Yang et al. [29] demonstrated a triboelectric energy harvester that could charge a lithium ion battery to power a biosensor for potential applications in water and harsh environments. The energy-harvesting device was fully enclosed in a spherical shell (Figure 4.8) and was shown to harvest energy from water waves. The spherically enclosed energy harvester produced voltages up to nearly 100V and a maximum current of about 8 μA when shaken by hand.

In 2013, Lin et al. [30] demonstrated water as a feasible triboelectric layer for contact electrification mechanism-based energy harvesters. A patterned

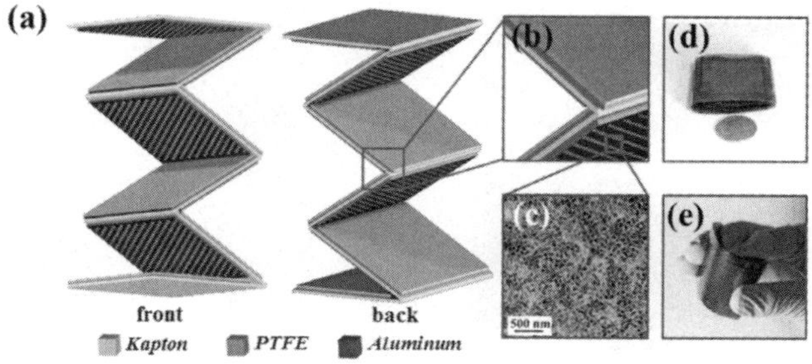

(i)

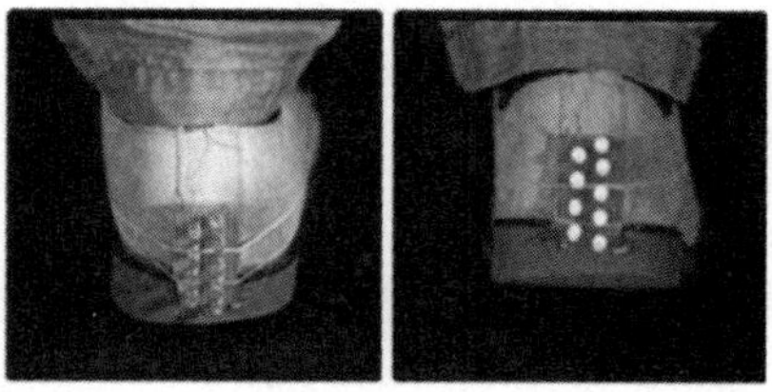

(ii)

Figure 4.7 (i) (a, b) Schematic diagram of the zigzag sgaped multilayered triboelectric energy harvester. (c) SEM micrograph of Al nanopores used one of the triboelectric layers. (d, e) Complete fabricated prototype of integrated multilayered triboelectric energy harvester. (ii) LED bulbs in the ON and OFF states during operation. (Adapted with permission from [22]. Copyright: American Chemical Society, 2013.)

PDMS film was used as the other triboelectric layer in the device. A schematic diagram of the operating mechanism of water-based triboelectric energy-harvesting device is shown in Figure 4.9. The device was shown to be able to light up to 60 LEDs. The energy harvester could generate up to a maximum open-circuit voltage of 52V and a short-circuit current density of 2.45 mA/m^2. Recently, another prototype has been reported to harvest electrostatic and mechanical energy from water [31]. The device utilized TiO_2 layer-based superhydrophobic surfaces to collect electrostatic energy from water, whereas a PTFE film and a SiO_2 layer were assembled to collect the mechanical energy from water. The overall power density of the device was observed to be 1.69 W/m^2.

Dhakar et al. [32] demonstrated a cantilever-based broadband triboelectric energy harvester as shown in Figure 4.10(a). The device utilized contact-separation mechanism to induce stiffening in the cantilever beam, which resulted into broadband characteristics of device. The maximum power output generated by the device was observed to be 0.91 μW at an acceleration of 1g. The maximum increase in the operating bandwidth of device was shown to be 284%. Different patterns were also studied using SU-8 micropillar arrays [Figure 4.10(b)]. It was shown that air void size between the SU-8 micropillars played an important role in determining the performance of the triboelectric

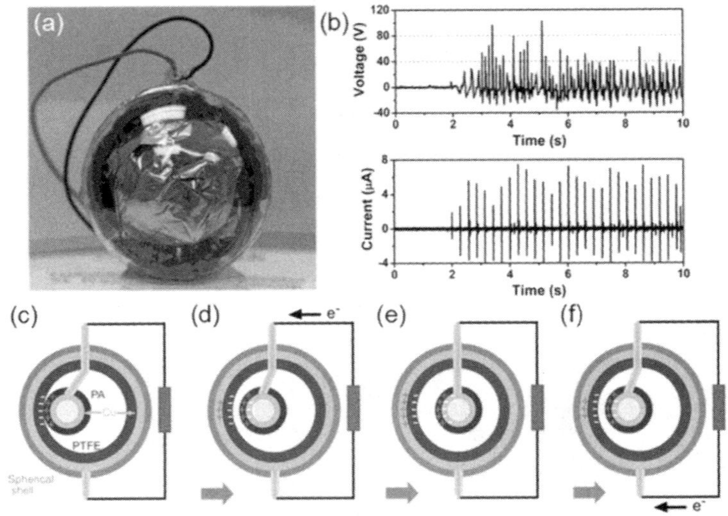

Figure 4.8 (a) A prototype of spherically enclosed energy harvester, (b) voltage and current generated by the device when shaken by hand, and (c–f) operating mechanisms of the energy-harvesting device [29].

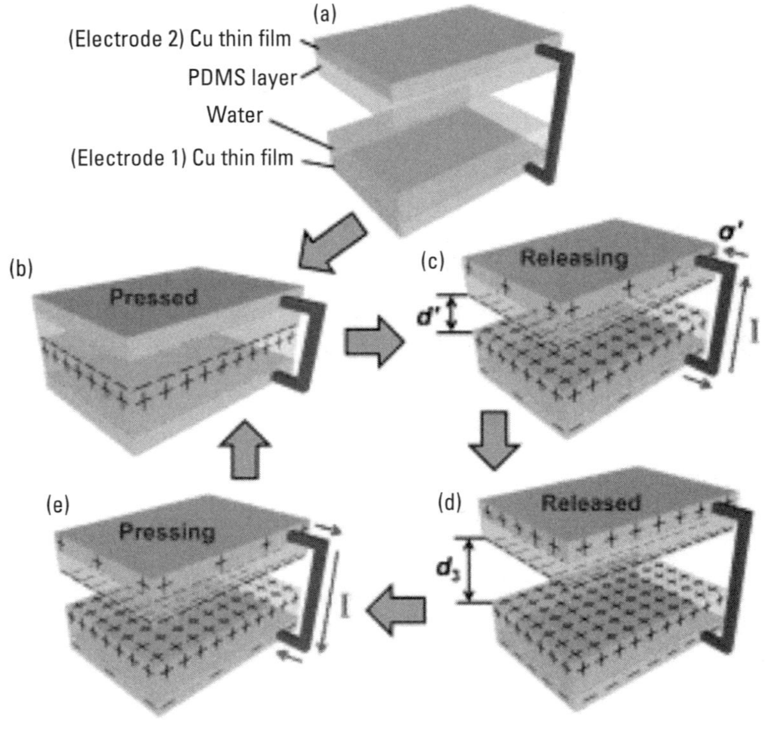

Figure 4.9 (a–e) Operating mechanism of triboelectric energy harvester using water as one of the triboelectric layers [30].

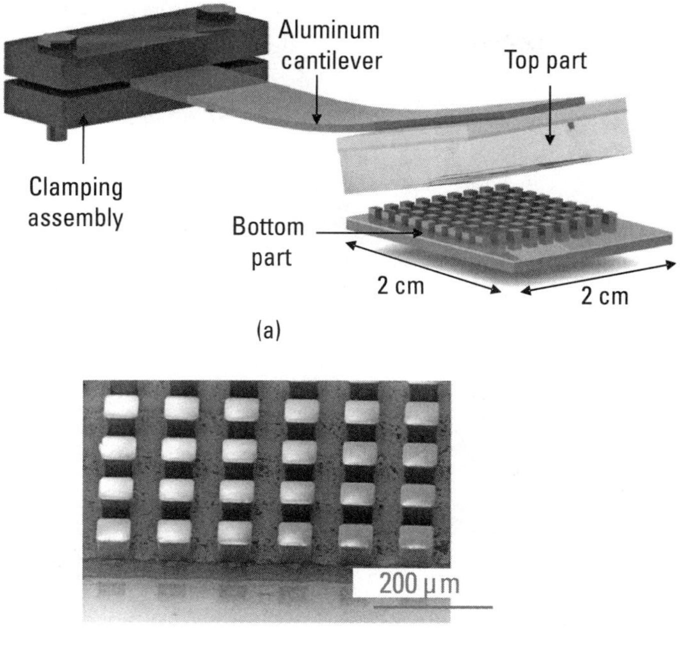

Figure 4.10 (a) Schematic drawing of the cantilever-based broadband triboelectric energy harvester. (b) SEM micrograph of SU-8 pillars used for patterning triboelectric layer [32].

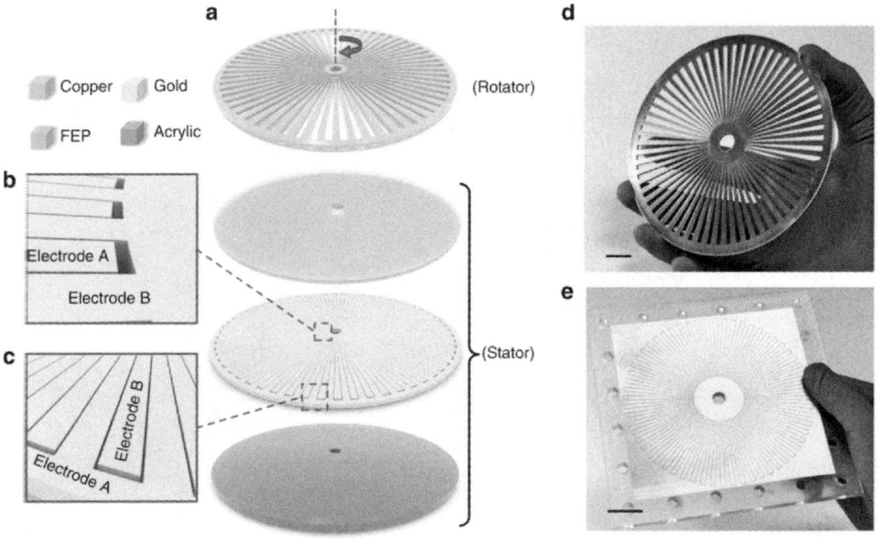

Figure 4.11 (a) Schematic design of the rotary mechanism-based triboelectric energy harvester, (b) the inner end of electrode, (c) the outer end of electrode, (d) a photograph of a rotator, and (e) a photograph of a stator [34].

energy harvester. Meng et al. [33] reported a triboelectric mechanism-based energy harvester that used the body as a conduit. It was demonstrated that using the human body as a conduit increased the output current by 39%. The device generated a maximum output voltage of 200V and a maximum output current density of 4.7 μA/cm^2.

Zhu et al. [34] demonstrated a rotary mechanism-based triboelectric energy harvester that used radial arrays of micro-sized sectors as contact surfaces. Figure 4.11 shows the schematic diagram and photograph of the prototype. The energy harvester was demonstrated to light LED bulbs, power an alarm system, power a digital clock, and charge a cellphone. The rotary motion in the triboelectric energy harvester was realized by using a small wind/water turbine that supplied torque to the device. The device was also shown to generate power using the human body motion. The prototype had a high-conversion efficiency of 24% and generated a maximum output power of 1.5W.

In conclusion, triboelectric energy harvesters present a novel mechanism to harvest mechanical energy apart from conventional piezoelectric and electromagnetic mechanisms. These devices have huge potential in the area of energy harvesting not just at the microscale or nanoscale but for a large-scale energy generation.

References

[1] Fan, F.-R., et al., "Transparent Triboelectric Nanogenerators and Self-Powered Pressure Sensors Based on Micropatterned Plastic Films," *Nano Letters,* Vol. 12, 2012, pp. 3109–3114.

[2] Fan, F.-R., Z.-Q. Tian, and Z. L. Wang, "Flexible Triboelectric Generator," *Nano Energy,* Vol. 1, 2012, pp. 328–334.

[3] Yang, Y., et al., "Self-Powered Magnetic Sensor Based on a Triboelectric Nanogenerator," *ACS Nano,* Vol. 6, 2012, pp. 10378–10383.

[4] Matsusaka, S., et al., "Triboelectric Charging of Powders: A Review," *Chemical Engineering Science,* Vol. 65, 2010, pp. 5781–5807.

[5] Harper, W., "The Volta Effect as a Cause of Static Electrification," *Proceedings of the Royal Society of London. Series A. Mathematical and Physical Sciences,* Vol. 205, 1951, pp. 83–103.

[6] Davies, D., "Charge Generation on Dielectric Surfaces," *Journal of Physics D: Applied Physics,* Vol. 2, 1969, p. 1533.

[7] Murata, Y., and S. Kittaka, "Evidence of Electron Transfer as the Mechanism of Static Charge Generation by Contact of Polymers with Metals," *Japanese Journal of Applied Physics,* Vol. 18, 1979, p. 421.

[8] Henniker, J., "Triboelectricity in Polymers," *Nature,* Vol. 196, pp 474–475, 1962.

[9] Diaz, A., and R. Felix-Navarro, "A Semi-Quantitative Tribo-Electric Series for Polymeric Materials: The Influence of Chemical Structure and Properties," *Journal of Electrostatics,* Vol. 62, 2004, pp. 277–290.

[10] Niu, S., et al., "Theoretical Study of Contact-Mode Triboelectric Nanogenerators as an Effective Power Source," *Energy & Environmental Science,* Vol. 6, 2013, pp. 3576–3583.

[11] Niu, S., et al., "Theory of Sliding-Mode Triboelectric Nanogenerators," *Advanced Materials,* Vol. 25, 2013, pp. 6184–6193.

[12] Matsusaka, S., M. Ghadiri, and H. Masuda, "Electrification of an Elastic Sphere by Repeated Impacts on a Metal Plate," *Journal of Physics D: Applied Physics,* Vol. 33, 2000, p. 2311.

[13] Lin, Z. -H., et al., "Enhanced Triboelectric Nanogenerators and Triboelectric Nanosensor Using Chemically Modified TiO2 Nanomaterials," *ACS Nano,* Vol. 7, 2013, pp. 4554–4560.

[14] Yang, W., et al., "Harvesting Energy from the Natural Vibration of Human Walking," *ACS Nano,* Vol. 7, 2013, pp. 11317–11324.

[15] Cheng, G., et al., "Pulsed Nanogenerator with Huge Instantaneous Output Power Density," *ACS Nano,* Vol. 7, 2013, pp. 7383–7391.

[16] Han, M., et al., "r-Shaped Hybrid Nanogenerator with Enhanced Piezoelectricity," *ACS Nano,* Vol. 7, 2013, pp. 8554–8560.

[17] Yang, Y., et al., "Silicon-Based Hybrid Energy Cell for Self-Powered Electrodegradation and Personal Electronics," *ACS Nano,* Vol. 7, 2013, pp. 2808–2813.

[18] Yang, Y., et al., "Single-Electrode-Based Sliding Triboelectric Nanogenerator for Self-Powered Displacement Vector Sensor System," *ACS Nano,* Vol. 7, 2013, pp. 7342–7351.

[19] Lin, L., et al., "Triboelectric Active Sensor Array for Self-Powered Static and Dynamic Pressure Detection and Tactile Imaging," *ACS Nano,* Vol. 7, 2013, pp. 8266–8274.

[20] Yang, Y., et al., "Triboelectric Nanogenerator for Harvesting Wind Energy and as Self-Powered Wind Vector Sensor System," *ACS Nano,* Vol. 7, 2013, pp. 9461–9468.

[21] Jing, Q., et al., "Case-Encapsulated Triboelectric Nanogenerator for Harvesting Energy from Reciprocating Sliding Motion," *ACS Nano,* Vol. 8, 2014, pp. 3836–3842.

[22] Bai, P., et al., "Integrated Multilayered Triboelectric Nanogenerator for Harvesting Biomechanical Energy from Human Motions," *ACS Nano,* Vol. 7, 2013, pp. 3713–3719.

[23] Du, W., et al., "A Three Dimensional Multi-Layered Sliding Triboelectric Nanogenerator," *Advanced Energy Materials,* Vol. 4, 2014.

[24] Guo, H., et al., "A Nanogenerator for Harvesting Airflow Energy and Light Energy," *J. Mater. Chem. A,* Vol. 2, 2014, pp. 2079–2087.

[25] Zhu, G., et al., "Triboelectric-Generator-Driven Pulse Electrodeposition for Micropatterning," *Nano Letters,* Vol. 12, 2012, pp. 4960–4965.

[26] Hou, T. -C., et al., "Triboelectric Nanogenerator Built Inside Shoe Insole for Harvesting Walking Energy," *Nano Energy,* Vol. 2, 2013, pp. 856–862.

[27] Zhu, G., et al., "Toward Large-Scale Energy Harvesting by a Nanoparticle-Enhanced Triboelectric Nanogenerator," *Nano Letters,* Vol. 13, 2013, pp. 847–853.

[28] Zhu, G., et al., "Power-Generating Shoe Insole Based on Triboelectric Nanogenerators for Self-Powered Consumer Electronics," *Nano Energy,* Vol. 2, 2013, pp. 688–692.

[29] Yang, Y., et al., "Fully Enclosed Triboelectric Nanogenerators for Applications in Water and Harsh Environments," *Advanced Energy Materials,* Vol. 3, 2013, pp. 1563–1568.

[30] Lin, Z. -H., et al., "Water–Solid Surface Contact Electrification and Its Use for Harvesting Liquid Wave Energy," *Angewandte Chemie International Edition,* Vol. 52, 2013, pp. 12545–12549.

[31] Lin, Z. -H., et al., "Dual-Mode Triboelectric Nanogenerator for Harvesting Water Energy and as a Self-Powered Ethanol Nanosensor," *ACS Nano,* Vol. 8, pp 6440–6448, 2014.

[32] Dhakar, L., F. E. H. Tay, and C. Lee, "Development of a Broadband Triboelectric Energy Harvester with SU-8 Micropillars," *Journal of Microelectromechanical Systems,* Vol. PP, No, 99, 2014.

[33] Meng, B., et al., "Single-Friction-Surface Triboelectric Generator with Human Body Conduit," *Applied Physics Letters,* Vol. 104, 2014, p. 103904.

[34] Zhu, G., et al., "Radial-Arrayed Rotary Electrification for High Performance Triboelectric Generator," *Nature Communications,* Vol. 5, 2014.

5

Strategies for High-Performance Vibration Energy Harvesters

Power consumption of some MEMS sensors is in the range of tens of microwatts to hundreds of microwatts. Within this range of power consumption, self-powered systems based on harvesting ambient energy become a viable solution. Using MEMS technology to realize tiny energy-harvesting devices will be a rational solution. It also allows us to integrate the sensor elements, ICs, wireless transmission part, and energy-harvesting device together into one standalone package or in a system-in-a-package (SiP) approach. Normally, the energy output per vibration cycle from MEMS energy harvester is less than several hundred nanojoules. Therefore, the output power is derived as 1 μW (i.e., 10 nJ times 100 Hz, or a few microwatts). It explains why the current reported resonant-mode MEMS energy harvesters can only provide output power in the range of 0.1 μW to 10 μW under the natural frequency.

To improve the output power of the MEMS energy harvesters, there are currently two major strategies to realize this. The first method is to employ hybrid conversion mechanisms of the energy harvester so that its power is the total one from each mechanism, for example, electromagnetic and piezoelectric mechanisms. The second method is to widen the bandwidth of the energy harvester, which can be realized by multifrequency scavenging, oscillator array, nonlinear mechanism, frequency-upconversion (FUC), and nonresonant methods. This chapter provides and discusses some existing strategies used to increase the output power of vibration energy harvesters and thus broaden their practical applications.

5.1 Hybrid Energy Conversion Strategies

5.1.1 Electromagnetic and Piezoelectric Hybrid Mechanisms

Harvesting energy from ambient vibrations to recharge a battery is a potential solution to realized handheld electronics. Typically, vibration-based energy harvesting devices employ one of the following energy transduction mechanisms: piezoelectric, electromagnetic, electrostatic, or triboelectric mechanisms. The transduction mechanism itself can generate electricity by exploiting the mechanical strain or relative displacement occurring with the system [1]. Mitcheson et al. compared the performance limits of the first three MEMS energy-harvesting mechanisms [2, 3]. The resonant quality factor of electromagnetic energy harvesters is proportional to resonant frequency. In contrast, the resonant quality factor of piezoelectric energy harvesters is proportional to the square of frequency. Thus, the optimum output power of piezoelectric mechanisms becomes rapidly diminished with increased frequency. Within the high-frequency range, the degeneration rate of output power of piezoelectric mechanisms is more rapid than the one in the electromagnetic case [2]. As a guideline, piezoelectric energy harvesters outperform electromagnetic energy harvesters at low frequency, and the electromagnetic mechanism is favorable in the high-frequency range. Piezoelectric energy harvesters usually produce high voltages and lower current. In comparison with piezoelectric energy harvesters, electromagnetic energy harvesters tend to produce relatively low AC voltage, and the voltage output is decreased when the size scales down [4]. A comparative study of electromagnetic and piezoelectric energy harvesters for portable devices has been presented [5]. It points out that piezoelectric energy harvesters present high power density and are more suitable for microsystem applications, while electromagnetic energy harvesters are good at applications of a relatively large size.

To overcome drawbacks of each individual type of the three energy-harvesting mechanisms, a creative energy harvester is proposed to harness energy from typing motions on a computer keyboard by combining electromagnetic (in terms of typing speed and frequency) and piezoelectric (using typing force) mechanisms [6]. Figure 5.1 shows 3-D view of a micromachined electromagnetic and piezoelectric power harvester. This energy harvester includes a 191-μm-thick layer of lead-zirconatetitanate piezoelectric transducer (PZT) ceramics coated with electrodes on double sides, and spiral coils of 13 turns with area of 1 cm^2 (i.e., 100-μm-wide aluminum coil electrode pattern with 100-μm spacing).

Based on the simplified electrical model of the electromagnetic power harvester, the output power given to the load can be expressed as

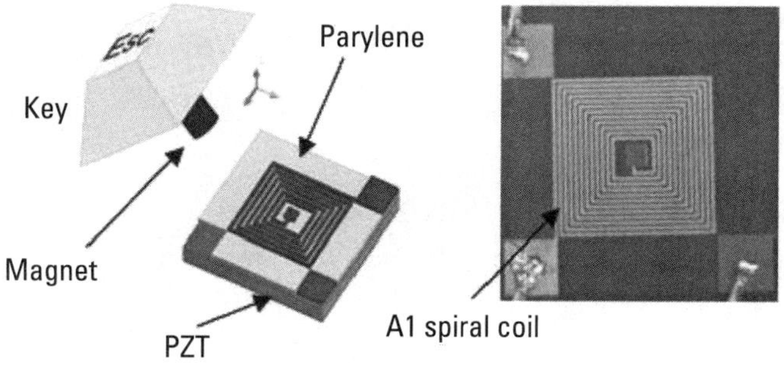

Figure 5.1 The 3-D view of the micromachined power harvester integrated microcoils on PZT and a picture of the fabricated device [6].

$$P_{EM} = \frac{V_S^2 R_L}{(R_S + R_L)^2 + (\omega L)^2} \tag{5.1}$$

where R_S is the internal resistance, L is internal inductance, and V_S and R_L are the generated voltage source and load resistance, respectively. Thus, the output power can be increased by adjusting R_S, L, and R_L. The effect of the coil inductance is negligible since $(R_S + R_L) \gg (\omega L)$ due to the small inductance and the low operating frequency. The maximum power is obtained when R_S and R_L are equal.

Thus, from an equivalent circuit model of the piezoelectric harvester, the output power will be

$$P_{PE} = \frac{V_S^2}{(1+\left(\frac{R_S + R_L}{\omega C_S R_S R_L}\right)^2)R_L} \tag{5.2}$$

where C_S is the internal capacitance. Since the typing motions are generally in a low and narrow frequency range (3–8 Hz), the frequency variance has a relatively small impact on the output power from the piezoelectric power harvester.

Figure 5.2 shows the experimental setup of the micromachined power harvester. This device includes a single key element integrated with a small magnet (5 mm long and 3mm in diameter), a piezoelectric element, a key-guiding structure, a rubber ring spacer, and a rubber cushion spring. Figure 5.3 shows the output performance of the piezoelectric and electromagnetic harvesters, respectively. The maximum power is about 40.8 μW from the PZT layer with a load resistance of 3 MΩ, while it is about 11.6 pW from the spiral coils with a 700Ω load. This energy harvester shows the potential of providing enough power to recharge batteries while the keyboard is in use.

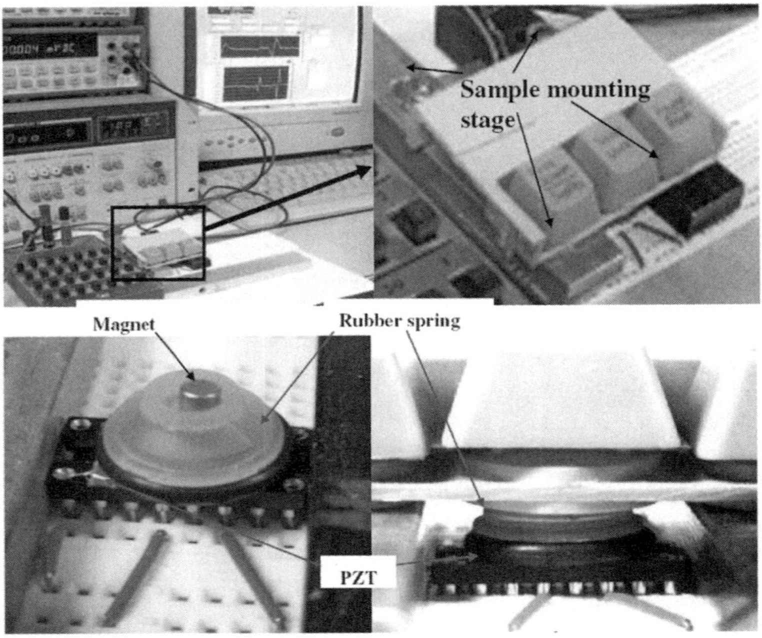

Figure 5.2 The experimental setup of power harvester for keyboard testing [6].

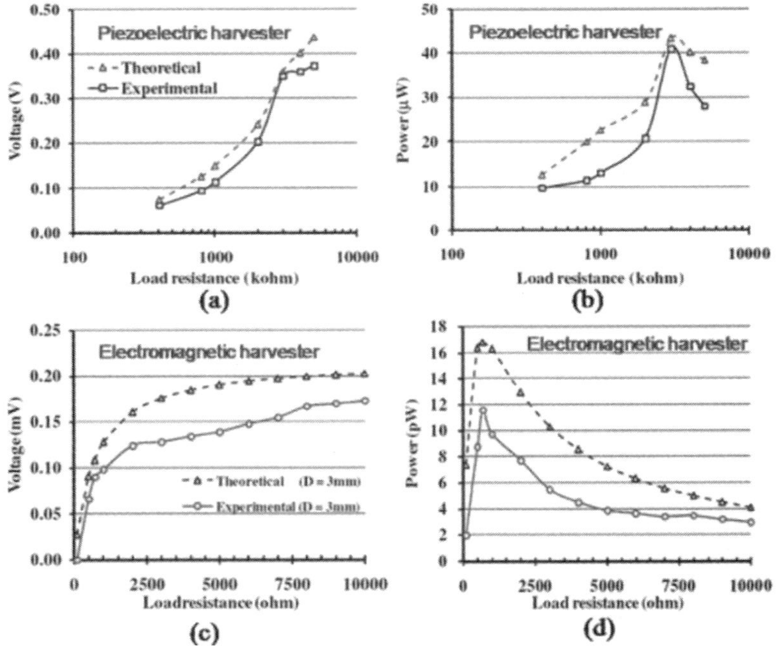

Figure 5.3 (a) The output voltage and (b) power from piezoelectric harvester and (c) the output voltage and (d) power from electromagnetic harvester [6].

The coupling of energy-harvesting mechanism is deployed to improve the output performance in comparison to the output of the single mechanism. Considering the air damping, as shown by Williams [7] and Challa [8], the total power output from the coupled mechanisms device (with two electrical dampers) based on lumped model can be written as

$$P_{coupled} = \frac{m_{struc}\left(\dfrac{\omega_s}{\omega_{struc}}\right)^3 \omega_s^{\,3}(\zeta_{e1}+\zeta_{e2})Y^2}{\left[\left[1-\left(\dfrac{\omega_s}{\omega_{struc}}\right)^2\right]+2\zeta_t\dfrac{\omega_s}{\omega_{struc}}\right]^2} \tag{5.3}$$

where ζ_t is the total damping ratio ($\zeta_t = \zeta_m + \zeta_{e1} + \zeta_{\varepsilon 2}$), $\zeta_m = b_m/2m_{struc}\omega_{struc}$ and are $\zeta_{ei} = b_{ei}/2m_{struc}\omega_{struc}$ are the mechanical and electrical damping ratios of each energy harvesting technique in the system, respectively, and ω_{struc} is the undamped natural frequency of the vibrating structure.

When the device is in resonance ($\omega_s = \omega_{struc}$), the output power is dependent on the resonance frequency, effective mass, amplitude of the vibration source, and damping components in the system. While the total electrical damping in the coupled system is the sum of the individual electrical damping components, the total power output of the coupled device is not simply the sum of the optimized, individual stand-alone energy harvesting techniques, as the increase in total damping in the system reduces the vibration amplitude (and hence the strain) of the piezoelectric cantilever beam, thereby reducing the stresses and the corresponding power output from the piezoelectric technique. Thus, from a design perspective, the energy harvested from the electromagnetic technique must be greater than the decrease in the energy from the piezoelectric technique to result in a net increase in power output of the coupled device. It is to be noted that the coupled device would have higher power output only for the case when the electrical damping from piezoelectric technique is less than the mechanical damping in the system [8].

To optimize the overall efficiency of the coupled device from the perspective of electrical damping, the coupling of a cantilever-based piezoelectric cantilever with the proof mass of magnet and electromagnetic techniques are proposed. Figure 5.4(a, b) shows the schematic and photograph of the coupled electromagnetic-piezoelectric energy harvester, respectively.

The performance comparison between the unimorph and bimorph cantilever design of PZT has been demonstrated by Wischke et al. [9]. The unimorph design generates more electromagnetic power because the lower resonance frequency results in the higher tip velocity. With the higher stiffness, the bimorph exhibits a much higher resonance frequency and, consequently, a

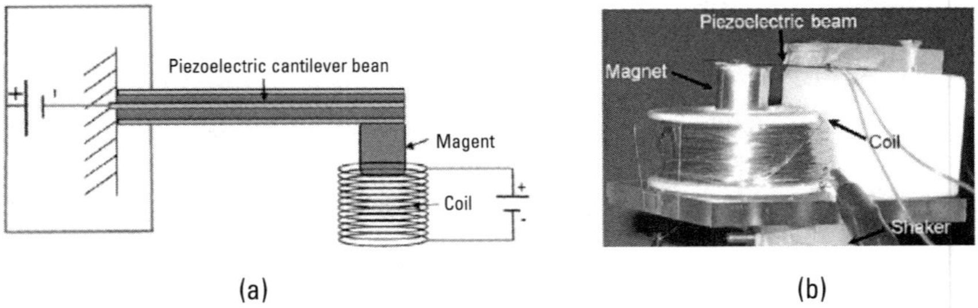

Figure 5.4 (a) Schematic of a coupled mechanism harvesters; and (b) photograph of the coupled electromagnetic-piezoelectric energy harvester [8].

much smaller tip velocity. Therefore, the electromagnetic power output is negligible compared to the increased piezoelectric output, which originates from the higher strain in the stiffer layout. Here a trade-off between the strain in the piezo and the velocity of the magnet is to increase the tip mass by using a larger magnet. Due to the fixed-free condition of the cantilever, its tip moves along a parabolic track. The magnet will therefore dip into the coil with a certain inclination that reduces the electromagnetic coupling. The power output of the hybrid piezoelectric-electromagnetic generator was increased up to 150% compared to a single unimorph piezoelectric cantilever. At the resonant frequency of 299 Hz, 60 μW, and 215 μW can be obtained with 10 m/s^2 stimulation and optimal load conditions from the electromagnetic and piezoelectric parts, respectively, as shown in Figure 5.5 [10]. The drawback of clamp-free cantilever is the inclination between magnet and coil, which results in the partial straining of piezoelectric layer.

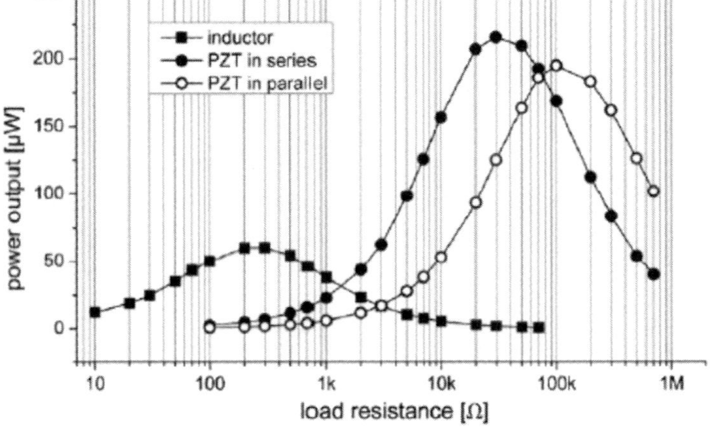

Figure 5.5 Output power from electromagnetic and piezoelectric part [10].

Researchers at the Illinois Institute of Technology have investigated the feasibility of hybrid mesoscale energy-scavenging topologies from linear motions of body for human-powered mobile electronics [11]. The structure of the hybrid PZT and permanent magnet energy scavenging is shown in Figure 5.6.

In this structure, the moving mass with an attached permanent magnet is connected to four serpentine piezoelectric beams on four sides, and the copper coil is fixed in the middle of the moving mass. With the human body's center of gravity up and down, both the piezoelectric beam and the copper coil will have power generated. The serpentine beam is 19.6 cm long, 0.4 mm wide, and 0.18 mm thick. The proof mass is made of tungsten alloy with 4.9 cm^3. The turn number of the copper coil is 350. The theoretical computation points out that, with a vibration frequency of 2 Hz and a vibration amplitude of 3 cm, 37 mW and 6 mW can be generated from the electromagnetic and piezoelectric parts, respectively.

As the energy harvester size scales down to the typical scales of MEMS structures, the proof mass vibration amplitude is limited if the ambient vibration amplitude is too large. So for constant acceleration vibration, mesoscale energy harvesters are suitable for low-frequency and large-amplitude applications while MEMS energy scavengers are more for the application of high-frequency and small-amplitude cases [12].

Printed circuit board (PCB) technology has been proposed as a low-cost approach for making millimeter-scale MEMS devices such as scanning mirrors [13] and electromagnetic energy harvester [14]. A novel hybrid mechanism harvester consists of a multilayer PZT cantilever, a PCB substrate of two-layer coils, and magnets as seismic mass [15]. This PZT cantilever is built up from a number of PZT layers with each layer of 30-μm thickness and a few micrometers thick screen-printed electrodes in between two layers of piezoelectric ceramic pattern. The length, width, and total thickness of the piezoelectric

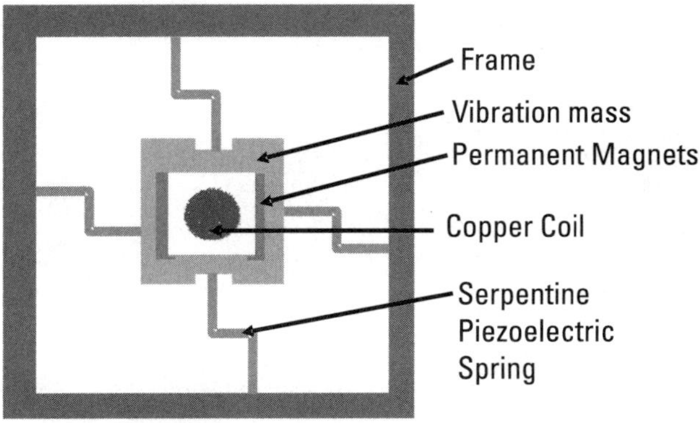

Figure 5.6 Proposed hybrid energy scavenging topology [11].

ceramic cantilever are 22 mm, 9.6 mm, and 0.65 mm, respectively. The bilayer copper coils are placed near the magnets, which are made of thin fire-resistant 4 (FR4) substrates based on a standard PCB technology. Three types of the hybrid energy harvesters are investigated to understand the influence of the relative position of PCB coils and magnets on output power, where PCB coils and magnets are manually assembled in the configurations, as shown in Figure 5.7.

The gap between the magnets and coils is kept at 0.5 mm for all types of devices, and it is adjusted by a 3-D nano-position controller. In the experimental setup platform, the dynamic signal analyzer is used to control the vibration amplitude of the shaker through power amplifier and the excited frequency, and to record the output voltages from PZT cantilever and coils. Figure 5.8(a, b) shows the output voltages of PZT cantilever and coils under the acceleration of 2.5g as a function of magnet numbers for devices of types I, II, and III, respectively. Because magnets are distributed symmetrically on the double sides of PZT cantilever end for types II and III, which results in almost the same output voltage from piezoelectric mechanism. However, the output voltage level is larger than the one of type I, especially for the case of more than four magnets, as shown in Figure 5.8(a). For type II and III devices, the PCB coils substrate is placed vertically against the end of cantilever, and this configuration results in lower damping ratio for a given input acceleration. When the poling direction of magnets is normal to the coils plane (e.g., the cases of type I and III), the output voltage from coils is much higher than the one of type II device (i.e., the poling direction of magnets parallel to the coils plane). Thus, the output voltage of type I devices ranges from 0.5 mV to 0.74 mV for magnets of 1 to 4, and the output voltage of 0.6 mV to 0.77 mV is derived for type III devices of 2, 4, and

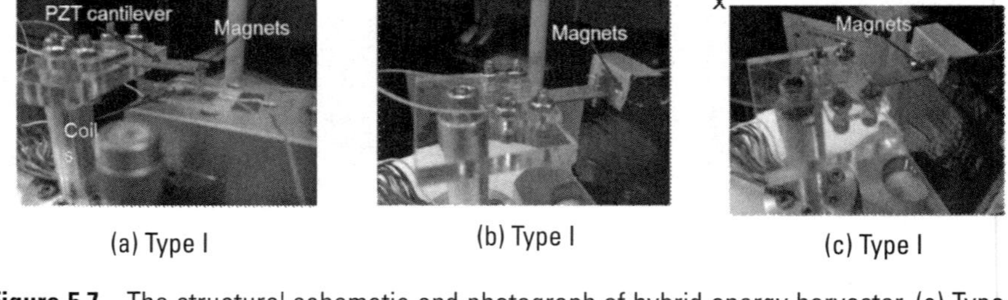

Figure 5.7 The structural schematic and photograph of hybrid energy harvester. (a) Type I: Magnets poling direction along z-axis, plane coils located in horizontal plane; (b) Type II: Magnets poling direction along z-axis, plane coils located in vertical plane; (c) Type III: Magnets poling direction along y-axis, plane coils located in vertical plane.

6 magnets. In contrast, the output voltage of 0.2 mV to 0.13 mV is derived for type II devices of 2, 4, and 6 magnets as shown in Figure 5.8(b).

Figure 5.8(c, d) shows the output power of PZT cantilever and PCB coils under different loading resistance for type I and III devices, respectively. When the loading resistance matched with the internal resistance of PZT cantilever, the maximum piezoelectric output power of types I and III is 107 μW and 176 μW under the vibrations of 2.5g acceleration at 350 Hz and 310 Hz, respectively. The maximum output power of types I and III from electromagnetic mechanism is 0.18 μW and 0.19 μW, respectively, when the loading resistance is matched with the internal resistance of coils. To evaluate the power density, we derived the overall volume of energy harvester including PZT cantilever and magnets. Thus, the volume of type I device with 4 magnets and type III device with 6 magnets is calculated as 194 mm^3 and 222 mm^3, respectively. The power densities of piezoelectric and electromagnetic mechanisms for type I and type III devices are derived as 550 $\mu W/cm^3$ and 0.93 $\mu W/cm^3$ and 790 $\mu W/cm^3$ and 0.85 $\mu W/cm^3$, respectively.

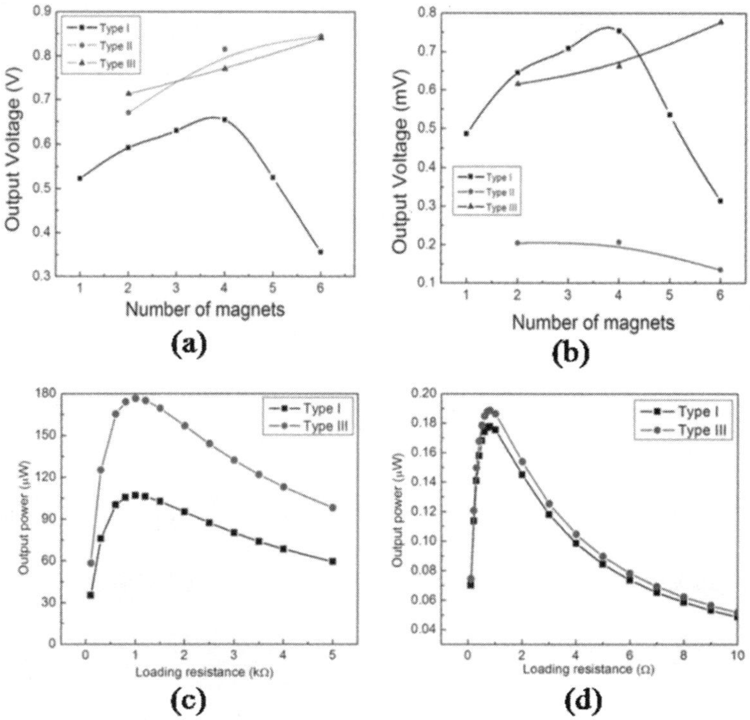

Figure 5.8 Measured open-circuit output voltages (Vrms) versus numbers of magnets under the cases of type I, type II, and type III. (a) Generated voltage from piezoelectric cantilever; (b) Generated voltage from electromagnetic coils; (c) output power generated power from piezoelectric cantilever; and (d) power generated power from electromagnetic coils.

5.1.2 Piezoelectric and Triboelectric Hybrid Mechanisms

A triboelectric nanogenerator (TENG) with the ultrahigh power density of 54,268 W/m^3 (i.e., 54.268 mW/cm^3) has been demonstrated, which proves the feasibility of powering microelectronic devices [16]. To effectively harvest vibration energy from the environment, low resonant frequency and wide bandwidth are the required characteristics of microharvesters. A novel flexible hybrid energy harvester based on a piezoelectric and triboelectric mechanism is shown in Figure 5.9 [17]. The resonant frequencies of the three cantilevers made of low Young's modulus polyvinylidene fluoride (PVDF) are designed to be 15, 30, and 45 Hz. With a 0.0225g PDMS proof mass, the length of each cantilever should be 25, 16, and 12 mm, respectively.

The piezoelectric energy harvester and TENG are assembled on the glass substrate to work with low-frequency vertical vibration input. Figure 5.10(a) shows the photo of the prototype. The microscope photo of the laser-etched PVDF thin film is also shown in Figure 5.10(b). In addition, scanning electron microscope (SEM) photos of the micro/nano dual-scale (MNDS) PDMS are shown in Figure 5.10(c, d). For the piezoelectric energy harvester, a sweep-frequency test from 1 to 60 Hz is carried out. Three voltage peaks are observed at 15, 32.5, and 47.5 Hz, with the value of 320, 288, and 264 mV, respectively. For the TENG, time-domain measurement is conducted at the first-order resonant frequency 15 Hz. The generated peak-peak voltage of TENG is as high as 20V and can instantaneously light up a commercial light emitting diode (LED). During the electrification process, the MNDS PDMS surface can greatly improve efficiency and output voltage of the TENG [18].

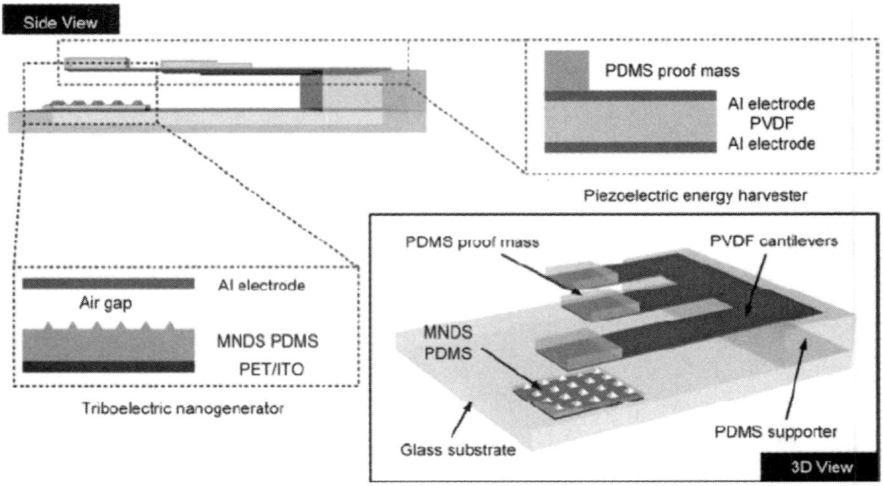

Figure 5.9 Side view and 3-D schematic of the device [17].

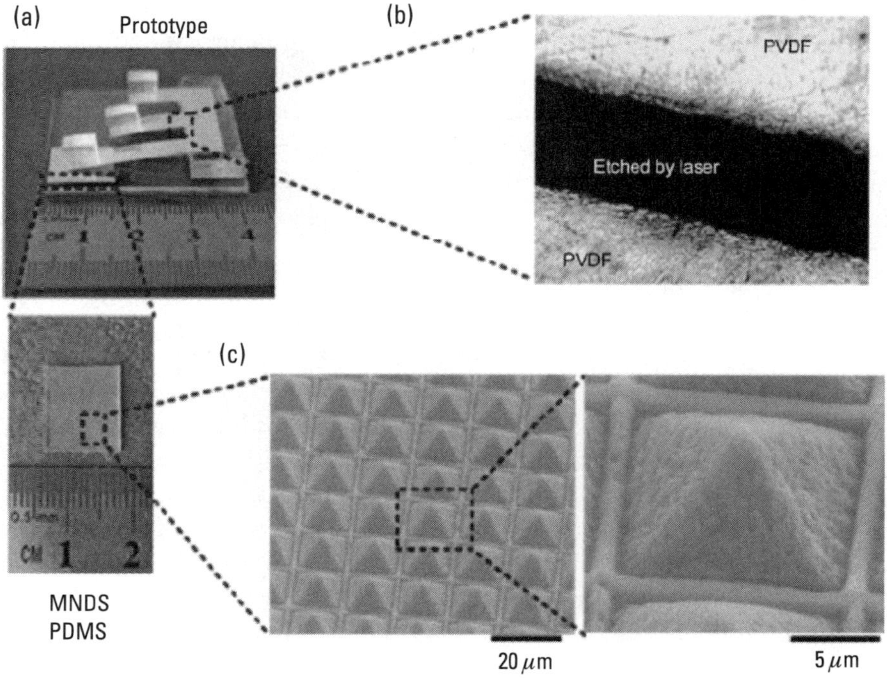

Figure 5.10 (a) Photo of the device and MNDS PDMS. (b) Microscope image of laser-etched PVDF thin film. (c) SEM images of the MNDS PDMS [17].

In recent years, ZnO nanowire harvesters on different substrate and distinct device structures have been reported [19–21]. The low output power of this ZnO harvester hinders their practical applications. Kim et al. [22] reported a novel harvest using both the piezoelectric effect from vertically aligned zinc oxide nanowire (ZnO NW) and the triboelectric effect from PDMS. Vertically aligned ZnO nanowires were grown on both sides of the flexible substrates by sol-gel spin-coating method and were sandwiched between the gold-coated vertically aligned ZnO nanowire electrode substrate using PDMS. This fabricated device successfully applied for biomechanical energy harvesting from human body activities like folding, stretching, and pressing. The generated average voltage and current value were 30V and 300 nA with output power density of 0.39 mW/cm^2 when folding the device by the fingers. The voltage and current values were 5V and 24V and 2 nA and 8 nA when stretching and pressing the fingers, respectively.

5.1.3 Piezoelectric and Electrostatic Hybrid Mechanisms

Electrostatic harvesters require either an asynchronous electrets configuration or dynamic operation using storage of a precharge and synchronization with the mechanical oscillation to operate properly [23]. It results in lower power densities due to the volume occupied by nontransducing components and power consumption by control electronics. To solve this problem, a system-based approach to couple both piezoelectric and electrostatic energy scavenging mechanisms simultaneously in a shared volume has been described at the design and simulation level. However, in-plane configurations are limited by trade-offs between piezoelectric coupling and material compliance as well as complexity of fabrication [24].

A piezoelectric resonator beam (PZT) is integrated with an electrostatic scavenger whose variable capacitor (CVAR) doubles as the proof mass as shown in Figure 5.11 [25]. The piezoelectric beam (PZT) forms the resonant beam simultaneously creating an AC voltage that provides displacement feedback signal and a rectified DC rail for control and load circuits. The output performance from electrostatic is lower than the generation from piezoelectric. However, the output performance from electrostatic can be improved by two methods. First, in terms of electrical area, it is imperative to move toward a monolithic ASIC that incorporates the required functionality with a focus on low power operation to minimize control circuit power consumption. Second, considering mechanical technology, the ability to dynamically and precisely control the electrostatic gap is paramount to obtaining efficient operation of the hybrid system. In addition, the enhancing of surface area at the minimum displacement would increase the maximum change in capacitance [26].

By employing higher conversion coefficient materials, higher values of piezoelectric damping could be obtained to more closely match the mechanical damping in the system to improve the power output of the stand-alone (original) piezoelectric device, but in terms of cost-effectiveness the use of a coupled device approach may be more desirable. Alternatively, another manner to significantly increase the power outputs from the energy-harvesting techniques is by reducing the mechanical damping in the system. In the future, to further increase the power output of the device will be realized through materials and device design [8].

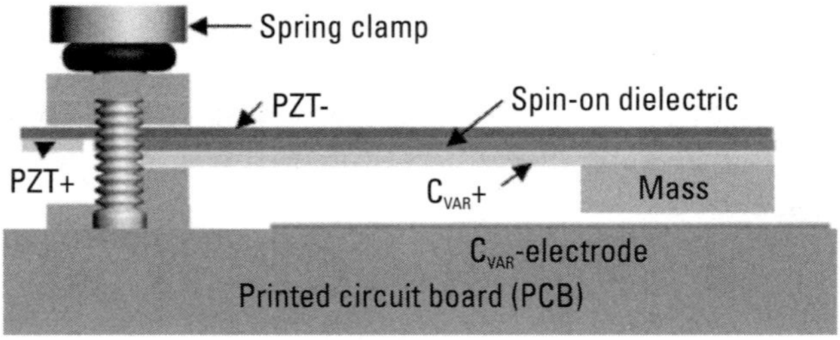

Figure 5.11 Diagram of hybrid scavenger cross section [25].

5.2 Frequency Broadening Strategies

The vibration-based energy-harvesting technique has been considered as a promising technology and has attracted noticeable research interests. No matter which transduction principle is exploited (i.e., piezoelectric, electromagnetic, or electrostatic), most of the MEMS energy harvesting research focuses on developing a linear spring-mass-damper system, in which the maximum output performance is achieved when the resonant frequency of the harvester matches the frequency of the ambient vibration. Any change of the ambient excitation frequency away from the resonant frequency of the harvester can result in a significant decrease in power generation. However, in ambient commonly occurring vibration sources, such as office buildings, manufacturing and industrial environments, military devices, transportations, household appliances, and human bodies, they are normally present random and irregular frequency peaks or contain various frequency peaks in different directions. These features have limited the output performance and applicability of traditional linear resonant energy harvesters. Broadening the bandwidth of vibration energy harvesters becomes one of the most challenging issues before their practical deployments.

To address the aforementioned frequency bandwidth limitation, the first solution is to develop a tunable energy harvester, such that the resonant frequency can be adjusted to match the frequency of ambient vibration at all times. The resonant frequency f_r of a spring-mass structure is given by

$$f_r = \frac{1}{2\pi}\sqrt{\frac{k}{m}} \tag{5.4}$$

where k is the spring stiffness and m is the inertial mass of the structure. Through the change of either the spring stiffness or the inertial mass, one can tune the resonant frequency of the spring-mass structure. As most reported vibration energy harvesters are based on a cantilever with a mass at the free end as shown in Figure 5.12, the spring stiffness depends on its materials and dimensions; thus, the resonant frequency f_{rc} is given by [27]

$$f_{rc} = \frac{1}{2\pi}\sqrt{\frac{Y\omega h^3}{4l^3\left(m+0.24m_c\right)}} \tag{5.5}$$

where Y is Young's modulus of the cantilever material; w, h, and l are the width, thickness, and length of the cantilever, respectively; and m_c and m are the mass of the cantilever and free end mass, respectively.

The resonant frequency can be tuned by changing the cantilever dimensions, varying the spring stiffness, straining the cantilever, and moving the

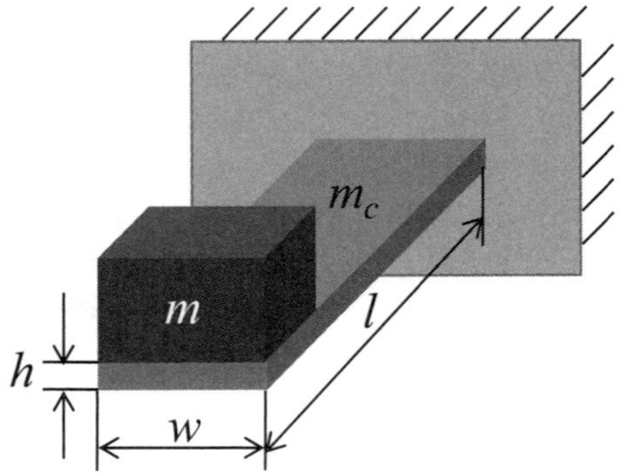

Figure 5.12 Spring-mass structure of a cantilever with mass.

center of gravity of the proof mass. There is a large number of frequency tuning solutions reported in the field of vibration energy harvesting. A comprehensive review and comparison was presented by Zhu et al. [28]. Changing the effective length [29] and the position of the gravity center [30] are potentially suitable for intermittent tuning. However, they are not suitable for in situ tuning or tuning with automatic control. The frequency can be tuned intermittently or continuously by changing the spring stiffness [31–33]. However, extra systems and energy are required or sometimes has to be adjusted manually. The electrical tuning method is much easier to implement than mechanical methods [34]. However, an extra closed loop system has to be introduced to control the tuning process. Among all these frequency tunable approaches reported in the field of vibration energy harvesting, unfortunately, most of the solutions can be achieved only in macroscale, due to the fabrication and structure assembly restrictions.

The other commonly used solution is to broaden the operating bandwidth of a vibration energy harvester. It is unlike a conventional energy harvester that the maximum power is generated only when its single resonant frequency matches with the dominant frequency of ambient vibration. A wideband energy harvester is capable of producing sufficient power in an increased bandwidth as long as the variant or random frequency of ambient vibration falls within its operating frequency range. Comparing to the frequency tunable approach, the frequency broadening strategy without requiring manual intervention has more application potential and thus attracted the attention of many researchers. This section mainly focuses on the effective strategies of the frequency broadening that are applicable in MEMS energy harvesting, including exploiting an oscillation system with multiple vibration modes, using a harvester array with different resonant frequencies, employing nonlinear spring structures, introducing amplitude limiters or mechanical stoppers, and adopting FUC mechanisms. In this section, a comprehensive review of the principles and operating mechanisms are summarized in detail with regard to their merits and applicability in different circumstances.

5.2.1 Multifrequency Harvesting Mechanisms

The multifrequency harvesting mechanism is designed to scavenge energy from vibration sources with multiple frequencies peaks along the same or different directions. Unlike a tunable or wideband energy harvester that increases the operating range of the first resonant mode in a continuous frequency range, a multifrequency energy harvester utilizes a spring-mass structure of multidegree-of-freedom (multi-DOF) to achieve multiple resonant modes at discrete frequencies. Each vibration mode of the device matches with a particular frequency peak of ambient vibration source to achieve more efficient vibration energy harvesting overall. In practice, the multi-DOF spring-mass structures are designed in three main configurations as shown in Figure 5.13. In configuration I, the oscillation structure contains multiple springs (k_1, k_2 ... k_i) and proof masses (m_1, m_2 ... m_i) connected in a series with a free end. The energy-harvesting transducers (f_1, f_2 ... f_i) can be implemented at any of the spring-mass hierarchy as shown in Figure 5.13(a). Similarly, configuration II consists of a series of spring-mass hierarchies but with a fixed-fixed end as shown in Figure 5.13(b). For configuration III [Figure 5.13(c)], it contains multiple spring structures connected with a single proof mass m. Most of the reported multifrequency energy harvesters can be categorized into these three configurations. Their differences are materials, geometries and constructions of the spring-mass structures, number of the spring-mass hierarchies, and energy conversion mechanisms. We will give some examples for each configuration in the following.

5.2.1.1 Configuration I

Dual-spring-mass energy-harvesting configurations consist of two masses connected in series by the energy conversion elements and springs. Theoretical investigations proved that the dual-spring-mass vibration energy harvester is able to provide multiple close and effective peaks in power response and significantly enhanced magnitude of power than the single-spring-mass configuration [35–37]. Yang et al. [38] reported a piezoelectric energy harvester with vibrations of multiple frequencies by incorporating configuration I. The total power of 46.22 μW generated at resonant frequencies of 89, 104, and 130 Hz, is obviously larger than the one of 28.35 μW for traditional single-spring-mass system. Another dual-spring-mass energy harvester based on configuration I was designed and microfabricated by Vullers et al. [39]. As shown in Figure 5.14(a), a packaged MEMS piezoelectric cantilever is attached at the tip of a metal beam to form a dual oscillation system. The generated power is significantly boosted

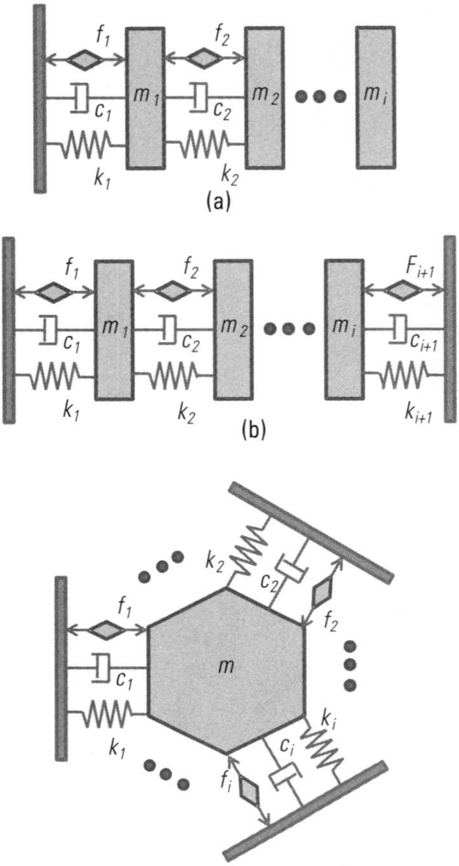

Figure 5.13 Configurations of the multi-DOF spring-mass structures: (a) configuration I; (b) configuration II; and (c) configuration III.

of more than 51 times due to the large packaged mass and spring ratios of the original cantilever and the secondary beam oscillator. Meanwhile, the bandwidth of the harvester is increased by the dual resonant peaks. El-Hebeary et al. [40] investigated multimodal vibration energy harvesting from V-shaped and open delta-shaped plates as shown in Figure 5.14(b). The plate structure with multiple magnets attached on it can be modeled as a series of spring-mass hierarchies. The device oscillates across the stationary coils to generate power at three natural frequencies in the range from 8 to 19 Hz.

5.2.1.2 Configuration II

Another type of multifrequency energy harvesters can be implemented by multiple bending modes of a continuous beam or by an array of cantilevers. Roundy et al. [41] proposed another single multi-DOF oscillator by incorporating

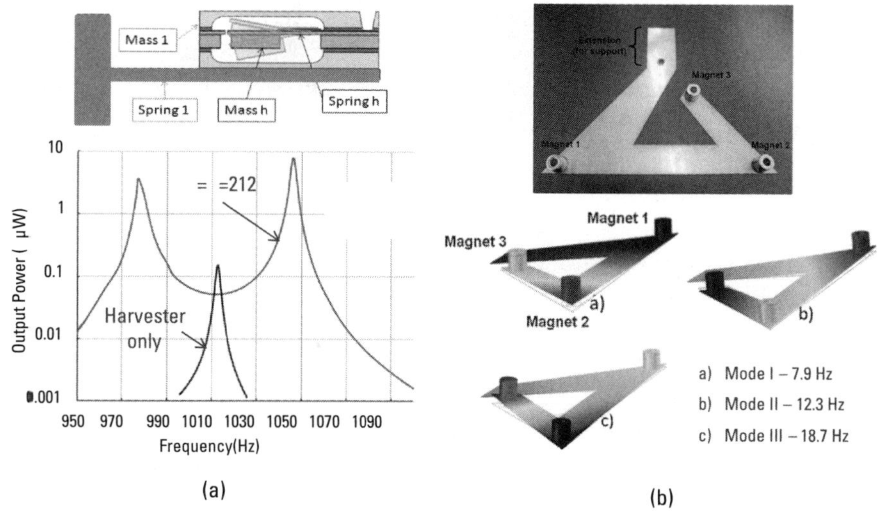

Figure 5.14 Multifrequency energy harvesters of configuration I by (a) Vullers [39] and (b) El-Hebeary [40].

multiple proof masses attached on a clamped–clamped beam to achieve multifrequency vibration harvesting. Based on a similar idea, Yang et al. [42] successfully fabricated an electromagnetic energy harvester with multiple frequencies as shown in Figure 5.15. The device consists of an acrylic supporting beam attached with three permanent magnets, and three sets of two-layer copper coils. The vibration energy under three resonant modes of 369, 938, and 1,184 Hz can be harvested, respectively. Qi et al. [43] explored a clamped–clamped

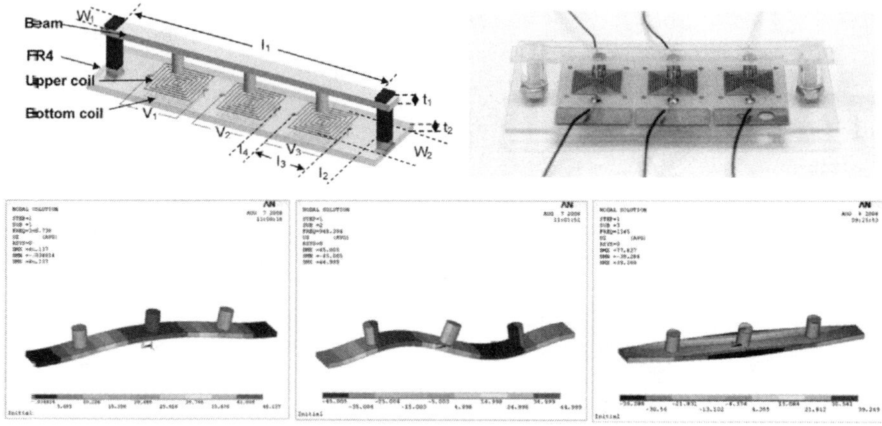

Figure 5.15 A multifrequency electromagnetic energy harvester of configuration II [42].

beam piezoelectric fiber composite (PFC) energy harvester with array of side mounted cantilevers. Those side-mounted cantilevers are added with different masses to be resonant at different frequencies and thus possess a multifrequency energy-harvesting characteristic. The above reviewed multi-DOF vibrations are achieved by exploiting multiple spring-mass oscillating structures. The dynamic behavior is governed by the geometry of the oscillating beam as well as the placement of the multiple magnets. Via a proper design of the system configuration, the spacing of the resonant modes can be determined so as to efficiently harvest vibration energy from a wide bandwidth.

5.2.1.3 Configuration III

Ching et al. [44] reported a multifrequency resonating energy harvester for wireless sensing as seen in Figure 5.16. The device consists of a NeFeB magnet attached on a spiral spring fabricated by laser-micromachining [Figure 5.16(a)]. Such a spring-mass structure exists in three resonating modes, which are out-of-plane vibration along the z-axis [Figure 5.16(c)], in-plane rotation along the x-axis [Figure 5.16(d)] and along 45 about x-axis [Figure 5.16(e)]. The electrical coil is fixed around the spring-mass structure such that the movement of the magnet results in varying amount of magnetic flux passing through the coil and

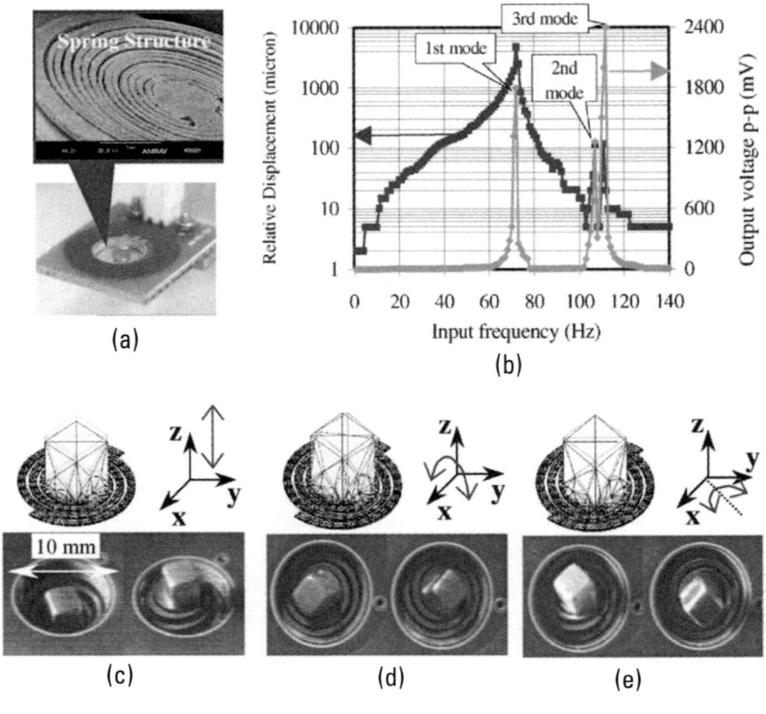

Figure 5.16 A laser-machined multifrequency energy harvester of configuration III [44].

a voltage is induced in the coil. Wang et al. [45] presented a similar multifrequency micro-energy harvester, which consists of a spring suspension structure attached with a permanent magnet and a stationary two-layer copper coil on glass substrate. The spring structure is composed of four copper square spiral spring beams and a platform on Si substrate. From the experimental results, the energy harvester device shows three resonant modes at 55, 121.25, and 122.2 Hz, respectively. Besides the circular and square spiral spring design, Liu et al. [46] developed a circular-rings and circular-mass system to form a 3-DOF micro-oscillator. The two-layer metal coils are patterned on the movable circular mass and a cylindrical permanent magnet is placed above the oscillator with a gap of 1 mm. The 3-D resonating modes of out-of-plane vibration along the z-axis (1,285 Hz) and in-plane vibration along 60 (1,470 Hz) and 150° (1,550 Hz) about x-axis are obtained. As the circular mass vibrates in-plane or out-of-plane, the moving coils cutting through the magnet field will induce an electrical current according to Faraday's law of induction. These three reported MEMS electromagnetic energy harvesters basically make use of the configuration III oscillator with multi-DOF to generate output power at three discrete resonant modes. Because of the symmetric configuration of the suspension structure, the natural frequencies of the second and third modes of such devices are very close to each other.

5.2.2 Broadband Harvesting of Oscillator Array

A straightforward concept of broadband energy harvesting is to integrate an array of single-DOF oscillators, each of which has different dimensions and hence different resonant frequencies. The power spectrum of the harvester array is a combination of the power spectra of each single harvester. Thus, the effective operating bandwidth can be thus essentially increased while the Q-factor does not decrease. Such types of single-DOF oscillator arrays have been proposed by several groups via piezoelectric transducers [47–49]. Xue et al. [49] reported a broadband piezoelectric harvester design by integrating multiple piezoelectric bimorphs with different aspect ratios. The bandwidth of the harvester can be shifted to the dominant frequency of ambient vibrations by increasing or decreasing the number of piezoelectric bimorphs in parallel. It is indicated that by the use of multiple bimorphs, the harvester system gains not only an increased frequency band from (97, 103) Hz to (87, 115) Hz, but also a much higher power output. Sari et al. [50] presented a MEMS electromagnetic energy harvester, which consists of a series of cantilevers patterned with double layer coils on top as shown in Figure 5.17. Each individual cantilever is designed to have different lengths resulting in an array of cantilevers with varying natural frequencies. The overall power spectrum of the device generated via the relative movement between a stationary magnet and the oscillating coils covers a

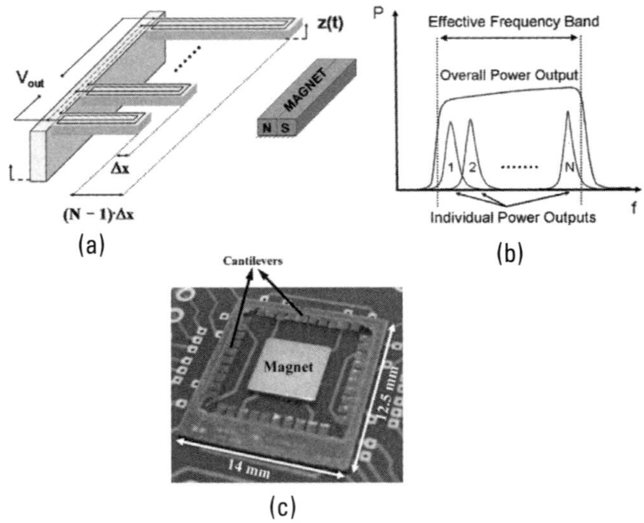

Figure 5.17 Broadband energy harvesters using oscillator array by Sari [50].

continuous frequency range from 4.2 kHz to 5 kHz. The drawback of this approach is the increased total volume and complexity of the fabricating of the harvester array. Moreover, it is volume inefficient since at a particular vibration frequency, only a single or a few individual harvesters contribute to the power output.

5.2.3 Broadband Harvesting of Nonlinear Mechanisms

The frequency broadening of an energy harvester can be realized by changing the effective spring stiffness of the oscillating system. The dynamic motion of a general oscillator can be expressed as

$$\ddot{x}+\gamma\dot{x}+\frac{dU(x)}{dx}=f(t) \tag{5.6}$$

where x represents the motion of the oscillator; γ is the viscous damping; $U(x)$ represents the potential function, and $f(t)$ is the excitation force of ambient vibration. For a nonlinear oscillator with a Duffing-type spring, the potential function $U(x)$ can be expressed in a quadratic form as

$$U(x)=\frac{1}{2}Ax^2+\frac{1}{4}Bx^4 \tag{5.7}$$

Figure 5.18 describes several types of potential functions for a Duffing spring [51]. Case I in the thin solid line represents a linear spring at $A > 0$ and $B = 0$;

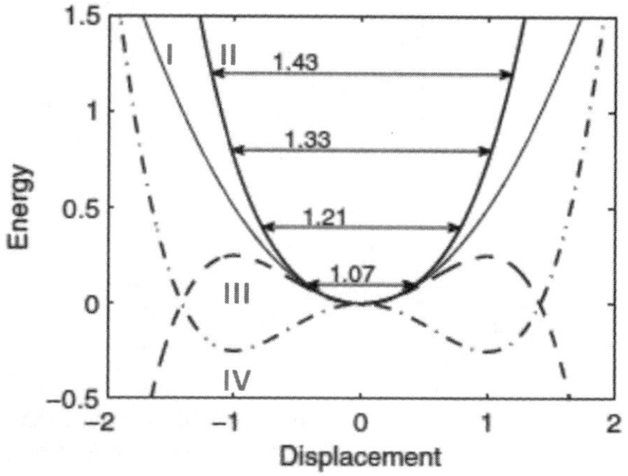

Figure 5.18 Different types of potential functions for a Duffing spring [51].

case II in the solid line indicates a spring stiffness hardening condition at $A > 0$ and $B > 0$; case III at $A > 0$ and $B < 0$ shows a spring softening curve in the dashed line where the stiffness decreases with displacement; and case IV at $A < 0$ and $B > 0$ results in a symmetric bistable potential in the dashed-dotted line.

From (5.7), the force-displacement relationship of a Duffing-type nonlinear spring is given by

$$F(x) = \frac{dU(x)}{dx} = Ax + Bx^3 \tag{5.8}$$

The spring force-displacement curves for the linear, hardening, and softening stiffness are represented in the thin solid line, the solid line, and the dashed line, respectively, as shown in Figure 5.19(a). Such nonlinearity of spring stiffness has the effect of shifting the resonant frequency. The shifted frequency response of a vibration-driven spring-mass-damper system with linear, hardening, and softening stiffness springs are shown in Figure 5.19(b).

It should be pointed out that the output power and bandwidth of the nonlinear harvester depend on the direction of approach of the vibration frequency to the resonant frequency. For a hard nonlinearity, the operating bandwidth can be increased when approaching the resonant frequency of the device from a lower frequency. For a soft nonlinearity, the bandwidth and output power will only produce an improvement when approaching the resonant frequency of the device from a higher frequency. This could be a restricted condition for nonlinear energy harvesters in a real application. Fortunately, as long as the vibration frequency of the device does not exceed the shifted resonant frequency, the output response will always fall onto the enhanced stable trace no matter whether the vibration frequency sweeps up or down. The broadened bandwidth

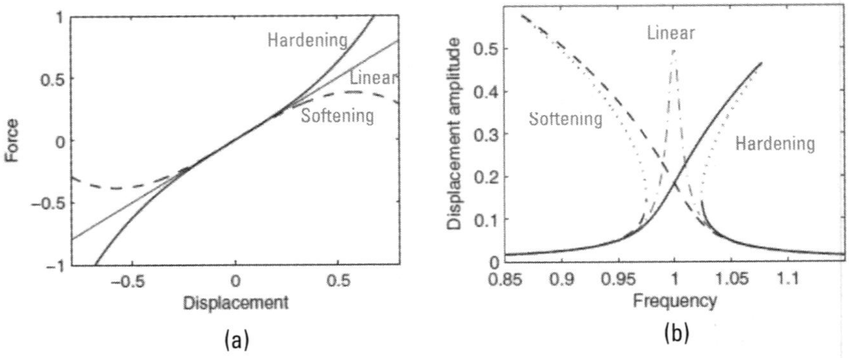

Figure 5.19 (a) Spring force-displacement curves and (b) frequency response of a spring-mass-damper system with the linear, hardening and softening stiffness springs [51].

depends on the damping ratio, the nonlinearity, and the input acceleration. In this section, we review recent advances in designing of nonlinear stiffness by introducing magnetic force, mechanical stress or stretching, and mechanical stoppers.

5.2.3.1 Magnetic Force-Induced Nonlinearity

To verify the concept of using mechanical nonlinearities to expand the bandwidth for vibration energy harvesting, magnetic force has been implemented by two different approaches. The first approach adopts a spring-mass oscillating structure integrated with magnetic components on it. The permanent magnets are assembled in the neighborhood for introducing a magnetic field. As the magnetic spring-mass structure oscillates, the spring stiffness will be hardened or softened by the surrounding repulsive or attractive magnetic force. The spring hardening approach induced by magnetic force could only broaden the frequency response in the right direction, while the spring-softening approach could broaden the frequency response in the left direction. In Figure 5.20(a), Stanton et al. [52] demonstrated a piezoelectric power-generating beam with a permanent magnet end mass interacting with the field of oppositely poled stationary magnets. Both the hardening and softening response can be realized by tuning the nonlinear magnetic interactions. For a constant distance h of 32 mm, the hardening and softening displacement responses are apparent at a distance d of 5 mm in Figure 5.20(b) and −2 mm in Figure 5.20(c). Moon and Holmes [53] proposed a magnetoelastic structure that contains a ferromagnetic cantilever with two permanent magnets located symmetrically near the free end. By changing the magnet spacing, the ferromagnetic cantilever may have three, two, or one stable equilibrium positions. Erturk et al. [54] developed a phenomenologically similar device with two piezoceramic layers attached to the root of the cantilever so as to obtain bimorph power generation.

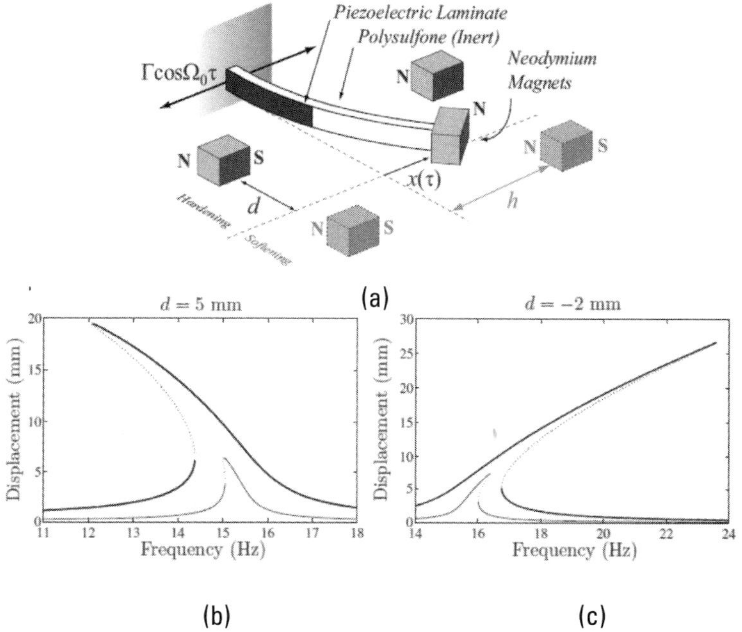

Figure 5.20 Magnetic force induced nonlinear energy harvester by Stanton [52].

Without the use of a mechanical spring structure, the other approach employs directly magnetic repulsive force to levitate a magnetic proof mass. Such prototype generally contains two permanent magnets fixed at the two ends of a tube. A magnetic proof mass is suspended in the middle of the tube by the magnetic repulsive forces from the top and bottom magnets. Such magnetic force induces nonlinearity as the middle magnet moves towards either the top or bottom magnet. Mann et al. [55] investigated the nonlinearity of an oscillator by magnetic levitation as shown in Figure 5.21(a). The device contains two outer magnets fixed at both ends of a Teflon tube and a center magnet suspended in the middle of the tube. As the center magnet is excited to oscillate, the variation of the magnetic repulsive forces between the center magnet and the two fixed magnets will induce a spring hardening of the restoring force. The nonlinear resonance can be tuned by varying the excitation amplitude or changing the spacing between the two fixed magnets as seen in Figure 5.21(b, c). Saha et al. [56] reported a similar electromagnetic energy harvesting prototype with coils wrapped around the outside of the tube. In the test, a power of 300 μW to 2.5 mW can be generated in the coils as the middle magnet vibrates up and down during human body walking and running. Foisal et al. [57] further developed an array of four generators placed side by side in model A and one above the other in model B. Such nonlinear electromagnetic energy harvesters are easy to be implemented in a very low-cost way. They are applicable to harvest energy

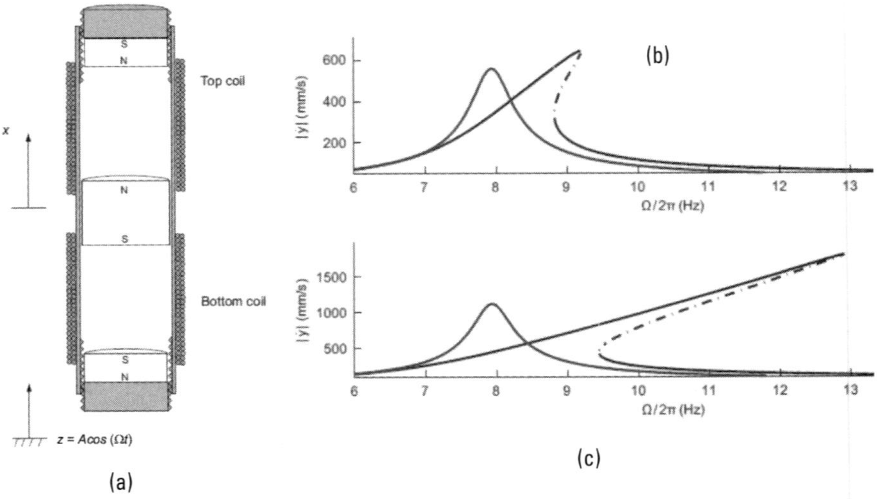

Figure 5.21 Nonlinear oscillator of magnetic levitation by Mann [55].

from vibrations of low-frequency and large amplitude, such as hand-shaking, walking, or running.

5.2.3.2 Mechanical Stress or Stretching-Induced Nonlinearity

The hardening Duffing of the spring structure can be simply achieved by a clamped-clamped beam or membrane that experiences mechanically stress or stretching at sufficient deflections or displacements as demonstrated in Figure 5.22. This type of approach can be implemented within MEMS devices and requires no additional power source. Shearwood et al. [58] and William et al. [59] first reported a micromachined electromagnetic power generator which consists of a permanent magnet attached to a polyimide membrane and a planar Au coil attached to the case. A large displacement of the membrane will cause it to be stretched and result in a spring-hardening effect. The mechanical pre-stress-induced nonlinear behavior was also found in the clamped-clamped beam structure proposed by Marzencki et al. [60]. The double-clamped beam is deposited with AlN piezoelectric thin film layer and a big seismic mass is located at the center. The nonlinear behavior can be obtained by carefully designing interlayer stresses in the MEMS structure. They design a high p+ doping of the top silicon layer to introduce a tensile shear stress on the interface with the AlN layer. The broadband phenomenon appears as the frequency sweeps up through the high built-in stresses in the beam. The frequency adaptability of over 36% is experimentally verified at an input acceleration of 2g. Hajati et al. [61] exploited an ultrawide bandwidth piezoelectric energy harvester by using a similar double-clamped PZT beam with center proof mass. The nonlinear resonator

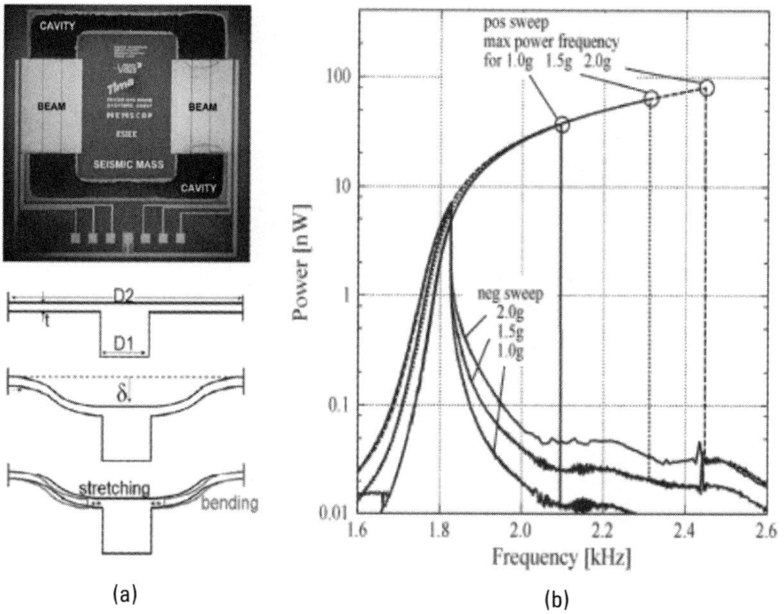

Figure 5.22 Mechanical stretching induced nonlinearity of a clamped-clamped beam [61].

shows a power bandwidth exceeding 50% of the center frequency and two orders of power magnitude improvement. The nonlinear stretching of fixed-fixed beams in an off-resonance mode was also demonstrated by Marinkovic and Koser [62, 63]. Besides spring hardening for out-of-plane vibration, Tvedt et al. [64] proposed a micromachined electrostatic energy harvester with in-plane motion. The four straight beams connected with the inertial mass have clamped support at the anchors so that they behave as hardening Duffing springs at sufficiently large displacement amplitudes.

It is promising to shape the spring with softening Duffing characteristics for the sake of increasing the operating bandwidth. Nguyen et al. proposed a MEMS electrostatic energy harvester with angled beam design [65, 66] as shown in Figure 5.23(a). It is found from the asymmetric force–displacement curve that the softening spring regime appears in the negative x-direction down to a displacement of about 90 μm, while in the opposite direction, the entire regime of displacement shows a spring-hardening behavior. The overall softening effect can be clearly seen in Figure 5.23(b), where the broadening of the operating bandwidth occurs as the frequency sweeps down. Later on, Nguyen et al. [67] demonstrated an extremely broadband energy harvester with an alternative curved beam design that displays an asymmetrical bistable behavior obtained purely through their geometrical design. The 3-dB bandwidth of the harvester is about 587 Hz at 0.208g acceleration for frequency down-sweep.

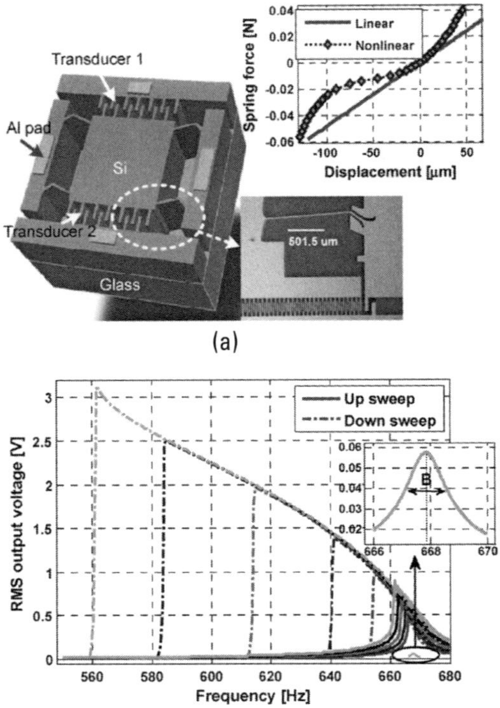

Figure 5.23 A nonlinear MEMS electrostatic energy harvester with angled beam design [66].

This approach offers a great freedom of spring geometry design according to the technology requirement and thus is particularly suitable for micromachined energy harvesters.

5.2.3.3 Mechanical Stopper-Induced Nonlinearity

In the literature, some researchers have leveraged mechanical stoppers to the energy-harvesting device for introducing another type of hard nonlinear stiffness, which is named as piecewise-linear stiffness. Figure 5.24 illustrates a piecewise linear model of the impact-based oscillating system with mechanical stoppers on two sides, which is modeled as a primary suspension system and consists of a proof mass m suspended by a spring k_0 and a damper c_0. The top and bottom mechanical stoppers are secondary suspension systems and are assumed to have spring stiffness of k_1 and k_2 and damping factors of c_1 and c_2, respectively. The distance between the proof mass and the top and bottom stoppers are d_1 and d_2, respectively, while it is assumed $d_1 < d_2$. The stoppers limit the relative movement of the mass and prevent the mass from excessive travel.

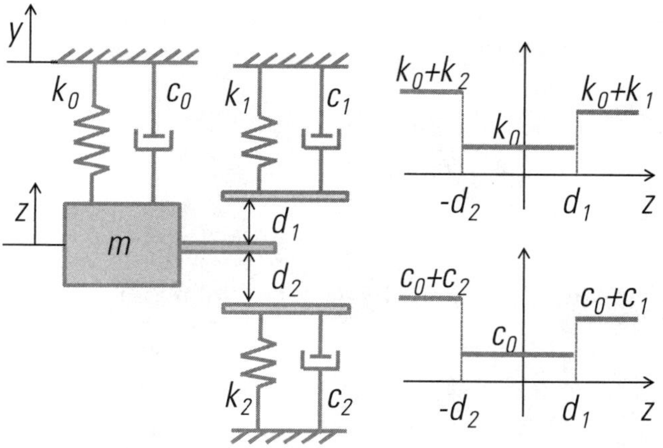

Figure 5.24 Piecewise linear model of an oscillating system with mechanical stoppers.

Narimani et al. [68] investigated the frequency response equation of a piecewise linear oscillation system to include both dual damping and stiffness behavior. The differential equation of motion of the impact-based oscillating system with stoppers engaged on two sides can be written as

$$\begin{cases} m_0\ddot{z}+(c_0+c_1)\dot{z}+(k_0+k_1)z-k_1d_1=-m_0\ddot{y} & (z\geq d_1) \\ m_0\ddot{z}+c_0\dot{z}+k_0z=-m_0\ddot{y} & (-d_2<z<d_1) \\ m_0\ddot{z}+(c_0+c_2)\dot{z}+(k_0+k_2)z+k_2d_2=-m_0\ddot{y} & (z\leq -d_2) \end{cases} \tag{5.9}$$

Equation (5.9) can be rearranged as follows:

$$\begin{cases} \ddot{z}+(2\xi_0\omega_0+2\xi_1\omega_1)\dot{z}+(\omega_0^2+\omega_1^2)z-\omega_1^2d_1=-\ddot{y} & (z\geq d_1) \\ \ddot{z}+2\xi_0\omega_0\dot{z}+\omega_0^2z=-\ddot{y} & (-d_2<z<d_1) \\ \ddot{z}+(2\xi_0\omega_0+2\xi_2\omega_2)\dot{z}+(\omega_0^2+\omega_2^2)z+\omega_2^2d_2=-\ddot{y} & (z\leq -d_2) \end{cases} \tag{5.10}$$

where $y(t) = Y\sin(\omega t)$, Y is the amplitude of the base excitation, ω is the excitation frequency; ξ_0 and ω_0 are the primary damping and frequency characteristics, ξ_1, ξ_2 and ω_1, ω_2 are the secondary damping and frequency characteristics, which can be further defined as

$$2\xi_0\omega_0=\frac{c_0}{m_0},\quad 2\xi_1\omega_1=\frac{c_1}{m_0},\quad 2\xi_2\omega_2=\frac{c_2}{m_0},\quad \omega_0^2=\frac{k_0}{m_0},\quad \omega_1^2=\frac{k_1}{m_0},\quad \omega_2^2=\frac{k_2}{m_0}.$$

To study the frequency response of the impact-based oscillating system, we use dimensionless variables

$$\tau = \omega_0 t\ ,\quad \rho = \frac{\omega}{\omega_0}\ ,\quad \rho_1 = \frac{\omega_1}{\omega_0}\ ,\quad \rho_2 = \frac{\omega_2}{\omega_0}\ ,\quad u = \frac{z}{Y}\ ,\quad v = \frac{y}{Y} = \sin(pt)\ ,\quad \delta_1 = \frac{d_1}{Y}\ ,\quad \delta_2 = \frac{d_2}{Y}$$

to obtain the following dimensionless equations of mass motion

$$\ddot{u} + 2\xi_0\dot{u} + u = \rho^2\sin(\rho\tau) + f(u,\dot{u}) \tag{5.11}$$

where

$$f(u,\dot{u}) = \begin{cases} -2\rho_1\xi_1\dot{u} - \rho_1^2\delta_1 + \rho_1^2\delta_1 & (u \geq \delta_1) \\ 0 & (-\delta_2 < u \leq -\delta_1) \\ -2\rho_2\xi_2\dot{u} - \rho_2^2\delta_1 + \rho_2^2\delta_2 & (u \leq -\delta_2) \end{cases} \tag{5.12}$$

The frequency response function, which describes the dimensionless amplitude a with frequency ρ, is obtained as

$$\pi^2\rho^4 = X_1^2 + X_2^2 \tag{5.13}$$

where

$$X_1 = -2\xi_0 a\rho\pi - \rho_1\xi_1 a\rho(\pi - 2\varphi_1 - \sin 2\varphi_1) - \rho_2\xi_2 a\rho(\pi - 2\varphi_2 - \sin 2\varphi_2) \tag{5.14}$$

$$X_2 = \pi a(1-\rho^2) - \begin{bmatrix} \frac{1}{2}\rho_1^2 a(2\varphi_1 - \sin 2\varphi_1 - \pi) + \frac{1}{2}\rho_2^2 a(2\varphi_2 - \sin 2\varphi_2 - \pi) \\ +2\rho_1^2\delta_1\cos\varphi_1 + 2\rho_2^2\delta_2\cos\varphi_2 \end{bmatrix} \tag{5.15}$$

$\varphi_1 = \sin^{-1}(\delta_1/a)$ and $\varphi_2 = \sin^{-1}(\delta_2/a)$ are the phase angles when the proof mass engages the top and bottom stoppers, respectively.

When a one-side stopper is involved in the impact-based oscillating system, the dimensionless differential equation of mass motion can be rewritten as

$$\ddot{u} + 2\xi_0\dot{u} + u = \rho^2\sin(\rho\tau) + f_i(u,\dot{u}) \tag{5.16}$$

where $i = 1$ or 2 represents the situation where the mass motion engages either the top or the bottom stopper, respectively.

$$f_1(u,\dot{u}) = \begin{cases} -2\rho_1\xi_1\dot{u} - \rho_1^2\delta_1 + \rho_1^2\delta_1 & (u \geq \delta_1) \\ 0 & (u < \delta_1) \end{cases} \tag{5.17}$$

$$f_2(u,\dot{u}) = \begin{cases} 0 & (u > \delta_2) \\ -2\rho_2\xi_2\dot{u} - \rho_2^2\delta_1 + \rho_2^2\delta_2 & (u \leq \delta_2) \end{cases} \tag{5.18}$$

An implicit equation for the amplitude ρ as a function of the excitation frequency ρ is given by

$$\pi^2\rho^4 = Z_1^2 + Z_2^2 \tag{5.19}$$

where

$$Z_1 = -2\xi_0 a\rho\pi - \rho_i\xi_i a\rho\left(\pi - 2\varphi_i - \sin 2\varphi_i\right) \tag{5.20}$$

$$Z_2 = \pi a\left(1-\rho^2\right) - \left[\frac{1}{2}\rho_i^2 a\left(2\varphi_i - \sin 2\varphi_i - \pi\right) + 2\rho_i^2 d_i \cos\varphi_i\right] \tag{5.21}$$

$\varphi_i = \sin^{-1}(\delta_i/a)$ is the phase angle when the proof mass engages the top or bottom stopper. In the case that the impact-based oscillating system employs a two-side or one-side stopper, the frequency response can be obtained analytically using (5.13) or (5.19). Liu et al. [69] investigated the key parameters for the frequency response, including base accelerations, damping ratios, frequency characteristics, and stopper distances based on the mathematical model. Through the numerical simulation and experimental measurements, they found that the damping ratio plays dominant role than stiffness in a piecewise linear vibration system. The frequency-broadening behavior is strengthened by a decrease in the damping and an increase in the stiffness of the stopper. In addition, a high base acceleration is also preferred to realize a better performance (wider operating bandwidth and higher power output). There is a trade-off for the stopper distance, since it affects the frequency bandwidth and mass motion amplitude with opposite trend. A lower stopper distance results in a wider frequency bandwidth at the expense of a reduction in the relative mass motion.

Soliman et al. [70, 71] presented an electromagnetic energy harvesting prototype with a piecewise-linear stiffness, which is realized by incorporating a mechanical stopper to vary the effective length of the beam. Numerical and experimental results demonstrated that the piece-wise linear prototype increases the up-sweep bandwidth of 240% wider than the traditional linear architecture at the half-power level, while maintaining the same bandwidth during down-sweep. Blystad et al. [72] and Moss et al. [73], respectively, investigated the use of a single-sided rigid end-stop and double-sided flexible piezoelectric bimorph-stops to construct impact-based broadband piezoelectric energy harvesters. For micro-scale piecewise-linear prototypes, Liu et al. [74] developed a piezoelectric MEMS energy harvester with low resonant frequency and wide operation bandwidth from 30 to 47 Hz by leveraging a metal carrier with a limited spacer as shown in Figure 5.25. Later on, a similar S-shaped MEMS PZT cantilever for energy harvesting from low-frequency vibrations below 30 Hz has been reported by the same group [75]. An in-plane MEMS nonlinear electret energy harvester has been developed by Suzuki's group [76, 77]. They employed a high-aspect-ratio in-plane nonlinear spring which is a hybrid of

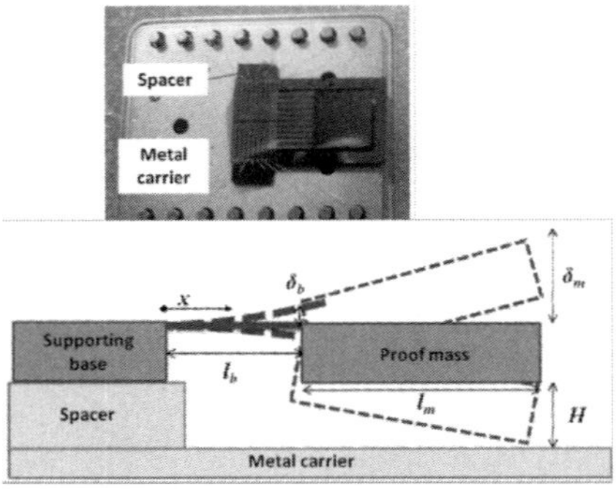

(a)

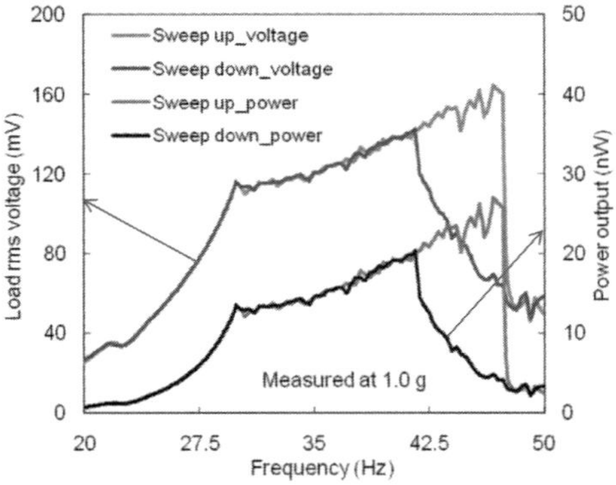

(b)

Figure 5.25 A wideband piezoelectric MEMS energy harvester with stopper by Liu [74].

main springs and fixed-free secondary springs to obtain a broad frequency range of 46 to 73 Hz. The operating bandwidth of such piecewise-linear harvesting system can be broadened as the stop distance decreases at the expense of lower output voltage and power.

5.2 4 Frequency-Upconversion Mechanisms

The maximum voltage and power generated from a vibration energy harvester is strongly dependent on the external vibration frequency and drops dramatically at low frequencies (1 to 10 Hz). Since most ambient vibration exists at low-frequency range, the vibration energy-harvesting with FUC mechanism was initially proposed for the purpose of upconverting low-frequency environmental vibrations to a high-frequency self-oscillation of the system and hence providing more efficient energy generation at low frequencies. On the basis of previous reviewed broadband harvesting approaches using nonlinear mechanisms, the FUC approaches have the advances of more efficient power generation at low frequencies and simultaneously possess a wide operating bandwidth. The FUC mechanisms can be achieved by utilizing mechanical impact, scrape-through, bistable bulking, and magnetic force approaches.

5.2.4.1 Mechanical Impact Approach

The impact-based FUC approach for energy harvesting was initially demonstrated by Umeda et al. [78, 79], who investigated the power transformation from mechanical impact energy to electric energy by the impact of a steel ball on a piezoelectric membrane. The operating mechanism is schematically shown in Figure 5.26. The steel ball with mass m at the height h firstly falls freely toward the center of the vibration under the effect of gravity. The piezoelectric membrane is then excited to generate bending vibration by the collision. An alternating voltage is supplied to the load resistance due to piezoelectric effect.

In general, the construction and mechanical model of an impact-based FUC energy harvester system assembled with oscillators 1 and 2 (termed as FUC-1 and FUC-2) are shown in Figure 5.27. The system contains an excitation oscillator with spring stiffness k_0, damping factor c_0, and proof mass m_0. FUC-1 with spring stiffness k_1, damping factor c_1, and proof mass m_1 is placed at a distance of x_1 above the proof mass m_0. Likewise, FUC-2 with stiffness k_2, damping factor c_2, and proof mass m_2 is placed at a distance of x_2 below the proof mass m_0. The resonant frequencies of the FUC oscillators (i.e., ω_1 and ω_2) are higher than that of the excitation oscillator ω_0. The operation sequence of the system is shown in Figure 5.27(c). The excitation oscillator operates in such a way that the proof mass moves upward and downward between the two FUC oscillators and impacts with them periodically. In stage A, the proof mass moves toward FUC-1. In stage B, the proof mass engages FUC-1 and they

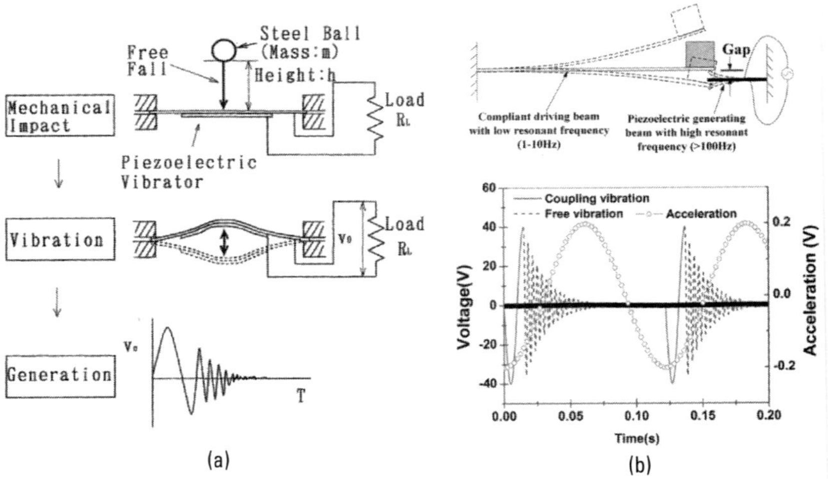

Figure 5.26 Operating mechanisms of the impact of a steel ball on a piezoelectric membrane by Umeda [78].

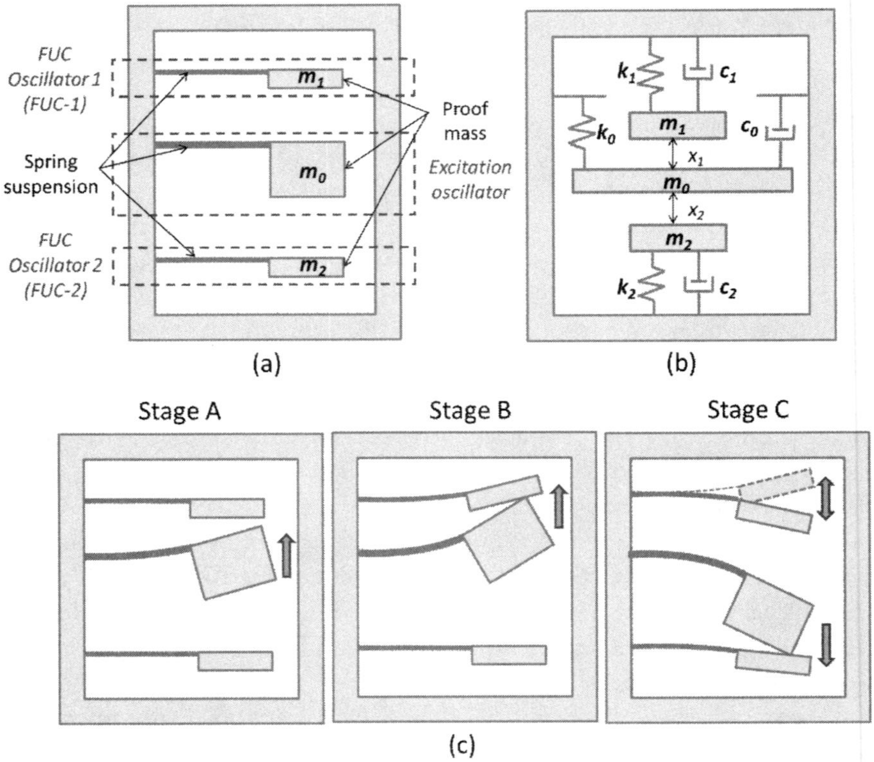

Figure 5.27 (a) Architecture, (b) mechanical model, and (c) operation illustration of a vibration-based energy harvester system with FUC oscillators (FUC-1 and FUC-2).

move together until the position where FUC-1 is released from the proof mass. In stage C, the proof mass continues its downward movement to FUC-2, during when FUC-1 self-oscillates at a high frequency, converting the mechanical energy in the spring to electrical energy. This process is then repeated in the opposite direction. When the excitation oscillator engages the FUC oscillator, there would be a sudden increase in the effective stiffness. The increase in stiffness would broaden the operating frequency range as described in the previous section. Thus, the FUC behavior and bandwidth broadening are achieved simultaneously. In the system, the main design consideration is the minimum acceleration and gap distances at which the FUC oscillators will begin operation. In addition, the spring stiffness and weight of the proof mass are also important factors for achieving optimized performance.

Based on the upconversion principle, Renaud et al. [80] and Manla et al. [81] demonstrated nonresonant energy harvesters driven by the repeated impact of a free ball on two piezoelectric plates. Gu et al. [82, 83] proposed an impact-driven FUC energy harvesting prototype that is demonstrated to be especially suitable for MEMS implementation. The FUC behavior is realized by the periodic impact between a compliant driving beam and the piezoelectric generating beams, through which the improved power output and broadened operating bandwidth can be gained simultaneously. Afterward, Liu et al. [84] presented a real MEMS-based FUC piezoelectric energy harvester as illustrated in Figure 5.28. By using a straight PZT cantilever as the FUC stopper, the operating bandwidth is extended from 12 to 26 Hz and the power density of the harvester is significantly improved to be 159.4 μW/cm^3.

5.2.4.2 Scrape-Through Approach

Similar to the impact-based FUC approach, a scrape-through energy-harvesting mechanism normally comprises a low-frequency (LF) oscillator and a high-frequency (HF) oscillator assembled with a predetermined gap. The energy conversion mechanism such as piezoelectric, electromagnetic, or electrostatic generating element is integrated with the HF oscillator. Taking the commonly used piezoelectric cantilever structure as an example, the operating mechanism of scrape-through process is illustrated in Figure 5.29. The HF cantilever acting as a top-stopper is arranged above the proof mass of the LF cantilever with a distance d and a lateral overlapping length l. As long as the vibration amplitude of LF cantilever is sufficiently large, the proof mass will periodically impact the HF cantilever. As a result, the operating bandwidth of LF cantilever will be broadened, at the expense of retardation of vibration amplitude. In the meantime, the overlapping distance is adjusted such that the proof mass of LF cantilever is able to scrape through the beam tip of HF cantilever during impact. Consider a particular instant of an oscillation cycle when the proof mass is at its lowest point at position ① and starts to move to position ② where the

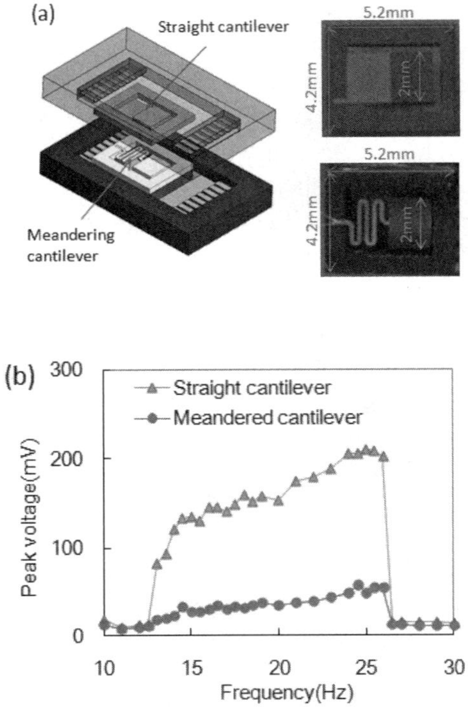

Figure 5.28 A MEMS-based FUC piezoelectric energy harvester demonstrated by Liu [84].

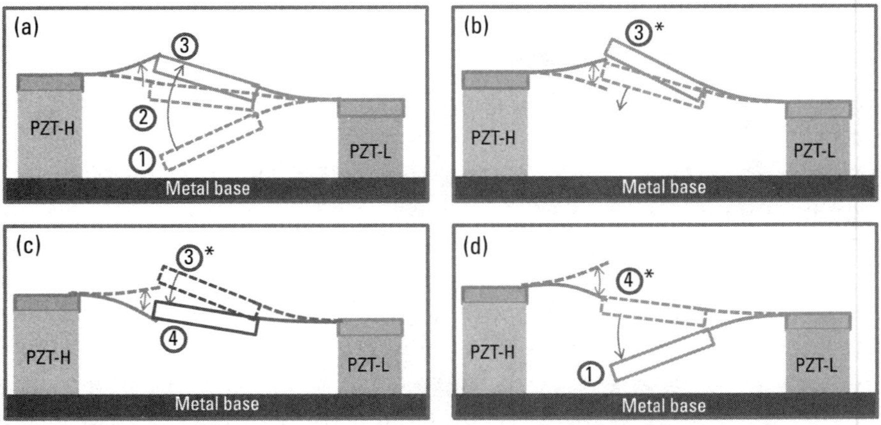

Figure 5.29 Operating mechanism of the scrape-through process.

proof mass starts to impact the beam of HF cantilever. Then the proof mass and the beam continue bending upward to position ③. During this movement, the overlapping distance between the proof mass and the beam decreases continuously until it reaches critical position ③ when the HF cantilever is just scraped through the proof mass and is released to self-oscillate at its resonant frequency. Thereafter, the proof mass continues its upward movement until it reaches its maximum amplitude at position ③* and starts to move downward until it reaches another critical position ④, where the overlapping distance between the proof mass and beam is again reduced to zero. From position ④, the proof mass scrapes through again to position ④* and the beam is released to self-oscillate again while the proof mass moves back to position ①. The cycle is repeated as the proof mass moves toward position ② again. During each vibration cycle, the LF cantilever impacts, scrapes through and releases the HF cantilever twice, resulting in oscillations at its high resonant frequency. The cyclic deformation of the PZT layer of LF and HF cantilevers will be transformed into electricity due to piezoelectric effect.

In the literature, Lee et al. [85] demonstrated a FUC approach realized by mechanical scrape-through contact between the tip of a piezoelectric harvesting beam and a set of superelastic ridges that slide past the tip of the harvesting beam [Figure 5.30(a)]. The amount of a generated voltage depends on the depth of a ridge and the rectification of FUC is a function of the ridge spacing. Liu et al. [86] demonstrated a MEMS-based piezoelectric energy harvester with frequency broadband and upconversion behaviors, which is achieved by a scrape-though process driven by ambient low-frequency vibrations. It consists of a low-frequency piezoelectric PZT cantilever (PZT-L cantilever) and a high-frequency piezoelectric PZT cantilever (PZT-H cantilever) assembled with a predetermined gap as shown in Figure 5.30(b). The resonant frequencies of the PZT-L and PZT-H cantilevers are 36 Hz and 618 Hz, respectively. During each

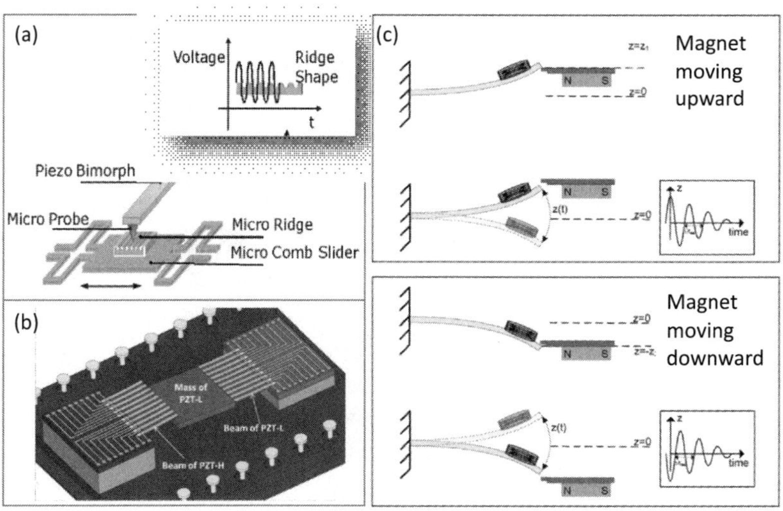

Figure 5.30 Schematic illustrations of scrape-through upconversion approaches by (a) Lee [85], (b) Liu [86], and (c) Zorlu [88].

vibration cycle, the PZT-L cantilever impacts, scrapes through, and releases the PZT-H cantilever twice, resulting in a self-oscillation of the PZT-H cantilever at its high resonant frequency of 618 Hz. Meanwhile, the operating frequency bandwidth is further increased from 33 to 43 Hz, while the output rms voltage is increased from 76 to 95 mV for an input acceleration of 0.6g. Zorlu et al. [87, 88] presented a FUC electromagnetic energy harvester using a mechanical scrape-through method from low-frequency vibrations as shown in Figure 5.30(c). The prototype consists of a polystyrene cantilever carrying a pickup coil and a magnet placed on a diaphragm. The vibrating diaphragm would scrape through the tip of the cantilever and upconvert to the cantilever's high-frequency vibration. Such approaches have the potential to offer the benefit of resonance, but with the difficulty of fine adjusting the overlapping distance.

5.2.4.3 Bistable Buckling Approach

A bistable buckling phenomenon (also known as snap-through buckling) is adapted to achieve highly efficient energy harvesting even at off-resonance conditions. Hence, it is more applicable for vibration sources of low frequencies and high accelerations, such as human activities. When the bulked clamped-clamped beam is pushed down from the initial position, it will first deflect and then snap to the second stable position as shown in Figure 5.31(a). For a bistable FUC harvesting device, the buckled beam is subjected to a low-frequency vibration above the threshold acceleration; it will snap through and activate the attached oscillating structure to resonate at high frequencies. Meanwhile, the bistable buckled structure can fully swing into another state at various vibration frequencies and thus has a large bandwidth particularly in low-frequency range, provided that the input frequency is sufficiently low enough for complete transition and the acceleration of the given input is sufficiently high to obtain a snap-through motion.

Jung and Yun [89] demonstrated a bistable buckling upconversion approach as shown in Figure 5.31(b). The prototype consists of two prebuckled slender bridges and four cantilever beams at the center of the bridges. A piezoelectric polyvinylidene fluoride (PVDF) layer is attached on top of each beam to convert the induced strain into electrical charge. When the excitation acceleration applied to the buckled bridge exceeds a threshold value, it immediately snaps through to the other equilibrium state. The rapid transition between the two equilibrium states generates a highly accelerated impulse like excitation and thereby caused the attached cantilever beams to vibrate freely at their resonant frequencies independently. The important characteristic of the proposed device is its wide operating frequency range. It was observed from Figure 5.31(b) that the bandwidths of Type 1 and 2 devices are approximately 22 and 26 Hz, respectively, which are 157% and 186% wider than that of a simple cantilever structure (i.e., 14 Hz).

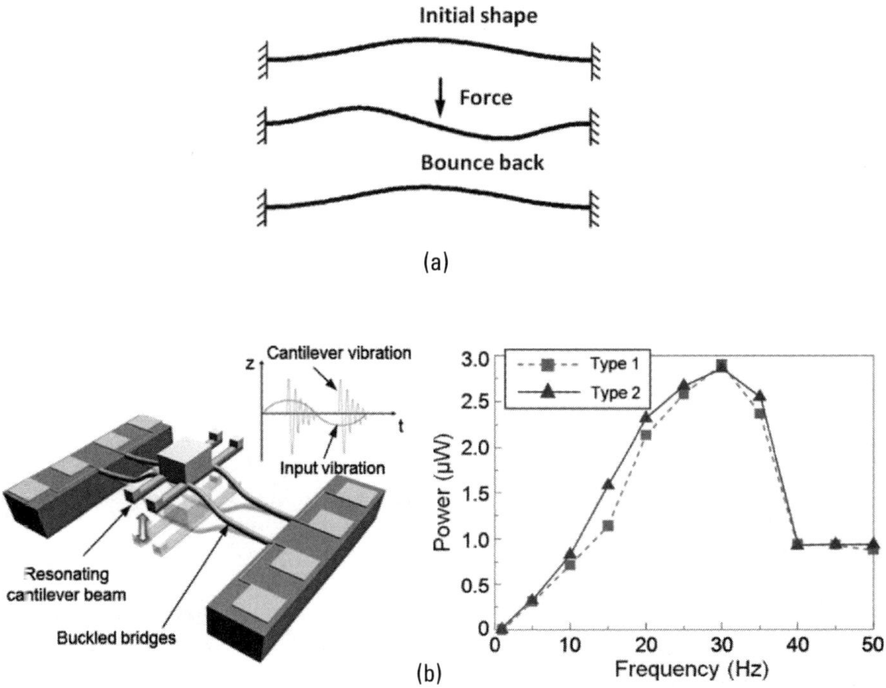

Figure 5.31 Schematic illustration of a bistable buckling FUC approach by Jung and Yun [89].

5.2.4.4 Magnetic Force Approach

Kulah and Najafi [90] demonstrated a FUC energy harvester prototype by utilizing magnetic attractive force as schematically shown in Figure 5.32. The prototype consists of two resonating structures. The top one is an NdFeB magnet mass suspended by a soft diaphragm and has a low resonant frequency to match the target scenario. The bottom ones are an array of cantilever beams coated with coils and have higher resonant frequencies. As the magnet mass resonates in response to external vibration, it moves closer to the cantilevers underneath, catches it at a certain point, pulls it up, and releases it at another point. The released cantilever starts to resonate at its high resonance frequency. A similar MEMS-based electromagnetic FUC harvester was fabricated by Sari et al. [91]. The important feature benefited from the FUC mechanism is a wide operating bandwidth from 50 to 150 Hz, which is effective as long as the input displacement amplitude is enough to activate the high-frequency oscillator.

Galchev et al. [92, 93] successfully fabricated a micro-frequency-increased generator (FIG) system based on the magnetic attractive force. The operation of the system is outlined in Figure 5.33(a). The driving FIG utilizes a large inertial mass to couple kinetic energy from the ambient into the structure and snaps

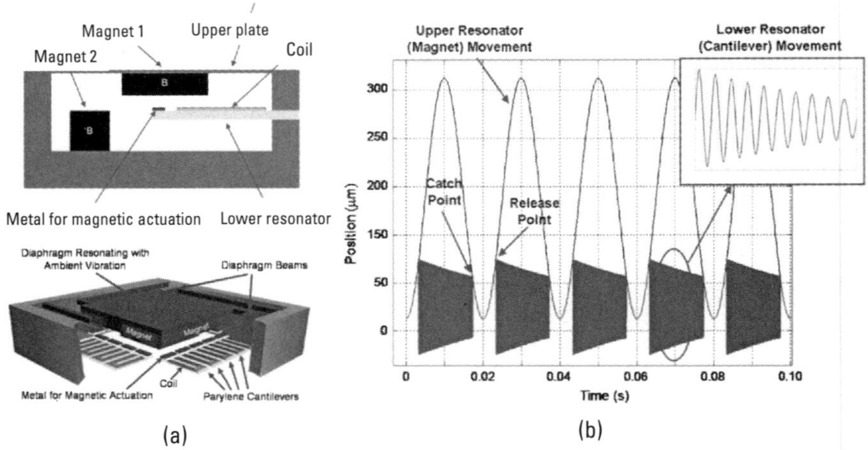

Figure 5.32 (a) Cross-section and 3-D view of an electromagnetic FUC harvester. (b) Movement of the low-frequency and high-frequency resonators [90].

back and forth between the two generating FIGs. As the inertial mass moves, it pulls and releases the generating FIG due to the magnetic attractive force. The freed FIG resonates at its high natural frequency, converting the stored mechanical energy into electrical energy. This process is then repeated in the opposite direction. Experimentally, the FIG system can generate a peak power of 57 μW and an average power of 2.3 μW from an input acceleration of 0.54g at only 2 Hz, and it is capable of providing a wide operating bandwidth (up to 30 Hz) and versatility without any modifications or tuning as shown in Figure 5.33(b).

Beside the employment of magnetic attractive force, Tang and Li [94] realized a bistable FUC piezoelectric energy harvester utilizing the magnetic repulsive force. The harvester consists of a longer and a shorter PVDF cantilever face to face with a certain distance. Permanent magnets and additional masses are mounted on each tip of the cantilever. The longer cantilever acts as a low-frequency (~10 Hz) primary resonator-generator (RG), while the short cantilever acts as a high-frequency (~102 Hz) secondary resonator-generator (RG). As the primary RG starts to vibrate and moves towards the tip magnet of the secondary RG, it will bend due to the magnetic repulsive force. Once the restoring force exceeds the repulsive force, the secondary RG is suddenly released and starts self-oscillation. The counteraction of the primary RG makes it vibrate in a bistable manner, thereby achieving wideband low-frequency generating capability with a small size. The device is demonstrated to generate over 1.5 μW of average power at the frequency range of 9 Hz to 19 Hz. This approach avoids mechanical collision and wear for long-term working durability. A similar concept was reported by the same group in [95].

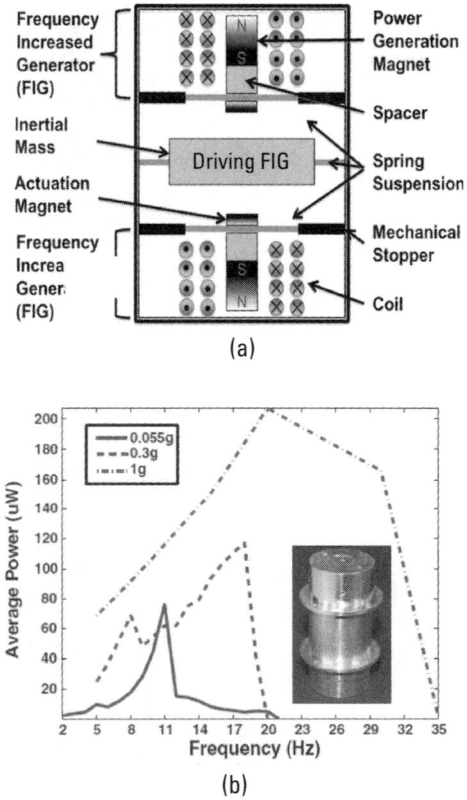

Figure 5.33 An electromagnetic FUC energy harvesting system using magnetic force demonstrated by Galchev et al. [93].

5.2.5 Nonresonant Mechanisms

To obtain the maximum output power, the vibration frequency of energy harvester should match this eigenfrequency of the vibration sources. However, the vibration frequency of ambient available vibration sources is random and varies from one case to the others. Human motions, wind-induced movements of the branches of the trees, and vehicle motions are examples for such vibrations that the operation frequency may change in a wide range [96–98]. Resonant-mode energy harvesters are designed to be connected to a vibrating base through a spring-mass-damper system, and they rely on a mechanical resonance within a limited bandwidth. Therefore, the resonant-mode harvesters are very difficult to adapt to the vibration source due to the limitation of their mass and spring stiffness. Thus, the energy of higher or lower vibration modes cannot be scavenged. Meanwhile, the vibration direction should match the sensitive direction of the energy harvester because some resonant harvesters have usually 1 DOF.

Nonresonant conversion mechanism works without a spring and does not have resonant behavior.

Lo and Tai [99] developed an electrets power generator using charged parylene HT-coated thin film polymer. As illustrated in Figure 3.15(b), when the parylene-coated rotor slides relative to the metal stator, the induced charges on the electrodes vary with the position of the rotor. Current flow is thus generated. The maximum power output, 17.98 μW, is obtained at 50 Hz with an external load of 80 MΩ. Due to the nonresonant in-plane movement of the rotor in the container, the power spectrum exhibits a relatively wide frequency range. For low frequencies, the generator can harvest 7.7 μW at 10 Hz and 8.23 μW at 20 Hz. At frequencies higher than 50 Hz, the rotor only partially moves in its supposed short distance and hence the generator produces less power. Yang et al. proposed a nonresonant out-of-plane electromagnetic energy harvester [100]. Unlike the conventional resonant vibration-based energy harvesters that

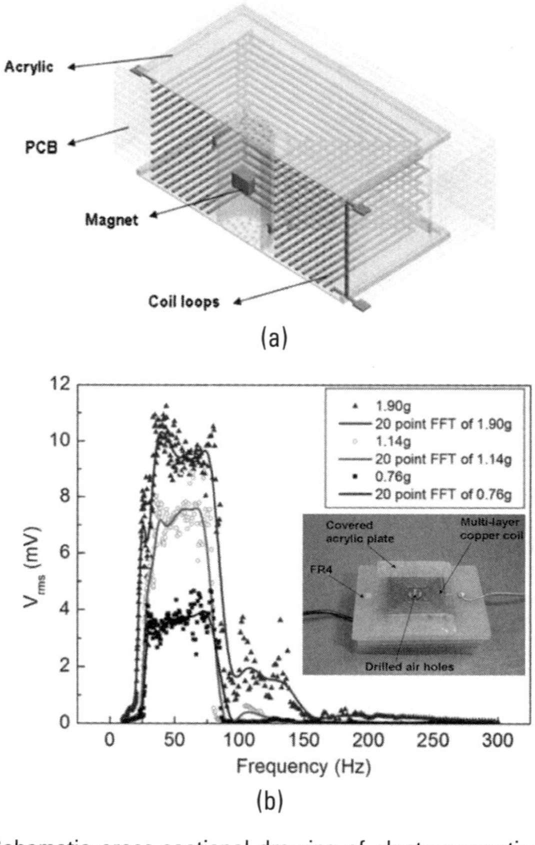

Figure 5.34 (a) Schematic cross-sectional drawing of electromagnetic wideband energy harvester. (b) Measured output voltage versus frequency under various accelerations of 0.76g, 1.14g, and 1.90g [100].

have a movable mass with springs linked to a stationary frame packaged in a house, there is no mechanical structure to physically connect magnet with *Fire Resistant 4* (FR4) substrates. Figure 5.34(a) shows the schematic cross-section graph of electromagnetic harvester, which includes up-down movable Neodymium (Nd) cylinder magnets inside a 4-mm-wide middle hole, multilayer coils, and covered acrylic plates. The poling NS direction of the magnet is along the central axis direction of cylinder magnet. The generated electrical energy per vibration cycle depends on the strength of magnetic field and the number of turns of the coils. Five pieces of FR4 PCB substrates with multilayer copper coils were stacked together to form the nonresonant-based energy harvester. To weaken the air damping effect, the middle hole is sealed by two acrylic plates with drilled air holes. Figure 5.34(b) shows the output voltage against frequency under 0.76g, 1.14g, and 1.9g, respectively. An average output voltage of 9 mV with 40 Hz bandwidth (i.e., from 40 to 80 Hz) is generated.

Another similar nonresonant harvester was proposed by Bowers et al. [101], which utilized a unidirectionally magnetized (NdFeB) permanent magnet ball that rolls and translates inside a spherical housing when subjected to

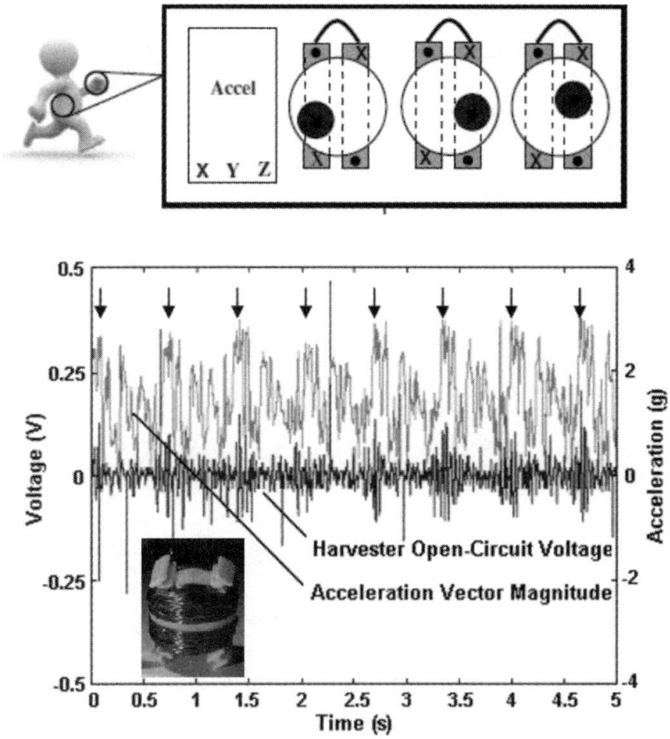

Figure 5.35 The input acceleration and open-circuit voltage versus time for the harvester placed in the pocket while running [101].

human motion. The sphere housing is wrapped with copper coils. When the human body walks or runs, the surrounding coils sense a time-varying magnetic flux induced by the rotation and translation of the magnetic ball and thus generate voltage as show in Figure 5.35. This particular design approach is promising because the ball will move under a large frequency range of nonspecific motions. The generated rms voltage ranges from ~80 mV to 700 mV and time-averaged power density is up to 0.5 mW/cm^3. Additionally, the rolling action is more favorable than previously reviewed sliding and bouncing harvesters, since sliding and bouncing dissipates more energy than rolling [99, 100].

References

[1] Beeby, S. P., M. J.Tudor, and N. M. White, "Energy Harvesting Vibration Sources for Microsystems Applications," *Measurement Science and Technology*, Vol. 17, 2006, pp. R175–R195.

[2] Mitcheson, P. D., et al., "Performance Limits of the Three MEMS Inertial Energy Generator Transduction Types," *J. Micromech. Microeng.*, Vol. 17, 2007, pp. S211–216.

[3] Mitcheson, P. D., et al., "Energy Harvesting from Human and Machine Motion for Wireless Electronic Devices," *Proc. IEEE* Vol. 96, 2008, pp. 1457–1486.

[4] Roundy, S., "On the Effectiveness of Vibration-Based Energy Harvesting," *J. Intell. Mater. Syst. Struct.*, Vol. 16, 2005, pp. 809–823.

[5] Poulin, G., E. Sarraute, and F. Costa, "Generation of Electrical Energy for Portable Devices: Comparative Study of an Electromagnetic and a Piezoelectric System," *Sens. Actuators A*, Vol. 116, 2004, pp. 461–471.

[6] Wacharasindhu, T., and J. W. Kwon, "A Micromachined Energy Harvester from a Keyboard Using Combined Electromagnetic and Piezoelectric Conversion," *J. Micromech. Microeng.*, Vol. 18, 2008, p. 104016.

[7] Williams, C. B., and R. B. Yates, "Analysis of a Micro-Electric Generator for Microsystems," *Sensors Actuators A*, Vol. 52, 1996, pp. 8–11.

[8] Challa, V. R., M.G. Prasad and F. T. Fisher, "A Coupled Piezoelectric-Electromagnetic Energy Harvesting Technique for Achieving Increased Power Output Through Damping Matching," *Smart Material and Structures*, Vol. 18, 2009, p. 095029.

[9] Wischke, M., and P. Woias, "A Multi-Functional Beam for Energy Scavenging from Vibration," *Proc. of PowerMEMS*, Sendai, Japan, November 9–12, 2008, pp. 77–80.

[10] Wischke, M., et al., "Electromagnetic Vibration Harvester with Piezoelectrically Tunable Resonance Frequency," *J. Micromech. Microeng.*, Vol. 20, 2010, p. 035025.

[11] Khaligh, A., et al., "Hybrid Energy Harvesting Topology for Human-Powered Mobile Electronics," *Proc. 34th Conf. IEEE Ind. Electron. Soc.*, Orlando, FL, November 2008, pp. 448–453.

[12] Wu, X., A. Khaligh, and Y. Xu, "Modeling, Design and Optimization of Hybrid Electromagnetic and Piezoelectric MEMS Energy Scavengers," *IEEE 2008 Custom Integrated Circuits Conference*, 2008, pp. 177–180.

[13] Yalcinkaya, A. D., O. Ergeneman, and H. Urey, "Polymer Magnetic Scanners for Bar Code Applications," *Sensors Actuators A*, Vol. 17, 2007, pp. 236–243.

[14] Yang, B., et al., "Electromagnetic Energy Harvesting from Vibrations of Multiple Frequencies," *J. Micromech. Microeng.*, Vol. 19, 2009, p. 035001.

[15] Yang, B., et al., "Hybrid Energy Harvester Based on Piezoelectric and Electromagnetic Mechanisms," *Journal of Micro/Nanolithography, MEMS, and MOEMS (JM3)*, Vol. 9, No. 2, 2010, p. 023002.

[16] Zhu, G., et al., "Toward Large-Scale Energy Harvesting by Nanoparticle-Enhanced Triboelectric Nanogenerator," *Nano Lett.*, Vol. 12, 2013, pp. 847–853.

[17] Han, M. D., et al., "Low-Frequency Wide-Band Hybrid Energy Harvester Based on Piezoelectric and Triboelectric Mechanism," *Science China Technological Sciences*, 2013.

[18] Zhang, X. S., et al., "Frequency-Multiplication High-Output Triboelectric Nanogenerator for Sustainably Powering Biomedical Microsystems," *Nano Lett.*, Vol. 13, 2013, pp. 1168–1172.

[19] Yang, B., et al., "Modeling and Experimental Study of a Low-Frequency-Vibration-Based Power Generator Using ZnO Nanowire Arrays," *IEEE Journal of Microelectromechanical Systems*, Vol. 21, No. 4, 2012, pp. 776–778.

[20] Lu, M. P., et al., "Piezoelectric Nanogenerator Using p-Type ZnO Nanowire Arrays," *Nano Lett.*, Vol. 9, No. 3, March 2009, pp. 1223–1227.

[21] Wang, Z. L., and J. H. Song, "Piezoelectric Nanogenertors Based on Zinc Oxide Nanowire Arrays," *Science*, Vol. 312, April 2006, pp. 242–246.

[22] Saravanakumar, B., et al., "Fabrication of ZnO Nanogenerator for Eco-Friendly Biomechanical Energy Harvesting," *RSC Advances*, 2013.

[23] Roundy, S., *Energy Scavenging for Wireless Sensor Networks with Special Focus on Vibrations*, Boston, MA: Kluwer Academic Publishers, 2004.

[24] Khbeis, M., J. McGee, and R. Ghodssi, "Design of a Hybrid Ambient Low Frequency, Low Intensity Vibration Energy Scavenger," *PowerMEMS 2006 Conference*, Berkeley, CA, November 29–December 1, 2006, pp. 287–290.

[25] Khbeis, M., J. McGee, and R. Ghodssi, "Development of a Simplified Hybrid Ambient Low Frequency Low Intensity Vibration Energy Scavenger System," *Transducers 2009*, Denver, CO, June 21–25, 2009, pp. 525–528.

[26] Khbeis, M., J. McGee, and R. Ghodssi, "Characterization of a Hybrid Ambient Low Frequency, Low Intensity Vibration Energy Scavenger System," *PowerMEMS 2009 Conference*, Washington, D.C., December 1–4, 2009, pp. 17–20.

[27] Blevins, R. D., *Formulas for Natural Frequency and Mode Shape*, Malabar, FL: Krieger, 2001.

[28] Zhu, D. B., M. J. Tudor, and S. P. Beeby, "Strategies for Increasing the Operating Frequency Range of Vibration Energy Harvesters: A Review," *Meas. Sci. Technol.*, Vol. 21, No. 2, 2010, p. 022001.

[29] Gieras, J. F., et al., "Electromechanical Energy Harvesting System," *Patent Publication Number*: WO2007070022(A2), WO2007070022(A3), 2007.

[30] Wu, X., et al., "A Frequency Adjustable Vibration Energy Harvester," *Proceedings of PowerMEMS 2008 + microEMS 2008*, 2008, pp. 245–248.

[31] Challa, V. R., et al., "A Vibration Energy Harvesting Device with Bidirectional Resonance Frequency Tunability," *Smart Mater. Struct.*, Vol. 17, No. 1, 2008, p. 015035.

[32] Leland, E. S., and P. K. Wright, "Resonance Tuning of Piezoelectric Vibration Energy Scavenging Generators Using Compressive Axial Preload," *Smart Mater. Struct.*, Vol. 15, No. 5, 2006, pp. 1413–1420.

[33] Eichhorn, C., F. Goldschmidtboeing, and P. Woias, "A Frequency Piezoelectric Energy Convert Based on a Cantilever Beam," *Proceedings of PowerMEMS 2008+ microEMS2008*, 2008, pp. 309–312.

[34] Wu, W. -J., et al., "Tunable Resonant Frequency Power Harvesting Devices," *Proceedings of SPIE*, Vol. 6169, 2006, pp. 55–62.

[35] Tang, X., and L. Zuo, "Enhanced Vibration Energy Harvesting Using Dual-Mass Systems," *J. Sound Vib.*, Vol. 330, No. 21, 2011, pp. 5199–5209.

[36] Tang, X., and L. Zuo, "Vibration Energy Harvesting from Random Force and Motion Excitations," *Smart Mater. Struct.*, Vol. 21, No. 7, 2012, p. 075025.

[37] Tang, L., and Y. Yang, "A Multiple-Degree-of-Freedom Piezoelectric Energy Harvesting Model," *J. Intel. Mat. Syst. Str.*, Vol. 23, No. 14, 2012, pp. 1631–1647.

[38] Yang, B., et al., "A Generator with Nonlinear Spring Oscillator to Provide Vibrations of Multi-Frequency," *Appl. Phys. Lett.*, Vol. 99, No. 22, 2011, p. 223505.

[39] Vullers, R. J. M., et al., "Large Power Amplification of MEMS Harvester by a Secondary Spring and Mass Assembly," *PowerMEMS 2012*, 2012, p. 211.

[40] El-Hebeary, M. M. R., M. H. Arafa, and S. M. Megahed, "Modeling and Experimental Verification of Multi-Modal Vibration Energy Harvesting from Plate Structures," *Sensor. Actuat. A: Phys.*, Vol. 193, 2013, pp. 35–47.

[41] Roundy, S., et al., "Improving Power Output for Vibration-Based Energy Scavengers," *IEEE Pervasive Computing*, Vol. 4, January-March 2005, pp. 28–36.

[42] Yang, B., et al., "Electromagnetic Energy Harvesting from Vibrations of Multiple Frequencies," *J. Micromech. Microeng.*, Vol. 19, 2009, p. 035001.

[43] Qi, S., et al., "Design of a Multiresonant Beam for Broadband Piezoelectric Energy Harvesting," *Smart Mater. Struct.*, Vol. 19, 2010, p. 094009.

[44] Ching, N. N. H., et al., "A Laser-Micromachined Multi-Modal Resonating Power Transducer for Wireless Sensing Systems," *Sensor. Actuat. A: Phys.*, Vol. 97-98, April 2002, pp. 685–690.

[45] Wang, P. -H., et al., "Design, Fabrication and Performance of a New Vibration-Based Electromagnetic Micro Power Generator," *Microelectron. J.*, Vol. 38, 2007, pp. 1175–1180.

[46] Liu, H., et al., "Feasibility Study of a 3D Vibration-Driven Electromagnetic MEMS Energy Harvester with Multiple Vibration Modes," *J. Micromech. Microeng.*, Vol. 22, 2012, p. 125020.

[47] Ferrari, M., et al., "Piezoelectric Multifrequency Energy Converter for Power Harvesting in Autonomous Microsystems," *Sensor. Actuat. A: Phys.*, Vol. 142, 2008, pp. 329–335.

[48] Shahruz, S. M., "Design of Mechanical Band-Pass Filters for Energy Scavenging," *J. Sound Vib.*, Vol. 292, 2006, pp. 987–998.

[49] H. A. Xue, H. A., Y. T. Hu, and Q. M. Wang, "Broadband Piezoelectric Energy Harvesting Devices Using Multiple Bimorphs with Different Operating Frequencies," *IEEE Trans. on Ultrason. Ferr.*, Vol. 55, September 2008, pp. 2104–2108.

[50] Sari, I., T. Balkan, and H. Kulah, "An Electromagnetic Micro Power Generator for Wideband Environmental Vibrations," *Sensor. Actuat. A: Phys.*, Vol. 145-146, 2008, pp. 405–413.

[51] Elvin, N., and A. Erturk, *Advances in Energy Harvesting Methods*, New York: Springer Science+Business Media, 2013.

[52] Stanton, S. C., C. C. McGehee, and B. P. Mann, "Reversible Hysteresis for Broadband Magnetopiezoelastic Energy Harvesting," *Appl. Phys. Lett.*, Vol. 95, 2009, p. 174103.

[53] Moon, F. C., and P. J. Holmes, "A Magnetoelastic Strange Attractor," *J. Sound Vib.*, Vol. 65, 1979, p. 22.

[54] Erturk, A., J. Hoffmann, and D. J. Inman, "A Piezomagnetoelastic Structure for Broadband Vibration Energy Harvesting," *Appl. Phys. Lett.*, Vol. 94, 2009, p. 254102.

[55] Mann, B. P., and N. D. Sims, "Energy Harvesting from the Nonlinear Oscillations of Magnetic Levitation," *J. Sound Vib.*, Vol. 319, 2009, pp. 515–530.

[56] Saha, C. R., et al., "Electromagnetic Generator for Harvesting Energy from Human Motion," *Sensor. Actuat. A: Phys.*, Vol. 147, 2008, pp. 248–253.

[57] Foisal, A. R. M., C. Hong, and G. -S. Chung, "Multi-Frequency Electromagnetic Energy Harvester Using a Magnetic Spring Cantilever," *Sensor. Actuat. A: Phys.*, Vol. 182, 2012, pp. 106–113.

[58] Shearwood, C., and R. B. Yates, "Development of an Electromagnetic Microgenerator," *Electron. Lett.*, Vol. 33, 1997, pp. 1883–1884.

[59] Williams, C. B., and R. B. Yates, "Analysis of a Micro-Electric Generator for Microsystems," *Sensor. Actuat. A: Phys.*, Vol. 52, 1996.

[60] Marzencki, M., M. Defosseux, and S. Basrour, "MEMS Vibration Energy Harvesting Devices with Passive Resonance Frequency Adaptation Capability," *J. Microelectromech. Syst.*, Vol. 18, December 2009, pp. 1444–1453.

[61] Hajati, A., and S. -G. Kim, "Ultra-Wide Bandwidth Piezoelectric Energy Harvesting," *Appl.Phys. Lett.*, Vol. 99, 2011, p. 083105.

[62] Marinkovic, B., and H. Koser, "Smart Sand—A Wide Bandwidth Vibration Energy Harvesting Platform," *Appl. Phys. Lett.*, Vol. 94, 2009, p. 103505.

[63] Marinkovic, B., and H. Koser, "Demonstration of Wide Bandwidth Energy Harvesting from Vibrations," *Smart Mater. Struct.*, Vol. 21, 2012, p. 065006.

[64] Tvedt, L. G. W., D. S. Nguyen, and E. Halvorsen, "Nonlinear Behavior of an Electrostatic Energy Harvester Under Wide- and Narrowband Excitation," *J. Microelectromech. Syst.*, Vol. 19, April 2010, pp. 305–316.

[65] Nguyen, D. S., et al., "Fabrication and Characterization of a Wideband MEMS Energy Harvester Utilizing Nonlinear Springs," *J. Micromech. Microeng.*, Vol. 20, 2010, p. 125009.

[66] Nguyen, S. D., and E. Halvorsen, "Nonlinear Springs for Bandwidth-Tolerant Vibration Energy Harvesting," *J. Microelectromech. Syst.*, Vol. 20, December 2011, pp. 1225–1227.

[67] Nguyen, S. D., E. Halvorsen, and I. Paprotny, "Bistable Springs for Wideband Microelectromechanical Energy Harvesters," *Appl. Phys. Lett.*, Vol. 102, 2013, p. 023904.

[68] Narimani, A., M. E. Golnaraghi, and G. N. Jazar, "Frequency Response of a Piecewise Linear Vibration Isolator," *J. Vib. Control*, Vol. 10, December 2004, pp. 1775–1794.

[69] Liu, H., et al., "Investigation of a MEMS Piezoelectric Energy Harvester System with a Frequency-Widened-Bandwidth Mechanism Introduced by Mechanical Stoppers," *Smart Mater. Struct.*, Vol. 21, 2012, p. 035005.

[70] Soliman, M. S. M., et al., "A Wideband Vibration-Based Energy Harvester," *J. Micromech. Microeng.*, Vol. 18, 2008, p. 115021.

[71] Soliman, M. S. M., et al., "A Design Procedure for Wideband Micropower Generators," *J. Microelectromech. Syst.*, Vol. 18, December 2009, pp. 1288–1299.

[72] Blystad, L. -C. J., and E. Halvorsen, "A Piezoelectric Energy Harvester with a Mechanical End Stop on One Side," *Microsyst. Technol.*, Vol. 17, 2010, pp. 505–511.

[73] Moss, S., et al., "A Low Profile Vibro-Impacting Energy Harvester with Symmetrical Stops," *Appl. Phys. Lett.*, Vol. 97, 2010, p. 234101.

[74] Liu, H., et al., "Piezoelectric MEMS Energy Harvester for Low-Frequency Vibrations with Wideband Operation Range and Steadily Increased Output Power," *J. Microelectromech. Syst.*, Vol. 20, October 2011, pp. 1131–1142.

[75] Liu, H., et al., "A New S-Shaped MEMS PZT Cantilever for Energy Harvesting from Low Frequency Vibrations Below 30 Hz," *Microsyst. Technol.*, Vol. 18, 2012, pp. 497–506.

[76] Miki, D., et al., "Large-Amplitude MEMS Electret Generator with Nonlinear Spring," *MEMS 2010: 23rd IEEE International Conference on Micro Electro Mechanical Systems, Technical Digest*, 2010, pp. 176–179.

[77] Suzuki, Y., et al., "A MEMS Electret Generator with Electrostatic Levitation for Vibration-Driven Energy-Harvesting Applications," *J. Micromech. Microeng.*, Vol. 20, 2010, p. 104002.

[78] Umeda, M., K. Nakamura, and S. Ueha, "Analysis of the Transformation of Mechanical Impact Energy to Electric Energy Using Piezoelectric Vibrator," *Jpn. J. Appl. Phys. Part 1*, Vol. 35, May 1996, pp. 3267–3273.

[79] Umeda, M., K. Nakamura, and S. Ueha, "Energy Storage Characteristics of a Piezo-Generator Using Impact Induced Vibration," *Jpn. J. Appl. Phys. Part 1,* Vol. 36, May 1997, pp. 3146–3151.

[80] Renaud, M., et al., "Harvesting Energy from the Motion of Human Limbs: The Design and Analysis of an Impact-Based Piezoelectric Generator," *Smart Mater. Struct.,* Vol. 18, 2009, p. 035001.

[81] Manln, G., N. M. White, and J. Tudor, "Harvesting Energy from Vehicle Wheels," *IEEE Transducer '09,* 2009, p. 4.

[82] Gu, L., "Low-Frequency Piezoelectric Energy Harvesting Prototype Suitable for the MEMS Implementation," *Microelectron. J.,* Vol. 42, 2011, pp. 277–282.

[83] Gu, L., and C. Livermore, "Impact-Driven, Frequency Up-Converting Coupled Vibration Energy Harvesting Device for Low Frequency Operation," *Smart Mater. Struct.,* Vol. 20, 2011, p. 045004.

[84] Liu, H., et al., "Piezoelectric MEMS-Based Wideband Energy Harvesting Systems Using a Frequency-Up-Conversion Cantilever Stopper," *Sensor. Actuat. A: Phys.*, Vol. 186, October 2012, pp. 242–248.

[85] Lee, D. -G., et al., "Novel Micro Vibration Energy Harvesting Device Using Frequency Up Conversion," *IEEE Transducer '07,* 2007, p. 4.

[86] Liu, H., et al., "A Scrape-Through Piezoelectric MEMS Energy Harvester with Frequency Broadband and Up-Conversion Behaviors," *Microsyst. Technol.,* Vol. 17, 2011, pp. 1747–1754.

[87] Zorlu, O., E. T. Topal, and H. Kulah, "A Vibration-Based Electromagnetic Energy Harvester Using Mechanical Frequency Up-Conversion Method," *IEEE Sens. J.,* Vol. 11, February 2011, pp. 481–488.

[88] Zorlu, O., and H. Kulah, "A MEMS-Based Energy Harvester For Generating Energy from Non-Resonant Environmental Vibrations," *Sensor. Actuat. A: Phys.*, Vol. 202, 2013, pp. 124–134.

[89] Jung, S. -M., and K. -S. Yun, "Energy-Harvesting Device with Mechanical Frequency-Up Conversion Mechanism for Increased Power Efficiency and Wideband Operation," *Appl. Phys. Lett.,* Vol. 96, 2010, p. 111906.

[90] Kulah, H., and K. Najafi, "Energy Scavenging from Low-Frequency Vibrations by Using Frequency Up-Conversion for Wireless Sensor Applications," *IEEE Sens. J.,* Vol. 8, March-April 2008, pp. 261–268.

[91] Sari, I., T. Balkan, and H. Kulah, "An Electromagnetic Micro Power Generator for Low-Frequency Environmental Vibrations Based on the Frequency Upconversion Technique," *J. Microelectromech. Syst.,* Vol. 19, February 2010, pp. 14–27.

[92] Galchev, T., H. Kim, and K. Najafi, "Micro Power Generator for Harvesting Low-Frequency and Nonperiodic Vibrations," *J. Microelectromech. Syst.,* Vol. 20, August 2011, pp. 852–866.

[93] Galchev, T. V., et al., "Harvesting Traffic-Induced Vibrations for Structural Health Monitoring Of Bridges," *J. Micromech. Microeng.,* Vol. 21, 2011, p. 104005.

[94] Tang, Q. C., and X. Li, "A Widen-Band Piezoelectric Energy-Harvester for High-Efficiency Power Generation at Low Frequencies," *IEEE Transducers 2013*, June 2013, pp. 697–700.

[95] Tang, Q. C., Y. L. Yang, and X. Li, "Bi-Stable Frequency Up-Conversion Piezoelectric Energy Harvester Driven by Non-Contact Magnetic Repulsion," *Smart Mater. Struct.*, Vol. 20, No. 12, 2011, p. 125011.

[96] Roundy, S., P. K. Wright, and J. Rabaey, "A Study of Low Level Vibrations as a Power Source for Wireless Sensor Nodes," *Comput. Commun.*, Vol. 26, 2003, pp. 1131–1144.

[97] Khoshhal, K., et al., "Probabilistic LMA-Based Classification of Human Behaviour Understanding Using Power Spectrum Technique," *13th Inter. Conf. Information Fusion*, EICC Edinburgh, UK, July 2010.

[98] Hostens, I., and H. Ramon, "Descriptive Analysis of Combine Cabin Vibrations and Their Effect on the Human Body," *J. Sound Vib.*, Vol. 266, 2003, pp. 453–464.

[99] Lo, H. -W., and Y.-C. Tai, "Parylene HT for Electret Generators," *J. Micromech. Microeng.*, Vol. 18, No. 10, 2008, p. 104006.

[100] Yang, B., and C. Lee, "Non-Resonant Electromagnetic Wideband Energy Harvesting Mechanism for Low Frequency Vibrations," *Microsyst. Technol.*, Vol. 16, 2010, pp. 961–966.

[101] Bowers, B. J., and D. P. Arnold, "Spherical, Rolling Magnet Generators for Passive Energy Harvesting from Human Motion," *J. Micromech. Microeng.*, Vol. 19, No. 9, 2009, p. 094008.

6

Microelectronic Circuits for Vibration Energy Harvesting

Vibration-based MEMS energy-harvesting devices have been increasingly reported in the literature, and the number of active research groups in the field has grown steadily. This interest is driven by a desire to eliminate primary batteries in low power microelectronic devices, particularly wireless sensor nodes [1, 2]. Large reductions of the power consumption have not only made the wireless networks more attractive, but also made the energy-harvesting approach more realistic. Normally the load circuitry requires a steady dc voltage; however, a vibration harvester does not produce such stable power source. It is therefore necessary to form the power management circuit as the interface between the harvester device and the load. In earlier chapters, the majority of work concentrated on the design and fabrication of microscale vibration energy harvesters; this chapter discusses the microelectronic circuits.

6.1 Overview of Energy-Harvesting Electronics

The electronic circuits for vibration energy harvesting have undergone revolutions from a simple energy harvester with peak rectifiers to complex topologies and control and even integrated with ICs. A classic energy-harvesting system consists of an energy-harvesting transducer, regulation/storage/management electronics and load, such as wireless sensor nodes as shown in Figure 6.1. In the block diagram, the energy-harvesting transducer generates electrical power from the ambient vibrations, based on piezoelectric, electromagnetic, and electrostatic transduction mechanisms. The ac power from the harvester is input

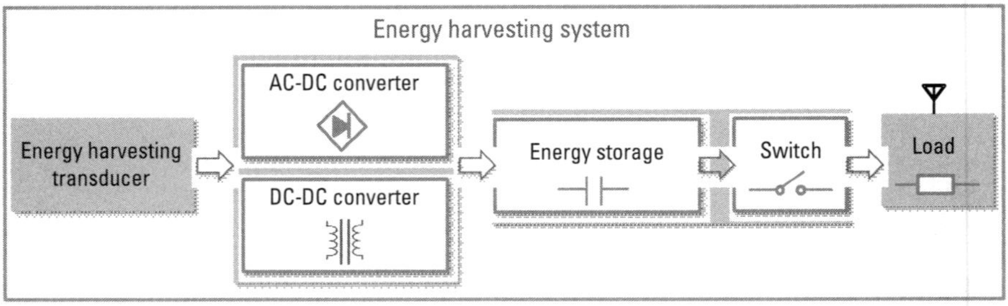

Figure 6.1 Classic energy-harvesting electronics.

to the converter, which converts the ac voltage to dc and steps up or down the dc voltage for different conditions. The converter is designed to accept instantaneous input voltages ranging from 0V to ±500V ac and input currents from 200 nA to 400 mA in either a steady stream of pulses or an intermittent and irregular manner with varying source impedances.

To achieve high power density, an impedance match between the energy-harvesting transducer and the electrical subsystem should be considered. This requires control of the input impedance of the circuit that interfaces to the transducer. However, the output voltage and current from the harvester are rarely compatible with load electronics and thus some manner of voltage regulation is required. Rectification, filtering, and voltage regulation are the basic requirements to convert the ac electrical output to dc from a vibration energy harvester useable in the majority of applications. To maximize the usable power from energy-harvesting transducer, researchers have been optimizing these subsystems through the adoption of more complex implementations, such as synchronous rectifiers, switched-mode converters, and peak power trackers. Early harvester electronics required a minimum of 4-V input to capture and store the energy from power generators. More recent designs feature a front-end voltage booster and claim to initiate capture and energy storage with voltage inputs at less than 100 mV.

6.1.1 Load Impedance Matching

To ensure that the energy produced in the harvester is efficiently transferred to the load, there is a fundamental requirement that the load impedance should be significantly larger than the internal impedance of the harvester. Although this arrangement achieves the maximum electrical efficiency, it could not obtain the maximum power transfer. It is well known that the maximum power transfer occurs when the load impedance is equal to the internal impedance. In the case of an ac energy source, the load should provide a conjugate match to the harvester. The internal impedance will be dependent upon the type of energy harvester and the conditions under which the harvester is operating.

To determine an optimal electrical load, a suitable equivalent circuit model of the harvester must be developed. A detailed explanation of the construction of the equivalent models can be found in [3], and therefore only a general explanation will be given here. Harvesting kinetic energy in vibrations may take one of several forms based on piezoelectric, electrostatic, and electromagnetic means. A piezoelectric device is typically a piezoelectric bending cantilever fixed on one end and attached to a free-moving mass on the other end. It produces a voltage in response to strain or deformation of the cantilever and is proportional to the mechanical-electrical coupling of the piezoelectric material and the stress applied. The equivalent circuit of the piezoelectric harvester is depicted as an open-circuit voltage (V_{PE}), a capacitor (C_{PE}), and a resistor (Rs) in series connection. The electromagnetic harvester harness kinetic energy by moving the coil through a magnetic field provided by a stationary magnet or moving a magnet across a stationary coil. Therefore, the voltage through the coil will be produced based on Faraday's induction law. The equivalent model contains an open-circuit voltage (V_{EM}) and the resistance of the coil (Rs) connected in series. The electrostatic harvester converts vibration to electrical energy by changing the gap distance or overlap area of the parallel plates of a capacitor. Under the charge-constrained condition, the model can be simplified to the equivalent circuit similar to that of an electromagnetic harvester. In some circumstances and operating modes, it may not be optimal to match the load impedance to that of the harvester due to some constraints; while this chapter considers the situation that there is always a defined load impedance that results in maximum power extraction from the transducer. It may therefore be more accurate to specify that the input impedance of the interface circuit to the transducer must be controllable, rather than always matched to the harvester.

6.1.2 Voltage Regulation Circuits

In an energy-harvesting system, the harvesting transducer, battery storage element, and load electronics all have different voltage and current characteristics. The voltage and current produced by the harvester may differ vastly from that of the battery and load elements, and hence must be processed before being connected to the load electronics. In addition, the voltage across the energy storage element may fluctuate depending on its state of charge. For example, the voltage across a thin-film Li Ion is around 2.7V when discharged and 4.2V when charged. However, the load electronics may not survive supply voltage variations exceeding 0.25V around a nominal value of about 1–1.5V. As a result, the ac-dc or dc-dc voltage regulator and trickle-charger circuits are used to transfer and condition energy from the harvester to the load through the battery.

6.1.2.1 AC-DC Converter

An AC-DC converter, also called an inverter or rectifier, manages AC voltage signal to dc by appropriately channeling charge into holding capacitor. Figure 6.2(a) shows a single diode-capacitor circuit for ac-dc rectifying. The basic principle is to transfer positive current from input voltage source V_{IN} into output capacitor C_O only when V_{IN} is positive with respect to ground. In this half-wave rectifier, the diode D holds positive voltage peak and the capacitor C_O holds it until the next positive cycle. However, this strategy only rectifies the positive phase of the voltage cycle. Thus, the capacitor C_O must be sufficiently large to sustain output current I_O without allowing the voltage across C_O to drop significantly. As an alternative of half-wave rectifier, Figure 6.2(b) illustrates a commonly used full-wave rectifier, also called a Wheatstone bridge, which is a combination of two half-wave rectifiers. This approach eases the holding requirements of C_O and consequently reduces the total variation of V_O. Considering the positive half-cycle that the voltage V_{IN} is positive with respect to ground, D_2 and C_O constitute a simple rectifying circuit, and D_4 is in series connection to complete the circuit loop. During this process, D_1 and D_3 are reversely biased and therefore off. Conversely, during the negative half-cycle, D_3 and C_O rectify the half-wave voltage and D_1 sources the current from ground. Hence, the circuits are completed and D_2 and D_4 are off during this time. The above asynchronous diodes can ease the circuit design but only at the expense of power losses. Approximately 0.6–0.8-V voltage drop across would exhibit every time they conduct current to V_O. Therefore, synchronous MOSFET switches are used as an alternative that drops considerably lower voltages, maybe on the order of 50–150 mV. The drawback to MOSFETs is that they require synchronizing signals to switch when desired, which makes the overall circuit more complex.

6.1.2.2 DC-DC Converter

The most straightforward method for converting high dc voltage to low is to use a linear regulator, which is actually an operational amplifier (op amp) in noninverting configuration. The most attractive features of a linear regulator are low noise and relatively fast response times. However, because the variable

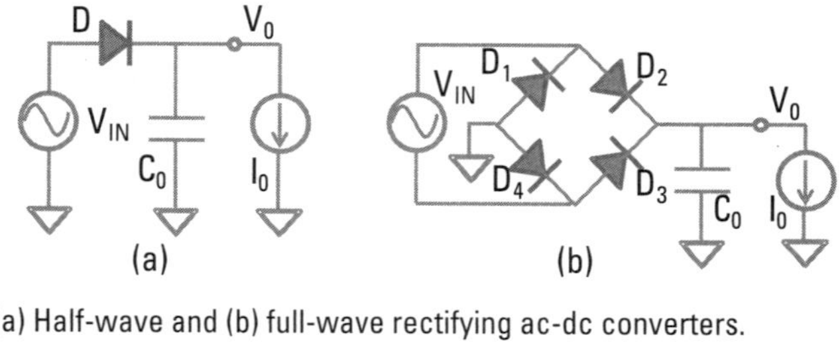

Figure 6.2 (a) Half-wave and (b) full-wave rectifying ac-dc converters.

and unpredictable load feature brings stringent stability requirements on the regulator, greatly different filter compensation strategies will be induced. In addition, the current requirement of a regulator is indeed larger than those of an op amp.

The switching dc-dc converters use boosting parallel-series or bucking series-parallel charge pumps to step-up or step-down voltages, which fall under two categories: electrostatic and magnetic. The electrostatic converter employs only capacitors to transfer energy and condition power, while a magnetic-switching converter uses capacitors only for storage and inductors for transferring energy. Basically speaking, switching power supplies enjoy higher power efficiency than linear regulators. This is because the voltages impressed across the power transistors (i.e., switches) are small in steady state (on the order of 50–150 mV), regardless of input voltage V_{IN} and output voltage V_O.

The basic strategy for step-up or boosting electrostatic-based charge pumps is to charge one or several capacitors to input voltage V_{IN} and then connect them in series to output voltage V_O as shown in Figure 6.3(a). Each capacitor holds an initial voltage equivalent to *VIN* and the overall *VIN* becomes $(N+1)\,V_{IN}$, where N refers to the numbers of pumping capacitors. The bucking or step-down strategy is similar but reversed as shown in Figure 6.3(b). A series of capacitors are charged to V_{IN}, initializing each capacitor to V_{IN}/N, and then all capacitors are connected in parallel with V_O, producing an initial voltage equivalent to V_{IN}/N at V_O. To increase the boosting amplification across N capacitors, several low-gain (i.e., two times) charge pumps in series are connected to use the boosted output of one to charge the next stage, and consequently achieving a higher overall gain of $2N\,V_{IN}$, as demonstrated in Figure 6.3(c).

6.1.3 Energy Storage Elements

Selection of the materials and design of an energy storage element can vary extensively depending on the lifetime requirements, size, cost, and so on. For autonomous microelectronic loads, the foremost microenergy storage devices being considered are microbatteries and electrochemical supercapacitors. This section details the design consideration, reviews existing technology and evaluates those microenergy storage devices.

6.1.3.1 Microbatteries

Batteries contain an ion-conducting electrolyte phase sandwiched by two chemically different electrodes. Once it is connected to a load, chemical reactions occur and ions travel across the electrolyte. A spontaneous flow of electrons occurs in the external circuit from the electrode with negative potential to the electrode with positive electrode, and can therefore be used to power the load electronics as seen in Figure 6.4(a). If the battery is rechargeable, the reverse

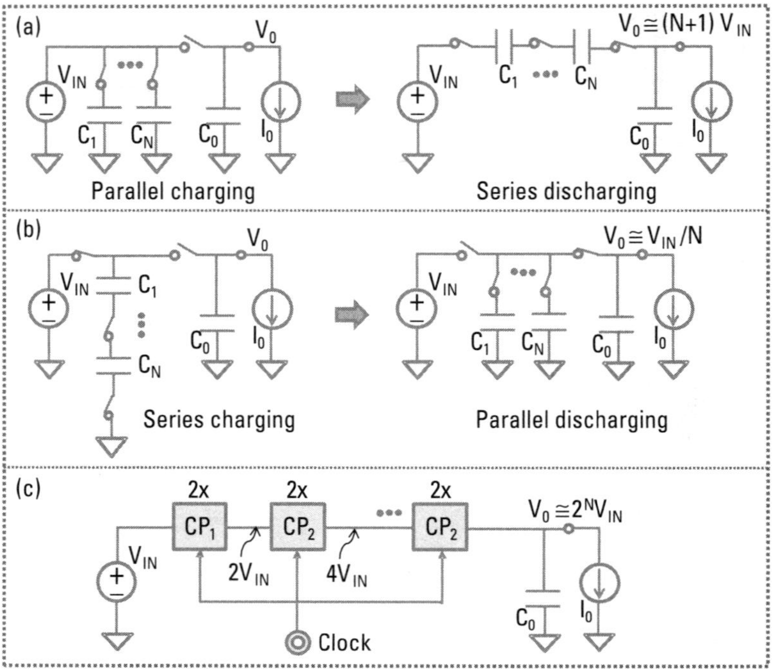

Figure 6.3 (a) Boosting (step-up) parallel-series and (b) bucking (step-down) series-parallel charge pumps. (c) Cascaded charge-pump configuration.

reactions can be induced. Construction of the first solid-state microbatteries was demonstrated in the early 1980s [4], as a result of substantial developments in solid-state ion conducting materials [5]. Decades of materials and processing innovations gave rise to a few successful microbattery implementations, with a variety of materials, geometries, and processing strategies. A few candidates of rechargeable microbattery chemistries have been summarized by Linden [6]. Most of those chemistries have been largely considered or even implemented in rechargeable micobatteries associated with varying materials compatibilities and fabrication processes.

The main components affecting the performance of a microbattery are the electrodes, electrolyte, current collectors, and a separator. Figure 6.4(c, d) illustrate the two-dimensional (2-D) microbatteries in planar or stacked configurations. The planar batteries are comprised of two electrodes positioned adjacent to each other on a substrate. This arrangement does not require a separator since the electrodes are physically detached. Normally, the planar cells are simply submerged in a liquid electrolyte or covered with a solid or polymer electrolyte film. The stacked batteries are fabricated by the consecutive deposition of films layer-by-layer and all films employ the same footprint area. The current distribution in a stacked cell runs perpendicular to the through thickness of the electrolyte/separator layer, resulting in uniform performance across both electrodes. In comparison to the planar microbatteries, it is more suitable for high-rate applications and could achieve better areal energy density. The 2-D microbatteries are categorized into two processing methods, that is, thin film and thick film deposition, and a variety of prototypes [7–18] are compared in Table 6.1.

For 2-D microbatteries, the amount of energy stored is determined by the electrodes and their thickness, while the rate of energy extracted depends on the contact area between electrodes and electrolytes, their through thickness, and the charge transport properties. This leads to a trade-off between the electrode and electrolyte dimensions. As an alternative, nontraditional three-dimensional (3-D) microbatteries have been implemented. By increasing the amount of electrode and electrolyte interfaces while maintain short ion transport distance between electrodes, 3-D microbatteries is able to improve the electrochemical performance dramatically. The most commonly adopted architectures for 3-D microbatteries are cells with interdigital cylinder or plate electrode array as shown in Figure 6.5(a, b) or positioning of concentric electrode and electrolyte

Table 6.1
Implementations and Performance Summary of 2-D Microbatteries

	Chemistry	**Geometry t—thickness f—footprint area**	**Operating Voltage (V)**	**Performance per 1 cm^2**
Thin film				
Stacked [7–11] Physical vapor deposition	Li-ion, glass electrolyte	Stacked t: 10–15 μm f: 1 mm^2–25 cm^2	3–4.2	10 μAh–1mAh; 50 μWh–5mWh; 10 mW
Thick film				
Stacked [12] Electrode polishing	Li-ion, ceramic electrolyte	t: 500 μm f: 1 cm^2	1.5–3.2	15 μAh; 42.5 μWh
Planar [13, 14] Microfabrication	Ni-Zn, liquid electrolyte	t: 50 μm f: 2 mm^2	1.2–1.85	305 μAh; 500 μWh; 150 mW
Planar, interdigital electrodes [15, 16] Dispenser printing	Li-ion, polymer electrolyte	t: 0.5–1 μm f: 6.6 cm^2	1.5–3	4.5 μAh–1mAh; 11 μWh
Stacked [15, 16]	$Zn\text{-}MnO_2$, gel electrolyte	t: 100 μm f: 0.25 cm^2	1–1.9	1.05 mAh; 0.37 μWh
Planar, interdigital electrodes [17, 18] Laser direct writer	$Zn\text{-}Ag_2O$, liquid electrolyte	t: 10–20 μm f: 4.9 mm^2	0.8–1.6	450 μAh; 0.6 mWh; 0.8 mW
Stacked, embedded in substrate [17]	Li-ion, gel electrolyte	t: 30 μm f: 9 mm^2	2.9–4.2	110 μAh; 1.32 mWh

Source: [1].

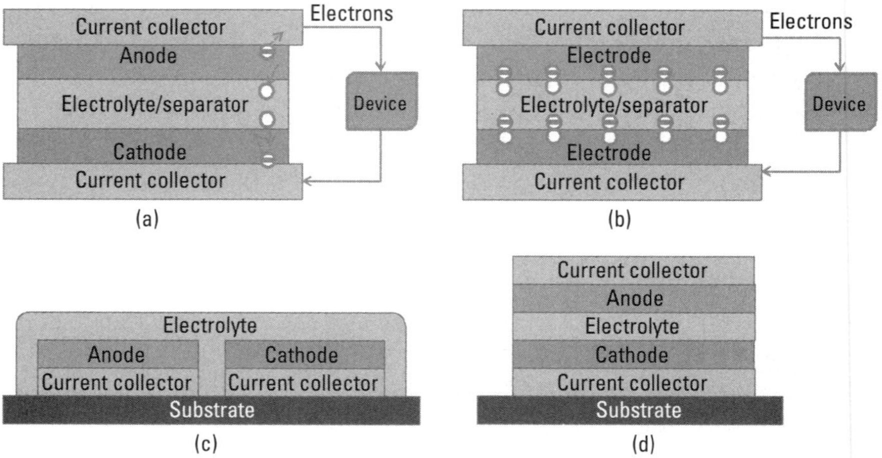

Figure 6.4 Schematics of (a) microbatteries, (b) microcapacitors, and (c) 2-D planar and (d) stacked microbattery configurations.

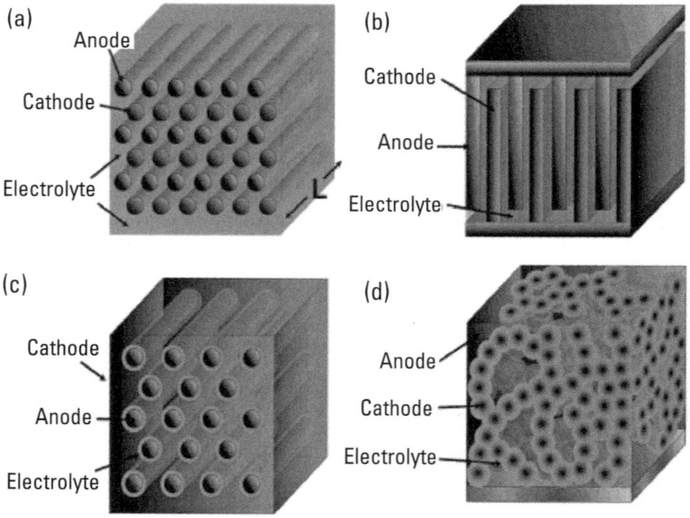

Figure 6.5 Architectures for 3-D microbatteries: (a) interdigital cylindrical array of electrodes; (b) interdigital plate array of electrodes; (c) an array of cylindrical electrodes coated with electrolyte and the surrounding filled with the second electrode material; and (d) an architecture similar to (c) with aperiodic interpenetration electrodes and electrolyte. (Copyright: ACS, 2004 [20].)

units surrounded by a matrix of the opposing electrode as shown in Figure 6.5(c, d). The development of 3-D architectures have been reviewed by Long et al. [19, 20], Rolison et al. [21], and Oudenhoven et al. [22]. The two main advantages are the increase in energy density of the device in a limited footprint and the decrease in the transport path of electrolyte ions. Promising research efforts in the 3-D microbatteries with one discontinuous element and continuous elements are summarized in Table 6.2 [1, 23–29]. New research efforts including processing, characterization, and modeling capabilities are required to deeply understand the 3-D architectures on the current and potential distributions within the cell, interfacial reaction kinetics, and overall device performance.

6.1.3.2 Microcapacitors

The electrochemical capacitors, also called supercapacitors or ultracapacitors, are expected to play an important role as high-power energy storage devices for miniaturized electronics. In comparison to batteries, supercapcitors operate within longer periods of time (e.g., millions of cycles) without the loss of their energy storage capacity. Supercapacitors can be classified into two categories: electromechanical double-layer capacitors (EDLCs) and pseudocapacitors or redox capacitors. As an electric field is applied across the capacitor, ions in the electrolyte can dissociate and travel towards oppositely charged electrode surfaces. The performance of microsupercapacitors is dependent on the intrinsic

Table 6.2
Implementations and Performance Summary of 3-D Microbatteries

	Chemistry	**Geometry**	**Operating Voltage (V)**	**Performance per 1 cm^2**
3D microbatteries with at least one discontinous element				
Removable silicon mold, electrodeposited rods [23]	Ni-Zn, liquid electrolyte	d: 50 μm t: 400 μm f: 0.25 cm^2	1.3–1.8	2.5 μAh; 4 μWh
Interlaced trenches in porous silicon mold [24]	Li-ion, liquid electrolyte	w: 50 μm; l: 290 μm; n: 8265; f: 1 cm^2	3–4	—
C-MEMS fabricated rods [25]	Li-ion, liquid electrolyte	d: 20 μm; l: 65 μm; n: 14400; f: 1 cm^2	0.7–3.5	10.6 μAh;
Super ink jet printed [26]	Zn-AgO, liquid electrolyte	d: 10 μm; l: 40 μm; n: 722; f: 0.9 cm^2	1.2–2	2.5 mAh; 3.95 mWh;
Thin film-coated concentric road array [27]	Li-ion, hybrid polymer electrolyte	d: 50 μm l: 0.5 μm f: 5.3 cm^2	1.3–2.2	2 mAh; 3 mWh
3D microbatteries with continuous elements				
Thin-film coated textured substrates [28]	Li-ion, glass electrolyte	w: 5 μm l: 130 μm f: 1 cm^2	3–4	15 mAh; 5 mWh; 50 mW
Templated ordered scaffold [29]	Li-ion, polymer electrolyte	t: 0.1343 cm f: 0.26 cm^2	1.6–3.3	25 μAh; 60 mWh

Source: [1].

properties of the electrode materials and the method of electrode fabrication. Carbon (EDLC) materials such as carbon nanotubes (CNTs), graphene and onion like carbon (OLC) [30], and pseudo-capacitive type electrode materials such as RuO_2, MnO_2, polypyrrole (Ppy), and polyaniline (PANI) have been used to fabricate microsupercapacitor electrodes. Due to its high specific surface area, good chemical stability and outstanding electrical properties, graphene is one of ideal candidates for energy storage [31, 32]. Graphene electrode can be used instead of carbon as an EDLC to filter ac. In addition, EDLCs from graphenes can minimize electronic and ionic resistance and produce capacitors with RC time constant of less than 200 ms, in contrast with ~1 second for typical EDLCs [33]. The fabrication techniques such as CVD, LBL, sputtering, electrochemical and electrophoretic deposition, spray deposition, and printing have been used to fabricate thin films of 2-D micro-supercapacitors [34]. Continued material improvements have diminished the gap of energy density between microbatteries and electrochemical capacitors.

6.2 Case Study of Energy-Harvesting Electronics

6.2.1 Interface Electronics for Piezoelectric Energy Harvesting

Piezoelectric energy harvesters convert the vibration energy into electrical energy based on the piezoelectric effect. Because the voltage generated across the piezoelectric element is ac signal, a basic ac-dc converter (or rectifier) is required to store the energy in a storage buffer. In addition, an impedance matching circuit is necessary to maximize the power generation. Oftentimes, an inductor can be induced to the circuit to compensate the contribution of the piezoelectric clamped capacitor. However, it cannot be adaptive to vibrations because of the large inductance at low frequency. Therefore, switching-type charging circuits have been proposed to overcome this drawback, where the switches are operated synchronously with the harvester vibration to optimize the power flow. Figure 6.6(a, b) illustrates two basic synchronized switching circuit (SSC) topologies [35]. The first type of switching circuit places the switches between piezoelectric element and rectifier. Typical examples are parallel SSHI (synchronized switching harvesting on an inductor) and series SSHI as shown in Figure 6.6(a) [36–42]. This technique is able to modify the generated voltage waveform to increase the collected power. The second type positions the switches between the rectifier and storage buffer, which is used to modify the charging current flowing into the storage buffer to speed up the charging and adapt the load [43, 44].

Although the SSC techniques can increase the collected energy or charging speed, they also require energy to supply active switches. It is hence of interest to achieve this control loop with totally self-powered electronics. As

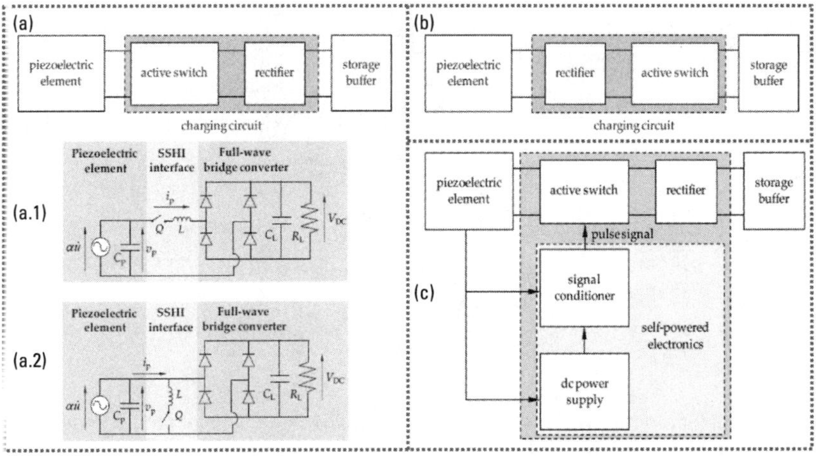

Figure 6.6 (a, b) Configurations of the piezoelectric element with synchronized switching circuit; and (c) the self-powered synchronized switching circuit; equivalent circuit of (a.1) a series SSHI converter and (a.2) a parallel SSHI converter [35].

demonstrated in Figure 6.6(c), the self-powered charging circuit comprises a synchronized switching converter, including active switches and rectifier, a signal conditioner to control the switching circuit, and a DC power supply for the signal control electronics. The low-voltage drop SSHI converter was developed by using the body diodes in MOSFET switches, which can diminish maximum 4 diodes drops. There are two main types of signal conditioners reported, that is, the peak detector [45–47] and the zero-crossing detector [41, 48]. The peak detector is more suitable for a series SSHI converter as the piezoelectric voltage is synchronized with the displacement. However, the phase delay for parallel SSHI is larger. The zero-crossing detector can be used for both series and parallel SSHI techniques. To avoid the phase compensation, an additional piezoelectric patch is required [41]. Ben-Yaakov's zero-crossing detector adapts the single piezoelectric patch, but it owns the large phase lag [48].

6.2.2 Interface Electronics for Electromagnetic Energy Harvesting

The general requirements of the interfacing electronics for an electromagnetic energy harvesting transducer are rectification, voltage step-up capability, and impedance matching control. A typical electrical interface as shown in Figure 6.7(a) is basically a step-up transformer that contains two Schottky diodes (D1 and D2) and a storage capacitor [49]. By using the step-up transformer, the low ac output voltage (tens or hundreds of millivolts) can be upconverted with appropriate transformer ratio. The diodes are used to rectify the stepped-up ac voltage, where D1 conducts during positive half voltage and D2 follows in the

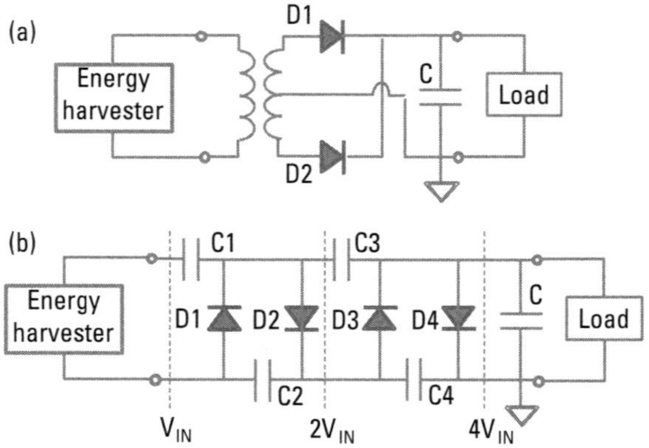

Figure 6.7 (a) Electrical interface circuit for rectification and voltage step-up; and (b) a Villard voltage multiplier for voltage boosting.

negative half [50–52]. Comparing with a standard diode bridge, this arrangement minimizes the effect of the diode voltage drop and is simple to be implemented. However, it may not transfer the maximum power to the load since an impedance match is not included in the configuration.

The voltage step-up capability can be achieved by using voltage multipliers such as the Villard multiplier as shown in Figure 6.7(b) and the Dickson multiplier. Cascading multiple stages of the Villard multiplier will boost the voltage greatly without using magnetic components [53, 54]. Each stage consists of two diodes and two capacitors. Mitcheson et al. [55] reported a dual-polarity boost converter as a potential solution to provide rectification, impedance match, and voltage step-up in one interface circuit for an electromagnetic harvester; meanwhile, minimizing diode voltage drops. Two boost converters are activated to rectify the positive and negative half-cycles of the ac voltage alternatively. The dual-polarity property eliminates the demand for a diode bridge rectifier. The authors also recommended introducing the synchronously switched MOSFETs for Schottky diodes to reduce the effects of power losses in the boost converter.

References

[1] Beeby, S., and N. White, *Energy Harvesting for Autonomous Systems*, Norwood, MA: Artech House, 2010.

[2] Priya, S., and D. J. Inman, *Energy Harvesting Technologies*, New York: Springer, 2009.

[3] D'Hulst, R. D., and J. Driesen, "Power Processing Circuits for Vibration-Based Energy Harvesters," *IEEE Power Electronics Specialists Conf.*, 2009, pp. 2556–2662.

[4] Creus, R., et al., "The Use of Ionic and Mixed Conductive Glasses in Microbatteries," *Mat. Sci. Eng. B*, Vol. 3, No. 1-2, 1989, pp. 109–112.

[5] Kennedy, J. H., "Thin Film Solid Electrolyte Systems," *Thin Solid Films*, Vol. 43, No. 1-2, 1977, pp. 41–92.

[6] Linden, H. D., *Handbook of Batteries*, 2nd ed., New York: McGraw-Hill, 1995.

[7] Bates, J. B., et al., "Thin-Film Lithium and Lithium-Ion Batteries," *Solid State Ionics,* Vol. 135, No. 1-4, 2000, pp. 33–45.

[8] Kushida, K., K. Kuriyama, and T. Nozaki, "Hundred-Micron-Sized All-Solid-State Li Secondary Battery Arrays Embedded in a Si Substrate," *Appl. Phys. Lett.*, Vol. 81, No. 26, 2002, pp. 5066–5068.

[9] West, W. C., et al., "Fabrication and Testing of All Solid-State Microscale Lithium Batteries for Microspacecraft Applications," *J. Micromech. Microeng.*, Vol. 12(a), 2002, pp. 58–62.

[10] Dudney, N. J., "Solid-State Thin-Film Rechargeable Batteries," *Mat. Sci. Eng. B*, Vol. 116, No. 3, 2005, pp. 245–249.

[11] Song, J., et al., "Solid-State Microscale Lithium Batteries Prepared with Microfabrication Processes," *J. Micromech. Microeng.*, Vol. 19, No. 4, 2009, p. 045004.

[12] Birke, P., et al., "A First Approach to a Monolithic All Solid State Inorganic Lithium Battery," *Solid State Ionics*, Vol. 118, No. 1, 1999, pp. 149–157.

[13] Humble, P. H., J. N. Harb, and R. LaFollette, "Microscopic Nickel-Zinc Batteries for Use in Autonomous Microsystems," *J. Electrochem. Soc.*, Vol. 148, No. 12, 2001, pp. A1357–A1361.

[14] Humble, P. H., and J. N. Harb, "Optimization of Nickel-Zinc Microbatteries for Hybrid Powered Microsensor Systems," *J. Electrochem. Soc.*, Vol. 150, No. 9, 2003, pp. A1182–A1187.

[15] Dokko, K., et al., "Sol-Gel Fabrication of Lithium-Ion Microarray Battery," *Electrochem. Commun.,* Vol. 9, No. 5, 2007, pp. 857–862.

[16] Ho, C. C., J. W. Evans, and P. K. Wright, "Direct Write Dispenser Printing of Zinc Microbatteries," *Proc. for PowerMEMS 2009*, December 2009, pp. 141–144.

[17] Arnold, C. B., H. Kim, and A. Pique, "Laser Direct Write of Planar Alkaline Microbatteries," *Appl. Phys. A,* Vol. 79, No. 3, 2004, pp. 417–420.

[18] Sutto, T. E., et al., " Laser Transferable Polymer-Ionic Liquid Separator/Electrolytes for Solid-State Rechargeable Lithium-Ion Microbatteries," *Electrochem. Solid-State Lett.*, Vol. 9, No. 2, 2006, pp. A69–A71.

[19] Long, J. W., and D. R. Rolison, "Architectural Design, Interior Decoration, and Three-Dimensional Plumbing En Route to Multifunctional Nanoarchitectures," *Acc. Chem. Res.*, Vol. 40, No. 9, 2007, pp. 854–862.

[20] Long, J. W., et al., "Three-Dimensional Battery Architectures," *Chem. Rev.*, Vol. 104, No. 10, 2004, pp. 4463–4492.

[21] Rolison, D. R., et al., "Multifunctional 3D Nanoarchitectures for Energy Storage and Conversion," *Chem. Soc. Rev.*, Vol. 38, No. 1, 2009, pp. 226–252.

[22] Oudenhoven, J. F. M., L. Baggetto and P. H. L. Notten, "All-Solid-State Lithium-Ion Microbatteries: A Review of Various Three-Dimensional Concepts," *Adv. Energy Mater.*, Vol. 1, No. 1, 2011, pp. 10–33.

[23] Chamran, F., et al., "Fabrication of High-Aspect-Ratio Electrode Arrays for Three-Dimensional Microbatteries," *J. Microelectromech. S.*, Vol. 16, No. 4, 2007, pp. 844–852.

[24] Ripenbein, T., et al., "Novel Porous-Silicon Structures for 3D-Interlaced Microbatteries," *Electrochimica Acta*, January 2009.

[25] Min, H., et al., "Fabrication and Properties of a Carbon/Polypyrrole Three-Dimensional Microbattery," *Journal of Power Sources*, Vol. 178, 2008, pp. 795–800.

[26] Ho, C. C., et al., "A Super Ink Jet Printed Zinc-Silver 3D Microbattery," *J. Micromech. Microeng.*, Vol. 19, 2009, p. 094013.

[27] Nathan, M., et al., "Three-Dimensional Thin-Film Li-Ion Microbatteries for Autonomous MEMS," *J. Micorelectromech. S.*, Vol. 14, No. 5, 2005, pp. 879–885.

[28] Niessen, R., F. Roozeboom, and P. Notten, "High Energy Density All-Solid-State Batteries: A Challenging Concept Towards 3D Integration," *Adv. Funct. Mater.*, Vol. 18, No. 7, 2008, pp. 1057–1066.

[29] Ergang, N. S., et al., "Fabrication of a Fully Infiltrated Three-Dimensional Solid-State Interpenetrating Electrochemical Cell," *J. Electrochem. Soc.*, Vol. 154, No. 12, 2007, pp. A1135–A1139.

[30] Simon, P., and Y. Gogotsi, "Capacitive Energy Storage in Nanostructured Carbon-Electrolyte Systems," *Acc. Chem. Res.*, Vol. 46, No. 5, 2012, pp. 1094–1103.

[31] Grande, L., et al., "Graphene for Energy Harvesting/Storage Devices and Printed Electronics," *Particuology*, Vol. 10, 2012, pp. 1–8.

[32] Choi, H. -J., et al., "Graphene for Energy Conversion and Storage in Fuel Cells and Supercapacitors," *Nano Energy*, Vol. 1, 2012, pp. 534–551.

[33] Miller, J. R., R. A. Outlaw, and B. C. Holloway, "Graphene Double-Layer Capacitor with AC Line-Filtering Performance," *Science*, Vol. 329, No. 5999, 2010, pp. 1637–1639.

[34] Beidaghi, M., and Y. Gogotsi, "Capacitive Energy Storage in Micro-Scale Devices: Recent Advances in Design and Fabrication of Microsupercapacitors," *Energy Environ. Sci.*, Vol. 7, No. 3, 2014, pp. 867–884.

[35] Liu, Y. -P., and D. Vasic, *Self-Powered Electronics for Piezoelectric Energy Harvesting Devices, in Small-Scale Energy Harvesting*, Intech, 2012.

[36] Lefeuvre, E., et al., "Piezoelectric Energy Harvesting Device Optimization by Synchronous Electric Charge Extraction," *J. Intel. Mat. Syst. Str.*, Vol. 16, No. 10, 2005, pp. 865–876.

[37] Guyomar, D., et al., "Toward Energy Harvesting Using Active Materials and Conversion Improvement by Nonlinear Processing," *IEEE Trans. on Ultrason. Ferr.*, Vol. 52, No. 4, 2005, pp. 584–595.

[38] Badel, A., et al., "Piezoelectric Energy Harvesting Using a Synchronized Switch Technique," *J. Intel. Mat. Syst. Str.*, Vol. 17, No. 8-9, 2006, pp. 831–839.

[39] Badel, A., et al., "Single Crystals and Nonlinear Process for Outstanding Vibration Powered Electrical Generators," *IEEE Trans on. Ultrason. Ferr.*, Vol. 53, No. 4, 2006, pp. 673–684.

[40] Minazara, E., et al., "Piezoelectric Diaphragm for Vibration Energy Harvesting," *Ultrasonics*, Vol. 44, 2006, pp. e699–e703.

[41] Chen, Y. Y., et al., "Self-Powered Piezoelectric Energy Harvesting," *IEEE IECON 2010*, 2010, pp. 7–10.

[42] Yang, Y. W., and L. H. Tang, "Equivalent Circuit Modeling of Piezoelectric Energy Harvesters," *J. Intel. Mat. Syst. Str.*, Vol. 20, No. 18, 2009, pp. 2223–2235.

[43] Wu, W. J., et al., "Modeling and Experimental Verification of Synchronized Discharging Techniques for Boosting Power Harvesting from Piezoelectric Transducers," *Smart Mater. Struc.*, Vol. 18, No. 5, 2009, p. 055012.

[44] Liu, Y. P., D. Vasic, and F. Costa, "Piezoelectric Energy Harvester Circuit for Capacitive Storage Buffer," *Electrimacs 2011*, June 6–8, 2011, Paris, France.

[45] Lallart, M., and D. Guyomar, "An Optimized Self-Powered Switching Circuit for Non-Linear Energy Harvesting with Low Voltage Output," *Smart Mater. Struc.*, Vol. 17, No. 3, 2008, p. 035030.

[46] Guyomar, D., and M. Lallart, "Nonlinear Conversion Enhancement for Efficient Piezoelectric Electrical Generators," *Ferroelectrics*, 2010.

[47] Lallart, M., Y. -C. Wu, and D. Guyomar, "Switching Delay Effects on Nonlinear Piezoelectric Energy Harvesting Techniques," *IEEE Trans. on Ind. Electron.*, Vol. 59, No. 1, 2012.

[48] Ben-Yaakov, S., and N. Krihely, "Resonant Rectifier for Piezoelectric Sources," *Appl. Power Electronics Conf. (APEC 2005)*, Vol. I, 2005, pp. 249–253.

[49] Wang, L., et al., "An Integrated Approach to Energy Harvester Modeling and Performance Optimization," *Behavioral Modeling and Simulation Workshop*, 2007, pp. 121–125.

[50] Shu, Y. C., and I. C. Lien, "Analysis of Power Output for Piezoelectric Energy Harvesting Systems," *Smart Mater. Struc.*, Vol. 15, No. 6, 2006, pp. 1499–1512.

[51] Lesieutre, G. A., G. K. Ottman, and H. F. Hofmann, "Damping as a Result of Piezoelectric Energy Harvesting," *J. Sound and Vib.*, Vol. 269, No. 3-5, 2004, pp. 991–1001.

[52] Ottman, G. K., et al., "Adaptive Piezoelectric Energy Harvesting Circuit for Wireless Remote Power Supply," *IEEE Trans. on Power Electr.*, Vol. 17, No. 5, 2002, pp. 669–676.

[53] Al-Lawati, M., M. Al-Busaidi, and Z. Nadir, "RF Energy Harvesting System Design for Wireless Sensors," *9th Inter. Multi-Conf. Systems, Signal and Devices*, 2012, pp. 1–4.

[54] Badel, A., et al., "Piezoelectric Vibration Control by Synchronized Switching on Adaptive Voltage Sources: Towards Wideband Semi-Active Damping," *J. Acoust. Soc. Am.*, Vol. 119, No. 5, 2006, pp. 2815–2825.

[55] Mitcheson, P. D., T. C. Green, and E. M. Yeatman, "Power Processing Circuits for Electromagnetic, Electrostatic and Piezoelectric Energy Scavengers," *Microsyst. Technol.*, Vol. 13, No. 11-12, 2007, pp. 1629–1635.

7

MEMS Acoustic Energy Harvesting

Acoustic noise is the result of the pressure waves generated by a vibration source. A sinusoidal wave is referred to as a tone, a combination of several tones is called a sound, and an irregular vibration is defined as noise. Acoustic power is the total amount of sound energy caused by a sound source over a given period of time. Acoustic energy is a clean, ubiquitous, and sustainable energy source, which is prevalent in our life ambient, for instance, airports, factory, and construction sites. Acoustic energy, in a manner similar to vibration energy, offers power densities on the order of 1 $\mu W/cm^2$ for a 100 dB acoustic signal [1], or approximately 964,000 $\mu W/cm^2$ at 160 dB, which is normally much higher than most signals. However, there are some applications where such high levels exist. Additionally, acoustic energy does not require the presence of sunlight. In recent years, many researchers have paid more attention to developing mechanisms to convert acoustic energy into electric energy. Acoustic energy harvesting has been demonstrated recently using a mesoscale Helmholtz resonator machined in aluminum, which can deliver 25 mW to a resistive load at a sound pressure level (SPL) of 152 dB [2, 3]. This acoustic energy could locally power a wireless active liner for suppression of engine noise in turbofan engine nacelles, where SPLs typically reach the maximum value of 150 dB [4].

7.1 Working Principle

Acoustic energy harvester converts acoustic energy into a usable form of electrical energy. There are two common transducer mechanisms, namely, piezoelectric and electrostatic methods. Piezoelectric transduction does not require an external power source other than an input signal and generates a high sensitivity with a potentially low noise floor. However, the integration is limited because the piezoelectric materials are difficult to fabricate by CMOS-MEMS process. The lightweight silicon membrane of electrostatic acoustic harvester can produce ultrasonic pressure waves in air more efficiently than a solid piezoelectric crystal. Compared with a piezoelectric harvester, the electrostatic harvester has an advantage in cost and output performance [5]. A Helmholtz resonator is a commonly used type of acoustic transducer. It consists of a cavity connected to the environment through a small neck. When excited by an acoustic input, a single resonance is seen, whereby the acoustic pressure inside the cavity is amplified to a level much greater than the incident acoustic signal. Horowitz et al. at the University of Florida introduced a micromachined piezoelectric acoustic energy harvester (AEH) using a Helmholtz resonator [6]. Figure 7.1 shows a schematic of the overall AEH system. This AEH system consists of a compliant piezoelectric back-plate Helmholtz resonator for energy conversion, power electronics for rectification and regulation, and battery for energy storage. Incident acoustic waves act on the AEH. The harvester scavenges the incident acoustic energy into an AC voltage, which is stored into batter by rectified circuit. This AEH employs a compliant piezoelectric back-plate Helmholtz resonator (HR) to create a coupled resonant system. The acoustical diaphragm compliance, C_{aD}, is found by placing a short circuit ($V = 0$) across the piezoelectric and subsequently computing the potential energy stored in the diaphragm for an applied pressure, P, and is thereby defined by [7]

$$C_{aD} = \left.\frac{\Delta Vol}{P}\right|_{V\to 0} = \frac{\int_0^{r2} w(r)|_{V\to 0}\, 2\pi\, dr}{P} \tag{7.1}$$

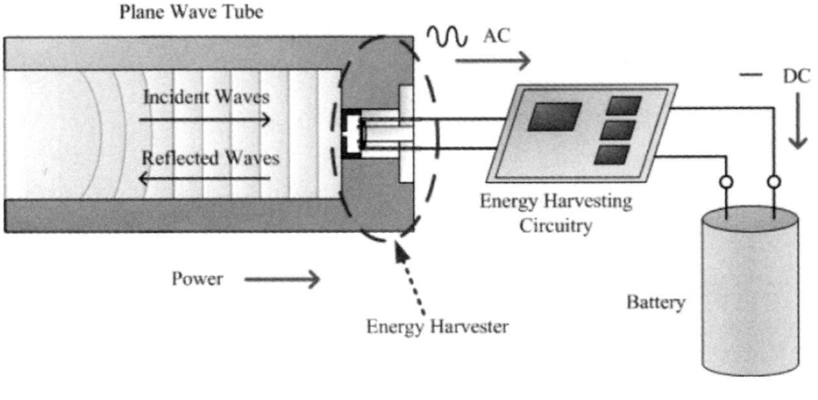

Figure 7.1 Schematic of overall acoustic energy harvester [5].

where ΔVol is the volumetric displacement and w(r) is the deflection as a function of the radius, *r*.

Physically, the coupling factor implies the fraction of energy that is ideally coupled between the acoustical and electrical energy domains. Here it is defined by

$$k_c^2 = \frac{d_A^2}{C_{ef} C_{aD}} \tag{7.2}$$

where C_{ef} is the free electrical capacitance of a parallel plate capacitor. d_A is the effective acoustic piezoelectric coefficient, which represents coupled energy, and the two elements that store potential energy associated with the transduction are C_{ef} and C_{aD}. The coupling factor is therefore related to the ideal ratio of the coupled energy to the stored energy, but does not account for losses [8].

7.2 Acoustic Microharvester

7.2.1 Electrostatic Conversion Method

The acoustic microharvester consists of some varied capacitors, which are charged by a DC voltage during operation. When this device is exposed to acoustic waves, the capacitors will engage and disengage at the same frequency of the incoming pressure field, which generates an output voltage proportional to the amplitude of the field. The main advantage of the electrostatic harvesters is the inherent impedance match between the harvester and the surrounding medium. This low-mechanical impedance of the membrane is usually negligible, which results in very efficient coupling of the sound waves into and from the sound-bearing medium [9]. Zhu et al. at the University of Newcastle proposed a novel 2-degree-of-freedom (2-DOF) microelectromechanical system (MEMS) device, which scavenges energy in arbitrary directions in a plane due to incoming ultrasonic waves, as shown in Figure 7.2(a) [10, 11]. The seismic mass moves along the *x*- and *y*-axes to absorb vibration energy in any in-plane direction. The variable capacitors of C_X and C_Y collect X and Y vibration contributions separately and flow the currents intro the load. The 100-V DC-voltage source is applied on the capacitor for polarization. However, the permanently charged electrets can replace an external bias voltage in a practical application for miniaturization [12, 13]. When the ultrasonic wave is applied on the converter, the DC voltage on the storage capacitor is charged up from 0.51V to 0.95V in 15 seconds with a 0.5-cm air gap. Thus, the energy stored on the storage capacitor changes from 0.13 to 0.451 μJ in 15 seconds, indicating that the power absorption capability of this energy converter is 21.4 nW. However, the nonuniform shaping of parallel comb fingers can significantly

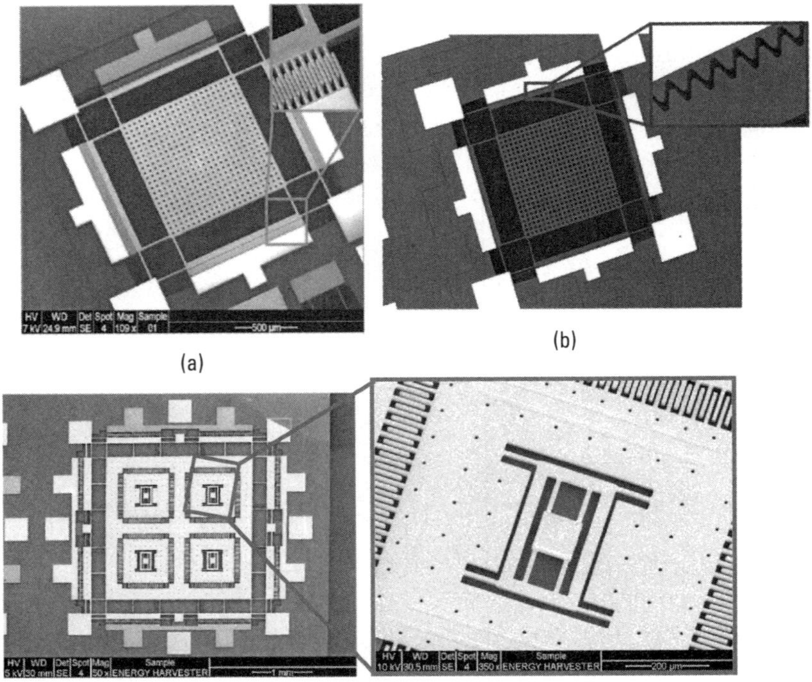

(a) (b) (c)

Figure 7.2 MEMS acoustic energy harvesters: (a) 2-DOF harvester with parallel combs [10, 11]; (b) 2-DOF harvester with triangular combs [16]; and (c) 3-DOF harvester [17].

affect the properties of electrostatic electrodes in both nanopositioning and energy-harvesting applications. In particular, an appropriately designed triangular electrode is able to provide both a larger total capacitance change compared with conventional gap-closing electrodes and a greater change in capacitance per unit displacement compared with overlap-varying electrodes [14, 15]. A triangular electrode structure may therefore result in a more effective transduction mechanism for electrostatic energy harvesting devices. Fowler et al. designed and fabricated 2-DOF ultrasonic harvester with triangular electrostatic electrodes, as shown in Figure 7.2(b) [16]. Compared with the harvester with parallel comb electrodes, this fabricated harvester has scavenged 27.6 nW of electrical power when the source of ultrasonic waves is applied at its mechanical resonance frequency of 38.95 Hz, which has an increase of 29% at the output power. In order to improve the output power, the same research group has proposed 3-DOF ultrasonic harvester, which includes in-plane x- and y-axes, as well as the out-of-plane z-axis, as shown in Figure 7.2(c) [17]. The experimental results show that the resonance frequency at three resonant modes is at approximately 36 kHz and this harvester can provide the electrical energy of 12.6 nW for charging of a load device through a rectifying circuit.

In order to increase the variable capacitor value and improve the output performance, Saboonchi et al. employed an electroplating method to form a higher aspect ratio geometry for an acoustic harvester [18]. The advantages of creating a thick microstructure layer include increasing the mass size without increasing the planar size to design a low-frequency transducer, which reduces the overall size of the transducer for the same spring coefficients, and designing a geometry with a higher spring coefficient to achieve a low resonance frequency, which affects the level of the dc bias voltage applied (note that the sensitivity of capacitive sensors is proportional to the level of DC). Figure 7.3(a) shows the process flow of MEMS harvesters. A 1.1-μm silicon oxide layer is deposited as a sacrificial layer. The 20-μm nickel and 0.5-μm gold layers are electroplating as the spring and mass system. This gold layer enhances the wire bonding procedure at the connection pads and reduces the corrosion potential of the nickel layer. The operational frequency can be varied over a wide range by changing the size and shape of the membrane. The capacitive harvester array cell physical properties are typically optimized based on the transducer application and the desired resonant frequency [19, 20]. COMSOL software is used to optimize the cell shapes and dimensions for an operating frequency of f= 10 MHz [21]. It is shown that the harvester directivity pattern is highly directional. The harvester beam pattern is shown to be steerable by applying a delay to the cell bias voltages, resulting in a smooth beam steering capability of the proposed harvester array. In order to tune the resonant frequency of harvester, two types of springs are designed and fabricated by changing the length of spring, as shown in Figure 7.3(b). The first modal frequencies of the harvester S1 and S2 in the z-direction are 51 kHz and 464 kHz, respectively. The MEMS harvesters exhibit resonance behavior at atmospheric pressure and are more sensitive at their design frequencies than similar frequency piezoelectric transducers. Considering the relative sizes of the two transducer types, this is a significant achievement.

Compared with this bulk micromachining process, the surface micromachining method can avoid the deep reactive ion etching with a high cost and reduce the size of the harvester. In 2006, Ozevin et al. developed capacitive-type resonant MEMS acoustic emission (AE) transducers that span the frequency range of 100 to 500 kHz using the PolyMUMPS (polysilicon multiuser MEMS process) method [22]. The AE method relies on the release of elastic waves caused by localized dynamic stress–strain fields in structures, which can originate from many source mechanisms, including crack growth, fiber breakage, and fretting [18]. The MEMS transducers were approximately 50 times less sensitive than conventional piezoelectric AE transducers under atmospheric pressure. Wright et al. optimized the properties of capacitive MEMS AE transducers to reach a sensitivity level comparable to conventional AE transducers. The MEMS device in this study had a reduced damping geometry to increase the resonant response at atmospheric pressure [23]. The research group at University of Colorado has

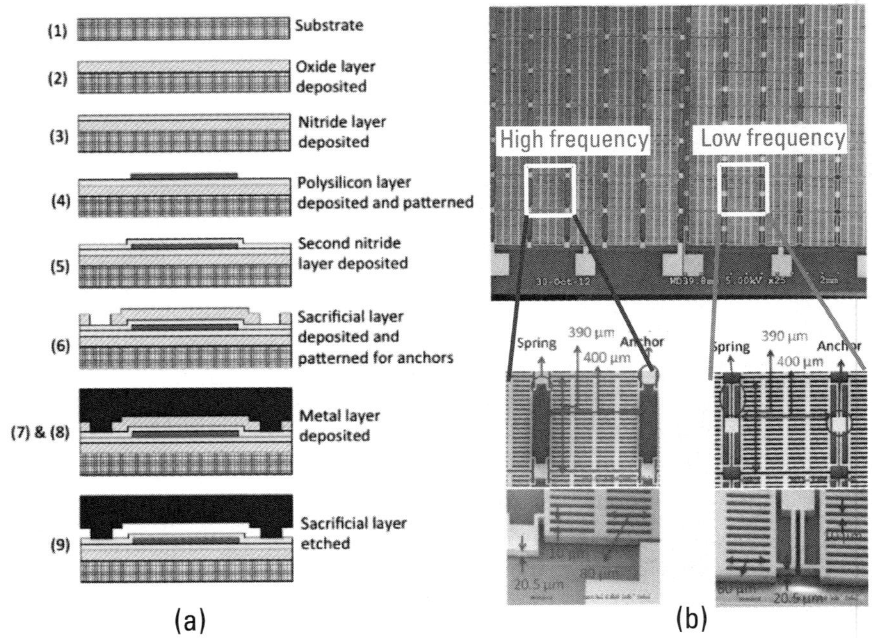

Figure 7.3 (a) The micromachining process flow; and (b) SEM images of high and low frequency design for an acoustic energy harvester [18].

analyzed the capability and limitations of MUMPs process for capacitive ultrasonic harvester [24]. The commercial MUMPs process that allows standardized device manufacture at a low cost is a possible solution. A full wafer costs a total of $4,000, but the process allows for building up to 25 shared designs on a single wafer so that 152 mm × 2 mm devices can be obtained for $160 to $180 (<12 per chip). However, there are several limitations on the MUMPs process that must be considered. One is the limitation of the microstructure thickness and the gap between the stationary layer and the freely moving microstructure layer. The process requirements limit any further design optimization. The second one is that the conductivity of polysilicon results in parasitic capacitance for similar inactive area. Another is that MUMPS also has no additional dielectric layer between the stationary and moving electrodes, other than air. The possibility of stiction reduces the reliability for long-term usage of the transducers for structural health monitoring (SHM) applications.

7.2.2 Piezoelectric Conversion Method

Piezoelectric acoustic harvester converts the acoustic energy into electric energy through the piezoelectric effect of piezoelectric materials, including PZT, ZnO, AlN, and PVDF. These functional films are formed by different micromachining processes, for example, sol-gel and hydrothermal. There are two types of structures for piezoelectric harvester, namely, a Helmholtz resonator and a sonic crystal. A Helmholtz resonator can effectively amplify sound pressure in the cavity and has been widely deployed for the acoustic energy harvester. Horowitz et al. introduced a micromachined acoustic harvester using a Helmholtz resonator with a lead zirconate titanate (PZT) piezoelectric composite diaphragm, as shown in Figure 7.4(a) [5]. Figure 7.4(b) shows the process flow of this device. The 270-nm-thick PZT was sol–gel deposited above a patterned 220 nm Ti/Pt layer. 100 nm thick TiO_2 layer acted as a diffusion barrier for the PZT was formed by evaporating a titanium layer and oxidizing at 650°C. This fabricated chip was flush mounted in a Lucite package and bonded with epoxy for testing. Preliminary energy-harvesting experiments indicate a power density of 0.34 $\mu W\ cm^{-2}$ for an acoustic input of 149 dB. An electromechanical Helmholtz resonator (EMHR) has been utilized to generate the resonant cavity pressure to deform a PZT piezoelectric back plate [25–27]. With an external sound with a 160-dB SPL at 2.6 kHz, the EMHR supplied a power of 30 mW. Most efforts are focused on improving the efficiency of the acoustic energy harvester (AEH) using a single HR at one vibration mode of the piezoelectric compliant plate or diaphragm that acts as the back plate of the cavity. To improve the conversion efficiency, Lallart et al. [28] used synchronized switch harvesting on an inductor (SSHI) with the PZT-membrane acoustic energy harvester and increased the power gain up to 10 and the bandwidth by 100%. An output power of ~60 μW with a power density of 0.7 $\mu W\ cm^{-2}$ have been obtained with an incident SPL of 100 dB at 240 Hz. Kimura et al. reported that PZT acoustic harvesters with no wet electrode fabrication processes exhibited further improved power density at the first resonance [29]. This energy harvester with the diaphragm diameter of 1.2 mm generated the highest energy density of 98 $\mu W/m^2$ at the

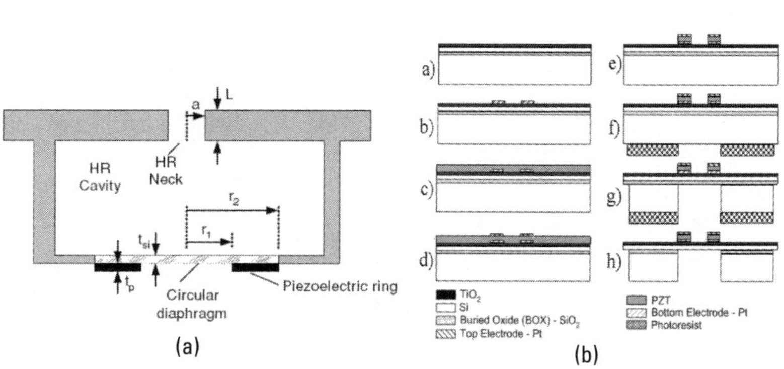

Figure 7.4 (a) The electromechanical Helmholtz resonator (HR) consists of a compliant piezoelectric-composite diaphragm that serves as a wall for the HR; (b) the micromachining process flow [5].

sound pressure level of 100 dB (0.01 W/m^2) at the first-resonance frequency of 16.7 kHz. However, these previous reports on the PZT acoustic energy harvesters did not consider the effect of the different polarity charges in the central part and the peripheral part of the diaphragms. Iizumi et al. reported on further improved performances of PZT acoustic energy harvesters that utilized different charge polarizations at the peripheral area and the central area of a PZT diaphragm with a diameter of 2 mm, using a combined device consisting of a peripheral harvester and central harvester, as shown in Figure 7.5 [30]. The deflected PZT capacitor generates a voltage difference between the two electrodes of Al and Pt/Ti owing to the polarization of PZT. This harvester, equipped with a Helmholtz sound collecting cone with a trumpet shape, generates a power of 1.7×10^{-13} W under a sound pressure level of 110 dB at the first resonance frequency of 6.28 kHz. The cone is confirmed to effectively increase the bandwidth of the energy harvester [31].

Apart from the sol-gel process for PZT film as functional diaphragm of acoustic energy harvester, some other processes and piezoelectric materials have been reported for the harvester. The hydrothermal method is used to deposit polycrystalline PZT thick film on Ti substrates [32] due to the limitation of PZT film realized by the sol-gel process. Starting materials for the reaction are lead nitrate ($Pb(NO_3)_2$ 99.0%), zirconium chloride oxide octahydrate ($ZrOCl_2 \cdot 8\,H_2O$ 99.0%), a powder of titanium oxide (TiO_2 99.0%) in potassium hydroxide (KOH) solution, and titanium metal for the substrate. The composition of the starting materials is determined as follows: 140 mL of 1 N aqueous $Pb(NO_3)_2$, 90 mL of 1 N aqueous $ZrOCl_2 \cdot 8\,H_2O$, 1 g TiO_2, and 300 mL of 4 N KOH solution. The temperature and a pressure of autoclave internal are 160°C and 0.55 MPa, respectively. The reaction time is 12 hours for each process. The fabricated hydrothermal PZT thick film is about 20 μm thick after 10 times of the deposition process. Although PZT materials have excellent

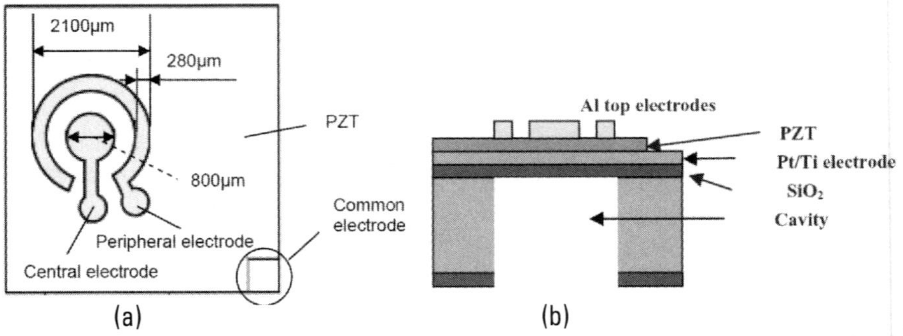

Figure 7.5 (a) PZT energy harvester with central and peripheral electrodes; (b) a schematic cross section [30, 31].

piezoelectric performance, many countries and organizations are restricting the usage of lead-based materials due to environmental pollution [33, 34]. The lead-free piezoelectric ceramic with formula of $(K_{0.5}Na_{0.5})_{0.97}Li_{0.03}(Nb_{0.9}Ta_{0.1})O_3$ (abbreviated as KNLNT-0.03/0.10) was reported and characterized [35]. This material has a clamped dielectric constant $\varepsilon_{33}{}^{S}/\varepsilon_0 = 890$, a piezoelectric coefficient $d_{33} = 245$ pC/N, an electromechanical coupling factor $k_t = 0.42$, and a Curie temperature of Tc > 300°C. The fabricated lead-free harvester had a bandwidth of 45% and a two-way insertion loss of −18 dB.

At high frequencies, acoustic attenuation is very high, so the focal point of the transducer must be located very close to the harvester surface. Thus, the rationale for employing strongly focused, high-frequency harvesters include two parts: obtaining a higher intensity at the focus and achieving a superior lateral resolution. Because the addition of a lens will generate additional attenuation, self-focused transducers are preferred in high-frequency applications. Feng et al. at the University of Southern California has reported a dome-shaped-diaphragm transducer (DSDT)-based piezoelectric ZnO film [36]. Figure 7.6(a) shows the fabrication process flow. The wax mold is formed by taking out the

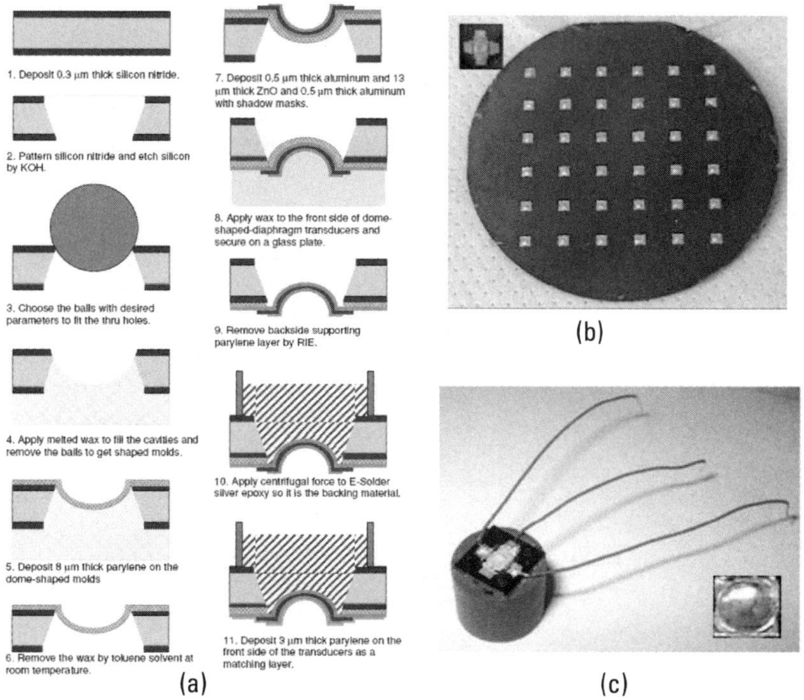

Figure 7.6 (a) The fabrication process of a dome-shaped transducer based on ZnO film; (b) the backside view of the diaphragm; (c) the fabricated transducer with wire bonding [36].

balls after the wax is cooled down and solidified. The c-axis oriented ZnO deposition is sputtered at 250°C, and the temperature will alter the shape of the parylene diaphragm due to the thermally induced stress in the parylene, if there is no other structurally rigid material in the diaphragm. Here the aluminum electrode as a structural layer provides enough stiffness to protect the designated spherical shape. This process can control the spherical surface construction. So the curvature and diameter of dome-shaped-diaphragm can be adjusted by modifying design parameters. Figure 7.6(b, c) shows the backside view and the single chip with the wire bonding of the fabricated device, respectively. Wang et al. developed a nanogenerator-based ZnO nanowire driven by an ultrasonic wave [37].

In addition to a Helmholtz resonator, a sonic crystal has been used to convert acoustic energy to electricity [38–40]. The sonic crystals have attracted a lot of attention in some possible applications, such as acoustic filters, vibrationless environments for high-precision mechanical systems, and transducers. A curved polyvinylidene fluoride (PVDF) piezoelectric beam is placed inside a defect region of a sonic crystal, which acts as a resonant cavity. The output power of ~35 nW is obtained with an incident SPL of 80–100 dB at 4.2 kHz. These reported acoustic energy harvester works at high frequencies (a few kilohertz). The straight tube acoustic resonator with PZT piezoelectric cantilever plates has been used to convert a low audible frequency sound to electricity [41]. With an incident sound pressure of 110 dB, an output voltage of 15.689V and a power of 12.697 mW are obtained at the first eigenmode frequency of 199 Hz. The corresponding areal and volume power densities are 0.635 mW cm^{-2} and 15.115 μW cm^{-3}, respectively. The same group reported a low-frequency acoustic harvester based on multiple PVDF beam arrays in quarter-wavelength acoustic resonator and its first operating frequency is 146 Hz [42]. To increase the operating bandwidth and improve the output power density, a wideband acoustic energy harvester with multimodal frequencies is proposed owing to the acoustic coupling of the vibrating plate and two cavities, as shown in Figure 7.7 [43]. The perforated plate serves as both the back plate of the front cavity and the faceplate of the back cavity. The length of the back cavity can be continuously changed by the mobile piston. A rectangular-shaped thin PZT-5H is placed in the radial direction of the annular brass plate to investigate the performance of the harvester. The experimental results show that the proximity of the multiple peaks exists in the frequency range of 1,100 to 1,400 Hz.

7.3 Application of Acoustic Energy Harvester

Acoustic energy commonly exists in a variety of everyday situations. Although the available acoustic energy is small, the energy requirement for certain

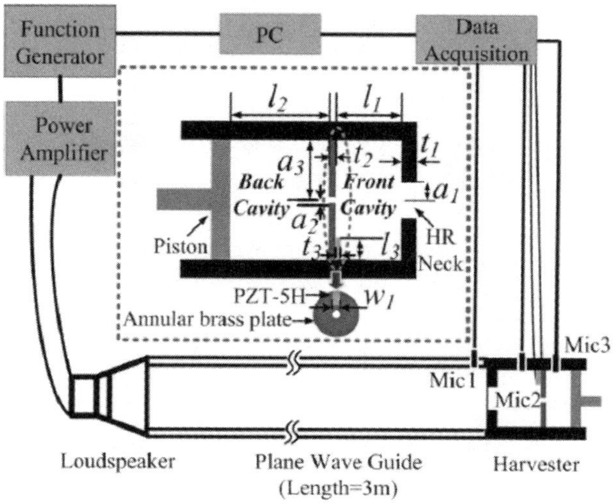

Figure 7.7 Illustration of the AEH with dual Helmholtz resonator and the experimental setup [43].

applications can be met. Currently, there are two main applications for acoustic energy: biomedical application and wireless sensor nodes. With the increasing interest in biomedical monitoring, the power supply of the implanted biomedical devices is becoming an important urgent issue. Implanted devices, such as embedded diagnostic sensors and pacemakers, continue to rely on batteries as their primary power source. Therefore, there is a continuing interest in developing methods of reliably and safely transmitting power to these devices [44]. Ultrasonic waves have been widely used in human organs for perspective testing and observation of fetal development in pregnant women [45]. As shown in [46], ultrasonic waves can effectively be used to mechanically excite an implanted MEMS device, meaning that a microscale energy harvester could be used to power an implanted system through harvesting of the resulting mechanical vibrations (Figure 7.8). The 200-MHz self-focused ultrasonic harvesters have been developed for specialized application in medical imaging and nondestructive evaluation [47], including thickness measurements of the layers of skin tumors, tissue characterization of diseased skin, eye tumor examination, and glaucoma research. The bandwidth is 28% centered at 200 MHz. This device can be used for a variety of tissue and cellular imaging needs and acoustic property measurement for a variety of substances, including biomolecular samples. Park et al. reported a multichannel chemical sensor based on capacitive micromachined ultrasonic transducer (CMUT), which is functionalized using various polymer layers, exhibited a mass sensitivity of 1 fg/Hz per a unit resonator and a volume sensitivity of 41.6 ppb/Hz to water vapor [48]. The same

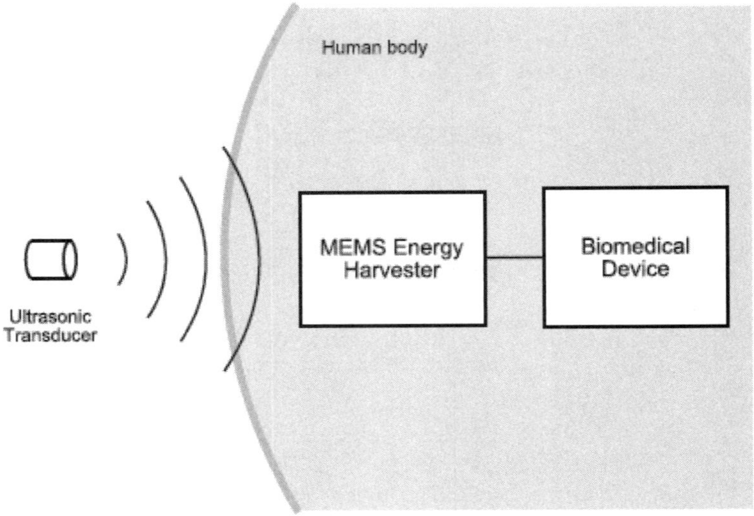

Figure 7.8 Overview of MEMS ultrasonic energy harvesting [46].

group also demonstrated a single-channel chemical sensor system for detecting sensitively dimethyl methylphosphonate (DMMP) [49]. Figure 7.9 shows the CMUT resonant structure with a vacuum cavity. When a dc bias voltage is applied between the plate (i.e., top electrode) and the substrate (i.e., bottom electrode), a high electric field built in the cavity provides sufficient electromechanical coupling between two electrodes. The circular plate is excited to vibrate by the AC voltage superimposed on the applied DC bias voltage. This

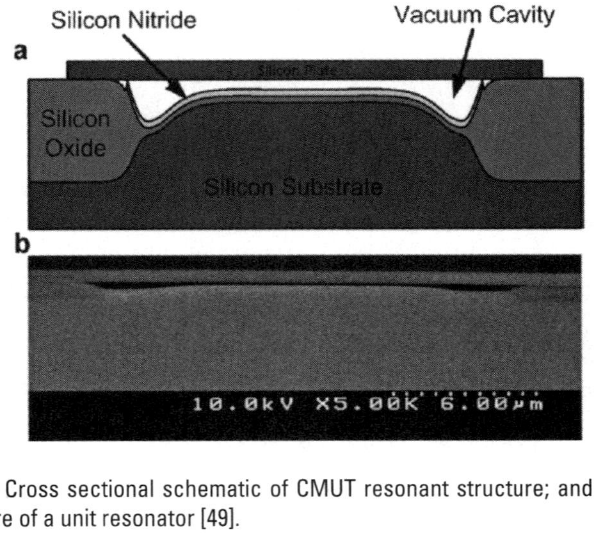

Figure 7.9 (a) Cross sectional schematic of CMUT resonant structure; and (b) SEM cross sectional picture of a unit resonator [49].

device can achieve a limitation of detection level of 56 ppb to DMMP. Chen et al. developed a capacitive micromachined ultrasonic transducer arrays for minimally invasive medical ultrasound [50]. This device can be micromachined into miniature platforms with various geometrical shapes, including needles, three-dimensional prisms, and other flexible-substrate configurations. High-frequency ultrasound may find other significant biomedical applications such as acoustic microscopy for noninvasive investigation of living cells (not only to visualize the cells but also examine their mechanical properties) [51, 52], a silicon-based acoustic lens up to the gigahertz range [53], and acoustic tweezers for noncontact manipulation of microparticles or cells based on the radiation force principle [54, 55].

Powering wireless sensor networks will be the other application of acoustic energy harvester. The PZT acoustic energy harvester integrated with CMOS IC chips could be a candidate power source for integrated smart systems [56]. Figure 7.10 shows a dream concept of an intelligent system. The piezoelectric acoustic energy harvester is expected to power MEMS sensors and actuators through acoustic waves from environmental noises. The harvesters with different diaphragm diameters of 1.5 and 2.0 mm generate powers of 1.1×10^{-11} and 5.1×10^{-12} W at their third resonance frequencies under the irradiation of 100-dB sound pressure, respectively. It is expected that the PZT energy harvester may be a candidate power source for intelligent integrated systems. In addition, PZT ultrasonic MEMS harvesters have been used for a structural health monitoring system [57], underwater communications [58], and powering a miniature wireless acoustic sensor nodes for monitor noise generated by aircrafts [59].

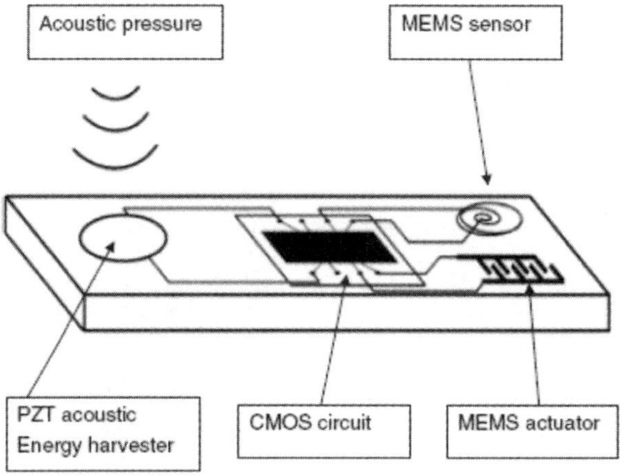

Figure 7.10 Concept of an integrated system with a piezoelectric acoustic energy harvester [56].

References

[1] Roundy, S., P. K. Wright, and J. Rabaey, "A Study of Low Level Vibrations as a Power Source for Wireless Sensor Nodes." *Computer Communications*, Vol. 26, 2003, pp. 1131–1144.

[2] Kadirvel, S., et al., "A Self-Powered Wireless Active Acoustic Liner," *12th AIAA/CEAS Aeroacoustics Conf.*, Cambridge, MA, May 2006.

[3] Horowitz, S., et al., "Technology Development for Self-Powered Sensors," *2002 AIAA 1st Flow Control Conf.*, AIAA paper 2002–2702, 2002, pp. 1–10.

[4] Taylor, R., et al., "Technology Development for Electromechanical Acoustic Liners," *ACTIVE '04*, Paper A04-093, 2004.

[5] Ladabaum, I., X. C. Jin, and B. T. Khuri-Yakub, "Miniature Drumheads: Microfabricated Ultrasonic Transducers," *Ultrasonics*, Vol. 36, 1998, pp. 25–29.

[6] Horowitz, S. B., et al., "A MEMS Acoustic Energy Harvester," *Journal of Micromechanics and Microengineering*, Vol. 16, 2006, pp. S174–S181.

[7] Prasad, S., et al., "Analytical Electroacoustic Model of a Piezoelectric Composite Circular Plate," *AIAA Journal*, Vol. 44, No. 10, 2006, pp. 2311–2318.

[8] Rossi, M., *Acoustics and Electroacoustics*, Norwood, MA: Artech House, 1988, pp. 245–308.

[9] Ergun, A. S., G. G. Yaralioglu, and B. T. Khuri-Yakub, "Capacitive Micromachined Ultrasonic Transducer: Theory and Technology," *Journal of Aerospace Engineering*, Vol. 16, No. 2, 2003, pp. 76–84.

[10] Zhu, Y., S. O. R. Mohermani, and M. R. Yue, "Ultrasonic Energy Transmission and Conversion Using a 2-D MEMS Resonator," *IEEE Electron Device Letters*, Vol. 31, No. 4, 2010, pp. 374–376.

[12] Zhu, Y., et al., "A 2-DOF MEMS Ultrasonic Energy Harvester," *IEEE Sensors Journal*, Vol. 11, No. 1, 2011, pp. 155–161.

[12] Peano, F., and T. Tambosso, "Design and Optimization of a MEMS Electrets Based Capacitive Energy Scavenger," *J. Microelectromech. Syst.*, Vol. 14, No. 3, 2005, pp. 429–435.

[13] Sterken, T., et al., "Harvesting Energy from Vibrations by a Micromachined Electret Generator," *Proc. Transducers*, 2007, pp. 129–132.

[14] Hoffmann, D., B. Folkmer, and Y. Manoli, "Analysis and Characterization of Triangular Electrode Structures for Electrostatic Energy Harvesting," *J. Micromech. Microeng.*, Vol. 21, No. 10, October 2011, pp. 104002-1–104002-10.

[15] Engelen, J. B. C., et al., "Improved Performance of Large Stroke Comb-Drive Actuators by Using a Stepped Finger Shape," *Proc. Int. Solid-State Sensors, Actuat. Microsyst. Conf.*, 2009, pp. 1762–1765.

[16] Fowler, A. G., S. O. R. Moheimani, and S. Behrens, "Design and Characterization of a-DOF MEMS Ultrasonic Energy Harvester with Triangular Electrostatic Electrodes," *IEEE Electron Device Letters*, Vol. 34, No. 11, 2013, pp. 1421–1423.

[17] Fowler, A. G., S. O. R. Moheimani, and S. Behrens S., "A 3-DOF MEMS Ultrasonic Energy Harvester," *IEEE Sensors*, October 29–31, 2012, Taipei, pp. 1–4.

[18] Saboonchi, H., and D. Ozevin, "MEMS Acoustic Emission Transducers Designed with High Aspect Ratio Geometry," *Smart Mater. Struct.*, Vol. 22, 2013, p. 095006.

[19] Hsu, T. R., *MEMS and Microsystems: Design, Manufacture, and Nanoscale Engineering*, 2nd ed., New York: John Wiley & Sons, 2008.

[20] Emadi, T. A., et al., "A MEMS Ultrasonic Transducer Imager Array, with Beam Width, Acoustic Power, and Frequency Modulations," *IEEE International Conference on Imaging Systems and Techniques*, Manchester, U.K., 2012.

[21] Emadi, T. A., and D. A. Buchanan, "Wide Range Beam Steering Capability of a 1-D MEMS Transducer Imager Array with Directional Beam Patter," *Sensors and Actuators A*, Vol. 202, 2013, pp. 193–196.

[22] Ozevin, D., et al., "Resonant Capacitive MEMS Acoustic Emission Transducers," *Smart Mater. Struct.,* Vol. 15, 2006, pp. 1863–1871.

[23] Wright, A., "A Multi-Axis Capacitance MEMS Sensor System for Acoustic Emission Sensing," Ph.D. dissertation, Carnegie Mellon University, 2009.

[24] Liu, J., C. Oakley, and R. Shanda, "Capacitive Micromachined Ultrasonic Transducers Using Commercial Multi-User MUMPs Process: Capability and Limitations," *Ultrasonics*, Vol. 49, 2009, pp. 765–773.

[25] Liu, F., et al., "Acoustic Energy Harvesting Using an Electromechanical Helmholtz Resonator," *J. Acoust. Soc. Am.*, Vol. 123, 2008, pp. 1983–1990.

[26] Liu, F., et al., "A Multiple Degree of Freedom Electromechanical Helmholtz Resonator," *J. Acoust. Soc. Am.*, Vol. 122, 2007, pp. 291–301.

[27] Phipps, A., et al., "Demonstration of a Wireless, Self-Powered, Electroacoustic Liner System," *J. Acoust. Soc. Am.*, Vol. 125, 2009, pp. 873–881.

[28] Lallart, M., et al., "Nonlinear Optimization of Acoustic Energy Harvesting Using Piezoelectric Devices," *J. Acoust. Soc. Am.*, Vol. 128, 2010, pp. 2739–2748.

[29] Kimura, S., et al., "Improved Performances of Acoustic Energy Harvester Fabricated Using Sol/Gel Lead Zirconate Titanate Thin Film," *Jpn. J. Appl. Phys.*, Vol. 50, 2011, pp. 06GM14 1–5.

[30] Iizumi, S., et al., "Lead Zirconate Titanate Acoustic Energy Harvesters Utilizing Different Polarizations on Diaphragm," *Procedia Engineering*, Vol. 25, 2011, pp. 187–190.

[31] Matsuda, T., et al., "Helmholtz Resonator for Lead Zirconate Titanate Acoustic Energy Harvester," *Journal of Physics: Conference Series*, Vol. 476, 2013, p. 012003.

[32] Ishikawa, M., et al., "Hydrothermal Polycrystalline PZT Thick Film Transducer for High Intensity Ultrasound Radiation at over 20MHz," *IEEE Ultrasonics Symposium*, 2006, pp. 2369–2372.

[33] Ringgaard, E., and T. Wurlitzer, "Lead-Free Piezoceramics Based on Alkali Niobates," *Journal of the European Ceramic Society*, Vol. 25, 2005, pp. 2701–2706.

[34] Saito, Y., et al., "Lead-Free Piezoceramics," *Nature*, Vol. 432, 2004, pp. 84–87.

[35] Wu, D. W., et al., "Lead-Free KNLNT Piezoelectric Ceramics for High-Frequency Ultrasonic Transducer Application," *Ultrasonics*, Vol. 49, 2009, pp. 395–398.

[36] Feng, G. H., et al., "Fabrication of MEMS ZnO Dome-Shaped-Diaphragm Transducers for High-Frequency Ultrasonic Imaging," *J. Micromech. Microeng.*, Vol. 15, 2005, pp. 586–590.

[37] Wang, X. D., et al., "Direct-Current Nanogenerator Driven by Ultrasonic Waves," *Science*, Vol. 316, 2007, pp. 102–105.

[38] Wang, W. C., et al., "Acoustic Energy Harvesting by Piezoelectric Curved Beams in the Cavity of a Sonic Crystal," *Smart Mater. Struct.*, Vol. 19, 2010, p. 045016.

[39] Wu, L. Y., L. W. Chen, and C. M. Liu, "Acoustic Energy Harvesting Using Resonant Cavity of a Sonic Crystal," *Appl. Phys. Lett.*, Vol. 95, 2009, p. 013506.

[40] Wu, L. Y., L. W. Chen, and C. M. Liu, "Acoustic Pressure in Cavity of Variously Sized Two-Dimensional Sonic Crystals with Various Filling Fractions," *Phys. Lett. A*, Vol. 373, 2009, pp. 1189–1195.

[41] Li, B., J. H. You, and Y. J. Kim, "Low Frequency Acoustic Energy Harvesting Using PZT Piezoelectric Plates in a Straight Tube Resonator," *Smart Mater. Struct.*, Vol. 22, 2013, p. 055013.

[42] Li, B., et al., "Harvesting Low-Frequency Acoustic Energy Using Multiple PVDF Beam Arrays in Quarter-Wavelength Acoustic Resonator," *Applied Acoustic*, Vol. 74, 2013, pp. 1271–1278.

[43] Peng, X., et al., "A Wideband Acoustic Energy Harvester Using a Three Degree-of-Freedom Architecture," *Applied Physics Letters*, Vol. 103, 2013, p. 164106.

[44] Arra, S., et al., "Ultrasonic Power and Data Link for Wireless Implantable Applications," *Proc. 2nd Int. Symp. Wireless Pervas. Comput.*, February 2007.

[45] Shimpf, S., and B. Schmidt, "MEMS Compatible PZT Arrays for Ultrasonic Applications on 3 Dimensional Carriers," *2008 IEEE International Frequency Control Symposium*, 2008, pp. 127–131.

[46] Denisov, A., and E. Yeatman, "Stepwise Microactuators Powered by Ultrasonic Transfer," *Procedia Eng.*, Vol. 25, 2011, pp. 685–688.

[47] Sharp, C. C., et al., "200 MHz Self-Focused ZnO MEMS Ultrasonic Transducers for Biomedical Imaging," *2004 IEEE International Ultrasonics, Ferroelectrics, and Frequency Control Joint 50th Anniversary Conference*, 2004, pp. 1946–1949.

[48] Park, K. K., et al., "Capacitive Micromachined Ultrasonic Transducers for Chemical Detection in Nitrogen," *Applied Physics Letters*, Vol. 91, 2007, p. 094102.

[49] Park, K. K., et al., "Capacitive Micromachined Ultrasonic Transducer (CMUT) as a Chemical Sensor for DMMP Detection," *Sensors and Actuators B: Chemical*, Vol. 160, 2011, pp. 1120–1127.

[50] Chen, J. K., "Capacitive Micromachined Ultrasonic Transducer Arrays for Minimally Invasive Medical Ultrasound," *J. Micromech. Microeng.*, Vol. 20, 2010, p. 023001.

[51] Weiss, E. C., et al., "Mechanical Properties of Single Cells by High-Frequency Time-Resolved Acoustic Microscopy," *IEEE Ultrason. Ferroelect. Freq. Contr.*, Vol. 54, No. 11, 2007, pp. 2257–2271.

[52] Jakob, A., et al., "P2E-5 Silicon-Based GHz Acoustic Lenses for Time-Resolved Acoustic Microscopy," *Proc. IEEE Ultrason. Symp.*, 2007, pp. 1605–1608.

[53] Jakob, A., et al., "Comparison of Different Piezoelectric Materials for GHz Acoustic Microscopy Transducers," *Proc. IEEE Ultrason. Symp.*, 2009, pp. 1722–1725.

[54] Lee, J. W., et al., "Single Beam Acoustic Trapping," *Appl. Phys. Lett.*, Vol. 95, 2009, p. 073701.

[55] Lee, J. W., et al., "Transverse Acoustic Trapping Using a Gaussian Focused Ultrasound," *Ultrasound Med. Biol.*, Vol. 36, 2010, pp. 350–355.

[56] Shinoda, S., et al., "Lead Zirconate Titanate Acoustic Energy Harvester Proposed for Microelectromechanical System/IC Integrated Systems," *Jpn. J. Appl. Phys.*, Vol. 49, 2010, pp. 04DL21 1–5.

[57] Guldiken, R. O., et al., "A Structural Health Monitoring System with Ultrasonic MEMS Transducers," *2011 SPIE Sensors and Smart Structures Technologies for Civil, Mechanical and Aerospace Systems*, Vol. 7981, 2011, pp. 79810F1–8.

[58] Martins, M., et al., "Optimization of Piezoelectric Ultrasound Emitter Transducers for Underwater Communications," *Sensors and Actuators A: Physical*, Vol. 184, 2012, pp. 141–148.

[59] Ralib, A. A. M., et al., "Design, Simulation and Fabrication of Piezoelectric Micro Generators for Aero Acoustic Applications," *Microsyst. Technol.*, Vol. 17, 2011, pp. 563–573.

8

MEMS Wind-Flow Energy Harvesting

Our ambient environment exhibits significant kinetic energy in the form of fluid flow such as wind, ocean currents, ventilation ducts, or water pipes. In the case that solar power, heat, or vibration energy is not available or enough, flow-driven energy harvesting could provide sufficient power source alternatively. It is reported that the mechanical power included by a cross-section of a flow is proportional to the third power of the flow speed. For instance, air flow at a speed of 1 m/s carries a mechanical power of about 600 mW/m^2. A large number of research works on wind turbines have been reported for high-power applications [1]. Most wind turbine generators are used in large-scale applications that generate kilowatts or megawatts of electric power based on electromagnetic transduction mechanism. However, reducing the size of a turbine disproportionally degrades its aerodynamic performance, increases its stall speed, and reduces its power output. Hence, it would not be suitable for small and low-power sensor nodes application. This chapter presents a few small-scale flow-driven energy harvesters, such as a small-scale windmill, the aeroelastic fluttering of a wind belt, and the vortex shedding of a flapping sheet.

8.1 Small-Scale Windmills for Energy Harvesting

A small-scale windmill transforms wind flow into rotational kinetic energy and further into electrical energy via electromagnetic [2], piezoelectric [3–5] and triboelectric [6] transducers. As shown in Figure 8.1(a, b), Priya [3] fabricated a piezoelectric windmill consisting of 12 bimorph transducers placing around the circumference of a center shaft. Each piezoelectric bimorph is assembled to just contact the tip of the triangular stopper. The wind flow causes the fan to rotate, and the torque generated from the rotation is transferred to the windmill shaft. The rotation of the windmill using camshaft mechanism results in the oscillatory motion of the bimorph due to the friction stoppers and eventually generation of electricity based on direct piezoelectric effect. The experimental results showed that the power output of about 5 mW was successfully achieved under wind speed of 10 to 12 miles/h. By changing the number of the bimorphs incorporated in the windmill, the power can be easily adjusted. This windmill provides a feasible solution to the remote powering of wireless sensors and communication devices. However, the size of this prototype is quite large and complicated. Researchers from the same group have addressed the drawbacks and developed an optimized small-scale windmill. The whole structure utilizing 18 piezoelectric bimorphs is made of plastic, which makes the design extremely low cost. At an average wind speed of 10 miles/h, the small-scale windmill is capable of providing 5-mW power output [4].

Recently, triboelectric energy harvesters (TriboEH) with the coupling of triboelectrification and electrostatic induction have been fully invented due to the advantages of being simple, cost-effective, and highly efficient [7, 8]. Xie et al. [6] developed a rotary TriboEH for harvesting wind energy available in the normal living environment. The device as illustrated in Figure 8.1(c, d) includes a framework, a shaft, a flexible rotor blade, and two aluminum (Al) stators. The soft rotor was structurally made of a polyester (PET) film with polymer nanowires on the surface. According to the triboelectric series [9], a polytetrafluoroethylene (PTFE) film was stuck to the PET blade as a triboelectric layer to get into contact with Al, while the Al blade acts not only as the counter triboelectric layer but also as an electrode. The top wind cup converts the wind flow into the rotation of the shaft, and the PET rotor blade will periodically sweep across the Al stators. Such consecutive face-to-face contact and separation between the rotor and the stator can introduce electricity. The rotary TioboEH successfully delivered a maximum power density of 39 W/m^2 at a wind speed of ~15 m/s, which has demonstrated to directly driving hundreds of commercial light-emitting diodes (LEDs) or rapidly charging capacitors. This work indicated great potential application in the future wind power technology.

8.2 Wind-Belt Fluttering for Energy Harvesting

A vibrating energy harvester can be invented by humdinger wind energy, known as the aerodynamic flutter phenomenon [10]. When a belt is exposed to the airflow at a certain angle, it vibrates in a direction perpendicular to the airflow. With an electromagnetic transducer, the micro wind-belt harvester (13 cm × 3 cm × 2.5 cm) has an output power of over 2 mW at airflow speeds higher than 5.5 m/s but its output power drops significantly (<200 μW) at low

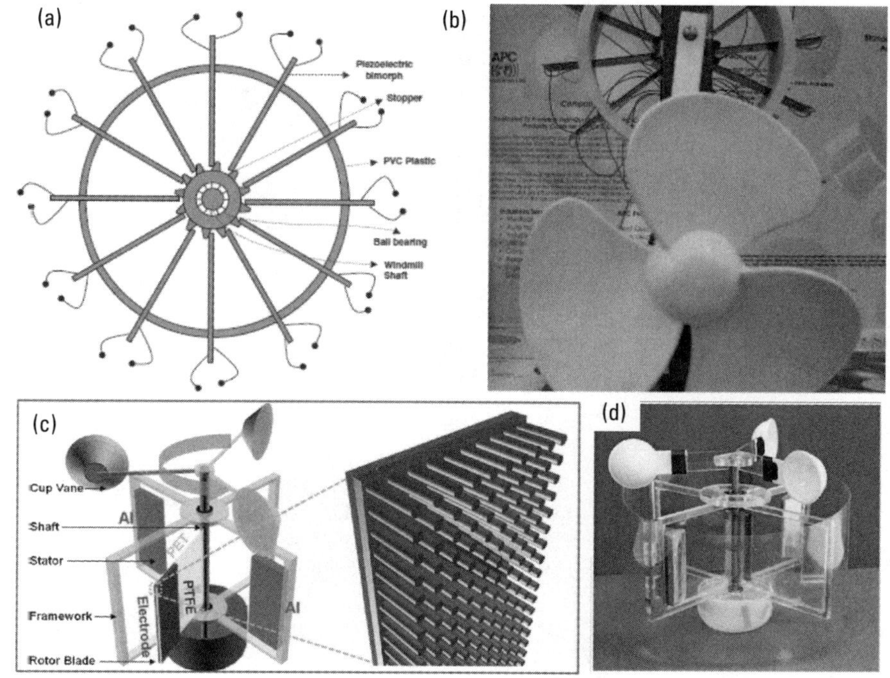

Figure 8.1 (a) The schematic diagram and (b) the photograph of the windmill using piezoelectric bimorph (Copyright: JSAP. [4]); and (c) the schematic drawing and (d) the photograph of the rotary triboelectric energy harvester from air flow with nanowire-like structures (Copyright: ACS, 2013. [6].)

air-flow speeds (<3.5 m/s). Fei et al. [11] implemented an electromagnetic harvester with copper coils and a permanent magnet to efficiently harvest energy from the wind-belt induced vibration as shown in Figure 8.2(a). Due to the fluid-structure interaction between the flexible belt and a random airflow, mechanical vibration of the belt can be introduced. Therefore, the wind harvesting system transfers wind-belt fluttering energy into electrical energy. For a 1-m-long belt, it is obtained at ~7-mW power at wind speed of 3 m/s. However, the dimension of the system is extremely large, and is not applicable to microscale energy harvesting.

Based on a similar fluttering phenomenon, Sun et al. [12] used piezoelectric polyvinylidene fluoride (PVDF) microbelt to extract energy from the low-speed air flow [Figure 8.2(b)]. The microfabricated PVDF thin film of a series of thickness from ~10 to 26 m was poled under an electric field of 70 MV/m for 5 minutes and cut into 20 mm × 2 mm belts with the length direction parallel to the β-phase forming tensile direction. Figure 8.2(c) shows the 3-D plot of the deduced maximum power output as functions of the PVDF film thickness

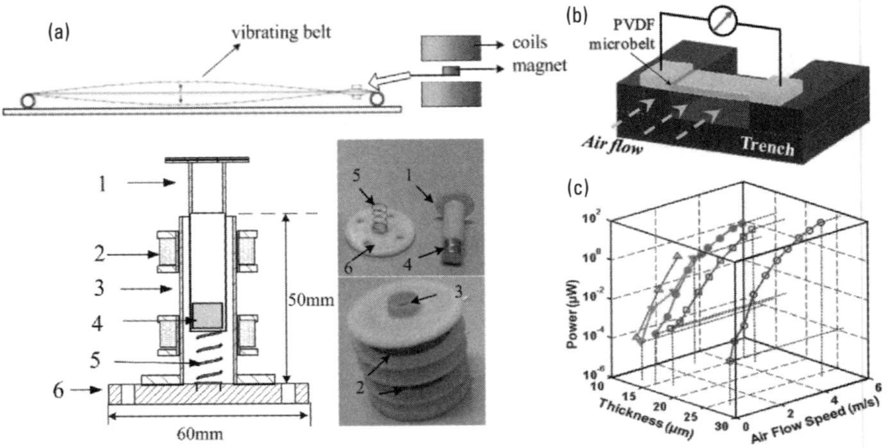

Figure 8.2 (a) Schematic illustration of the wind-belt fluttering integrated with a magnet-coil-based electromagnetic energy harvester (Copyright: Elsevier, 2012. [11]); (b) schematic drawing of a PVDF microbelt for harvesting energy from air flow; and (c) 3-D plot of the maximum power output as functions of the PVDF film thickness and air flow speed (Copyright: RSC, 2011. [12].)

and air flow speed. By connecting the PVDF microbelt (17 mm thick) to a capacitor through a rectifier bridge circuit, the maximum stored energy was calculated to be 8 μJ at 2m/s to 1.8 mJ at 4m/s, which is sufficient to power small electronic devices, such as a digital stopwatch. Their capability for harvesting energy from simulated respiration was also demonstrated.

8.3 Vortex-Induced Vibration for Energy Harvesting

Vortex-induced vibrations have been extensively investigated by researchers for small-scale piezoelectric wind harvesting [13]. The driving forces that oscillate the piezoelectric cantilever or "eel" can be generated in a flowing medium (air or water) by either an obstacle in the flow such as a bluff body or a von Karman's vortex street [14–18]. At present, two different design concepts of utilizing vortex-induced vibrations have been developed as shown in Figure 8.3. The first configuration concept is to attach a flexible piezoelectric cantilever at the trailing edge of a bluff body. Figure 8.3(a) shows a fluid flowing toward the leading edge of a bluff body [19, 20]. The high stagnation pressure near the edge impels flow around the bluff body. Two shear layers are formed that trail at the back side of the flow and bound the wake. The regular vortices pattern is called a vortex street. It is noted that the vortex shedding of a bluff body in a steady flow depends on the Reynolds number, Strouhal number [21, 22], smoothness of the structure, dimensions of the bluff body [23–25], and other factors [26].

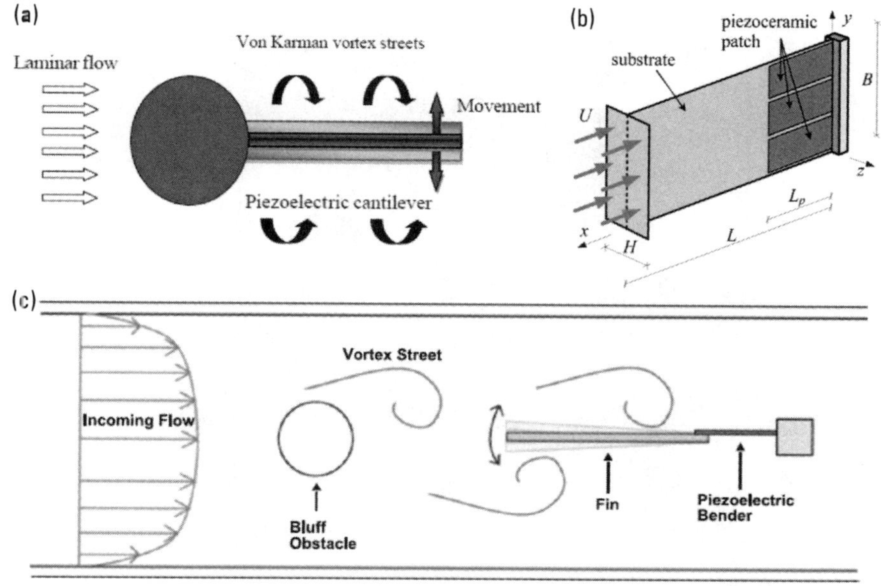

Figure 8.3 Schematic illustrations of (a) a vortex force-induced movement of the cantilever with a bluff body (Copyright: SPIE. [20]), (b) a T-shaped piezoelectric cantilever under a fluid flow (Copyright: AIP Publishing, LLC. [28]), and (c) a vortex shedding from a bluff obstacle in front of a piezoelectric bender (Copyright: IOP Publishing. [32].)

The Reynolds number is related to the free stream velocity U, characteristic dimension D of the bluff body, and the kinematic viscosity of the fluid υ as:

$$\mathrm{Re} = \frac{UD}{\nu} \tag{8.1}$$

To maximize the power generation from fluid-flow energy, vibration-based harvesting devices are designed to match one of their natural frequencies (typically the fundamental frequency) to the base excitation frequency [27]. Under such a lock-in condition, the fluttering structure oscillates at the same frequency as the undisturbed wake behind the bluff body, and consequently its wavelength and amplitude are similar to the undisturbed vortex street. Therefore, an efficient vortex-induced, wind-driven energy harvester should be designed to have a wideband lock-in region in case the flow speed is not constant.

Based on this design concept, Kwon [28] utilized a bimorph cantilever with a T-shaped bluff body that hastens the occurrence of a vortex-induced flutter at a low fluid speed as shown in Figure 8.3(b). The prototype can generate output power from a wind speed of 4 m/s and can provide a maximum power of 4.0 mW at a wind speed of 15 m/s. Some researchers also investigated novel tree-like piezoelectric flapping sheets [29, 30], as illustrated in Figure 8.4. To

Figure 8.4 Illustration of tree-like piezoelectric flapping sheets including (a) parallel-flow stalk-leaf design and (b) cross-flow stalk-leaf design (Copyright: AIP Publishing, LLC. [29].)

make sure that the device is flapping in the air flow, the parallel flow and cross-flow stalk leaves using flexible PVDF sheets are designed [29]. The dominant exciter of this flapping phenomenon is the vortex shedding from the bluff body [31]. The horizontal stalk leaf is driven by the pressure difference between the two sides of the leaf induced by the vortex shedding from the bluff body or the tailing edge. In contrast, the vertical stalk leaf is initially driven by the vortex shedding from the tailing edge. The torsion and bending deformation of the stalk could induce an attack angle and a sideslip angle between the harvester and the air flow, which further result in aerodynamic lift and sideslip forces that eventually trigger the flutter motion. Experimental studies have shown that the cross-flow stalk configuration produced significantly more power than the parallel-flow configuration, even without a bluff body. The device of an adhered double-layer stalk is observed to produce a peak power of approximately 615 μW at an 8-m/s wind speed on a 5-MΩ load. Compared with the rotary windmill and other piezoelectric vibration harvesters, this flapping PVDF device possesses the advantages of low-cost, biocompatible reconfigurable morphology and a broadband response to wind speeds and directions.

The second design strategy to achieve air-flow energy-harvesting couples vortex shedding from a bluff obstacle. The piezoelectric cantilever is arranged behind the bluff body as demonstrated by Weinstein et al. [32] in Figure 8.3(c). By attaching the fin to the cantilever, the deflection can be maximized and its response can be easily adjusted for various flow conditions. Ideally, this design provides relatively high performance over a wide range of flow speeds. The experimental results showed that the operating frequency of the device can be tuned to resonance at low speeds from 2 to 5 m/s by changing the position of the weights along the fin. In a 15-cm-diameter air duct, power generation of 200 μW and 3 mW were achieved for flow speeds of 2.5 and 5 m/s, respectively. The output performance could be sufficient to power a wireless sensor node for smart building monitoring. Zhu et al. [33] demonstrated an airflow energy harvester that consists of a wing attached to a cantilever. The cantilever was placed behind a fluff body in either a vertical or horizontal orientation. The operating principle was also based on the vortex shedding induced asymmetric lift force on the two sides of the cantilever. An electromagnetic transducer was employed to extract electrical energy from the airflow-induced vibrations. Experimental results showed that once the airflow speed is over 2 m/s, an average power generated can exceed 90 μW, which is sufficiently large to power some wireless sensor nodes.

8.4 Helmholtz Resonance for Energy Harvesting

The Helmholtz resonance phenomenon occurs when air blows across the neck of an empty chamber. As the external force pushes the air into the chamber, the pressure inside increases. Once the force is removed, the high-pressure air inside will flow out and the chamber pressure becomes slightly lower than the outside. As a result, the air will be blown back in. This process repeats, with the change of the pressure decreasing every time. As illustrated in Figure 8.5(a), a Helmholtz resonator can be simplified as a mass-spring system, where the mass is the volume of the air in the neck of the resonator and the spring is the volume of the air in the chamber of the resonator. The diameter of the port is related to the mass of the air and the volume of the chamber. A longer neck would make for a larger mass and vice versa. The resonant frequency depends on the volume of the chamber and the aperture as [34]:

$$f_H = \frac{v}{2\pi}\sqrt{\frac{A}{V_H L'}} \tag{8.2}$$

where v is the air flow speed, A and L' are the cross-sectional area and the apparent length of the neck, and V_H is the volume of the chamber.

As shown in Figure 8.5(b), Matova et al. [35] combined a piezoelectric energy converter with a Helmholtz resonator. The resonator converts air-flow energy to mechanical oscillations and the piezoelectric converter converts the vibration energy into electricity. The resonant frequency of the resonator was adjusted to match the complementing piezoelectric converter, so that the

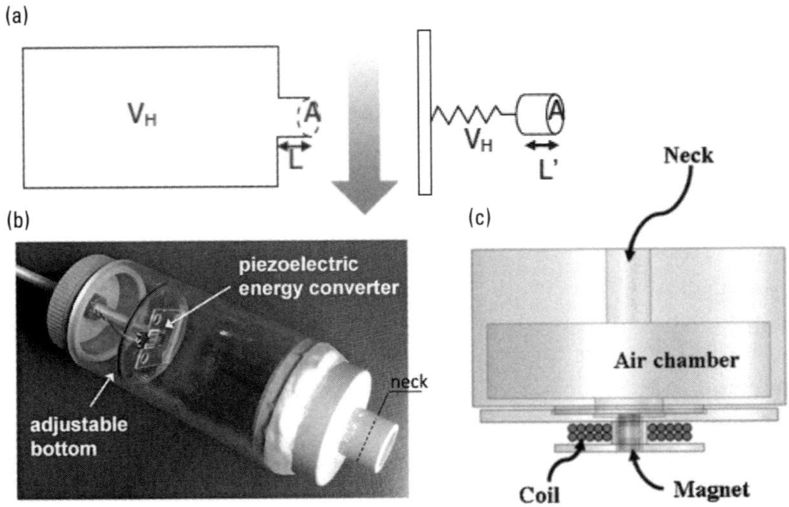

Figure 8.5 (a) Schematic diagram of a Helmholtz resonator with its simplified spring-mass model; (b) photograph of wind-flow energy harvesting device with a combination of piezoelectric converter and Helmholtz resonator; and (c) illustration of a Helmholtz resonator integrated with electromagnetic transducer. (Copyright: IOP Publishing. [35, 36].)

maximum energy was obtained as 42.2 μW at a flow speed of 20 m/s. Kim et al. [36] reported two types of airflow energy harvesters, which are based on windbelt induced oscillation and Helmholtz resonance of electromagnetic transducers. The Helmholtz resonator as shown in Figure 8.5(c) is simply a gas-filled chamber with an open neck, in which a standard second-order (i.e., spring and mass) fluidic oscillation occurs. An electromagnetic transducer consisting of a permanent magnet and coil system is attached to a diaphragm on the chamber and performs the mechanical-to-electrical conversion. The Helmholtz-resonator-based energy-harvesting device achieved a peak-to-peak output voltage of 4 mV at 1.4 kHz, from an input pressure of 0.2 kPa, which is equivalent to the wind speed of 5 m/s.

8.5 MEMS-Based Air-Flow Energy Harvesting

At present, most air-flow energy harvesters are achieved with a relatively large volume or implemented with additional large component such as supporting framework for a windmill rotation, bluff body for vortex shedding, chambers for Helmholtz resonance, and so on. For instance, St. Clair et al. [37] embedded a piezoelectric beam within a cavity for micro wind-driven energy harvesting as shown in Figure 8.6(a). The self-excited oscillation of the beam occurred as the flow rate of air through the beam is large enough such that the nonlin-

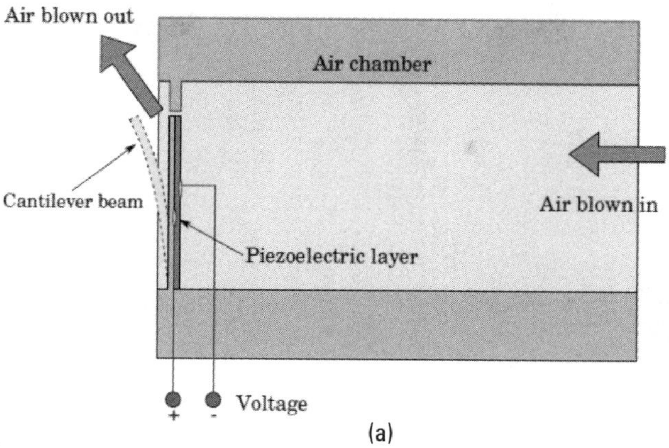

(a)

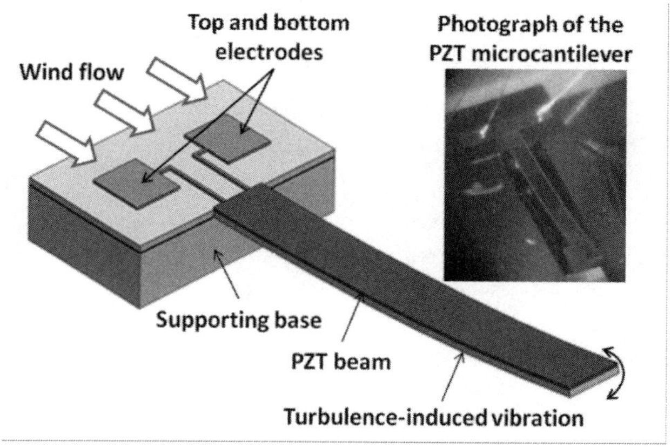

Figure 8.6 (a) Micro wind-flow energy harvesting with a piezoelectric beam assembled on one side of an air chamber (Copyright: AIP Publishing, LLC. [37].) (b) Schematic illustration and photograph of a piezoelectric PZT microcantilever immersed in a wind flow (Copyright: AIP Publishing, LLC. [40].)

ear air pressure force offsets the damping of the structure. The dimensions of the micro-PZT beam are only 12 mm in width, 13 mm in length, and 0.127 mm in thickness. However, an air chamber is additionally constructed using a polyvinyl chloride (PVC) pipe with an inside diameter of 76.2 mm, which makes the whole system relatively large. He and Gao [38] integrated a MEMS piezoelectric energy-harvesting element onto a metal sheet. As long as the wind speed increased to a critical speed, the impact of the metal sheet cantilever with the rigid stop caused the MEMS harvesting element to self-oscillate and the piezoelectric layer of the harvesting element to generate electrical energy. Later on, He et al. [39] from the same group employed a similar optimized AlN-based MEMS energy harvester fabricated with a bulk micromachining process to directly harvest wind energy. The maximum power output is about 2.27 μW for a wind speed of 16.3 m/s flowing from the proof mass to the fixed end of the cantilever.

Only a few air-flow energy harvesters have been realized by MEMS technology. Due to the microfabrication compatibility, piezoelectric thin films are commonly adopted for MEMS-based air-flow energy harvesting. Liu et al. [40, 41] developed two types of piezoelectric (PZT) microcantilevers with both flow sensing and energy-harvesting capability for the potential applications of self-sustained flow-sensing microsystems. The first PZT microcantilever (3 mm × 0.3 mm × 5 μm) was immersed in a wind flow along a parallel incidence direction with respect to the cantilever surface [Figure 8.6(b)], while the second PZT microcantilever (2 mm × 1.65 mm × 5 μm) with a Si proof mass (2 mm × 1.65 mm × 0.4 mm) connected at its free end was immersed in a wind flow perpendicular to the cantilever surface. When the PZT microcantilever is completely immersed within the air flow boundary layer, it will self-oscillate in a turbulence-induced vibration [42] and exhibit a modulated response nearby its resonant frequency. The experimental results have indicated a better performance for the second PZT microcantilever, of which the output voltage and optimized power density are obtained as 48.1 mV and 28.5 mW/cm^3 at flow velocity of 15.6 m/s, respectively. This work has elucidated the smart function of using PZT microcantilevers as flow sensors and wind-driven energy harvesters replying on the dragging force of air flow.

References

[1] Herberta, G. M. J., et al., "A Review of Wind Energy Technologies," *Renew. Sustain. Energy Rev.*, Vol. 11, No. 6, August 2007, pp. 1117–1145.

[2] Koutroulis, E., and K. Kalaitzakis, "Design of a Maximum Power Tracking System for Wind-Energy-Conversiton Applications," *IEEE Trans. on Ind. Electron.*, Vol. 53, No. 2, 2006.

[3] Priya, S., "Modeling of Electric Energy Harvesting Using Piezoelectric Windmill," *Appl. Phys. Lett.*, Vol. 87, No. 18, 2005, p. 184101.

[4] Priya, S., et al., "Piezoelectric Windmill: A Novel Solution to Remote Sensing," *Jpn J. Appl. Phys.*, Vol. 44, No. 3, 2005, pp. L104–L107.

[5] Myers, R., et al., "Small Scale Windmill," *Appl. Phys. Lett.*, Vol. 90, No. 5, 2007, p. 054106.

[6] Xie, Y., et al., "Rotary Triboelectric Nanogenerator Based on a Hybridized Mechanism for Harvesting Wind Energy," *ACS Nano*, Vol. 7, No. 8, 2013, pp. 7119–7125.

[7] Wang, S. H., L. Lin, and Z. L. Wang, "Nanoscale Triboelectric-Effect-Enabled Energy Conversion for Sustainably Powering Portable Electronics," *Nano Lett.*, Vol. 12, 2012, pp. 6339–6346.

[8] Zhu, G., et al., "Toward Large-Scale Energy Harvesting by a Nanoparticle-Enhanced Triboelectric Nanogenerator," *Nano Lett.*, Vol. 13, 2013, pp. 847–853.

[9] Diaz, A. F., and R. M. Felix-Navarro, "A Semi-Quantitative Tribo-Electric Series for Polymeric Materials: The Influence of Chemical Structure and Properties," *J. Electrost.*, Vol. 62, 2004, pp. 277–290.

[10] Humdinger Wind Energy, July 28, 2010, http://www.humdingerwind.com.

[11] Fei, F., J. D. Mai, and W. J. Li, "A Wind-Flutter Energy Converter for Powering Wireless Sensors," *Sensor. Actuat. A: Phys.*, Vol. 173, No. 1, 2012, pp. 163–171.

[12] Sun, C., et al., "PVDF Microbelts for Harvesting Energy from Respiration," *Energy Environ. Sci.,* Vol. 4, No. 11, 2011, p. 4508.

[13] Priya, S., and D. J. Inman, *Energy Harvesting Technologies,* New York: Springer Science+Business Media, 2009, pp. 3–76.

[14] Giacomello, A., and M. Porfiri, "Underwater Energy Harvesting from a Heavy Flag Hosting Ionic Polymer Metal Composites," *J. Appl. Phys.,* Vol. 109, No. 8, 2011, p. 084903.

[15] Taylor, G. W., et al., "The Energy Harvesting Eel: A Small Subsurface Ocean/River Power Generator," *IEEE J. of Oceanic Eng.,* Vol. 26, No. 4, October 2001, pp. 539–547.

[16] Allen, J. J., and A. J. Smits, "Energy Harvesting Eel," *Journal of Fluids and Struct.,* Vol. 15, No. 3-4, April-May 2001, pp. 629–640.

[17] Tan, Y. K., and S. K. Panda, "A Novel Piezoelectric Based Wind Energy Harvester for Low-Power Autonomous Wind Speed Sensor," *33rd Annual Conf. of the IEEE Industrial Electronics Society (IECON)*, 2007, pp. 2175–2180.

[18] Ovejas, V. J., and A. Cuadras, "Multimodal Piezoelectric Wind Energy Harvesters," *Smart Mater. Struct.*, Vol. 20, No. 8, 2011, p. 085030.

[19] Ghasemi-Nejhad, M. N., et al., "Piezoelectric Energy Harvester Operating in Flowing Water," *Proceeding of SPIE,* Vol. 7643, 2010, pp. 76432Z–76432Z-8.

[20] Ghasemi-Nejhad, M. N., V. Sivadas, and A. M. Wickenheiser, "A Study of Several Vortex-Induced Vibration Techniques for Piezoelectric Wind Energy Harvesting," *Proceeding of SPIE,* Vol. 7977, 2011, pp. 79770F–79770F-13.

[21] Shu, Y. C., I. C. Lien, and W. J. Wu, "An Improved Analysis of the SSHI Interface in Piezoelectric Energy Harvesting," *Smart Mater. Struct.*, Vol. 16, 2007, pp. 2253–2264.

[22] Lefeuvre, E., et al., "Piezoelectric Energy Harvesting Device Optimization by Synchronous Charge Extraction," *J. Intell. Mater. Syst. Struct.*, Vol. 16, 2005, pp. 865–876.

[23] Achenbach, E., and E. Heinecke, "On Vortex Shedding from Smooth and Rough Cylinders in the Range of Reynolds Number 6 Multiplied by 10^3 to 5 Multiplied by 10^6," *J. Fluid Mech.*, Vol. 109, 1981, pp. 239–251.

[24] Lienhard, J. H., and L. W. Liu, "Locked-In Vortex Shedding Behind Oscillating Circular Cylinders, with Application to Transmission Lines," *ASME-Paper ASME FE-24(67-FE-24)*, 1967.

[25] Roshko, A., "Wake and Drag of Bluff Bodies," *J. Aeronautical Sciences*, Vol. 22, No. 2, 1955, pp. 124–132.

[26] Blevins, R. D., *Flow-Induced Vibrations*, 2nd ed., Malabar, FL: Krieger Publishing Company, 1994.

[27] Wickenheiser, A. M., and E. Garcia, "Broadband Vibration-Based Energy Harvesting Improvement Through Frequency Up-Conversion by Magnetic Excitation," *Smart Mater. Struct.*, Vol. 19, No. 6, 2010, p. 065020.

[28] Kwon, S. -D.,"A T-Shaped Piezoelectric Cantilever for Fluid Energy Harvesting," *Appl. Phys. Lett.*, Vol. 97, No. 16, 2010, p. 164102.

[29] Li, S., J. Yuan, and H. Lipson, "Ambient Wind Energy Harvesting Using Cross-Flow Fluttering," *J. Appl. Phys.*, Vol. 109, No. 2, 2011, p. 026104.

[30] Bryant, M., E. Wolff, and E. Garcia, "Aeroelastic Flutter Energy Harvester Design: The Sensitivity of the Driving Instability to System Parameters," *Smart Mater. Struct.*, Vol. 20, No. 12, 2011, p. 125017.

[31] Williamson, C. H. K., and R. Govardhan, "Vortex Induced Vibrations," *Annu. Rev. Fluid Mech.*, Vol. 36, 2004, pp. 413–455.

[32] Weinstein, L. A., et al., "Vortex Shedding Induced Energy Harvesting from Piezoelectric Materials in Heating, Ventilation and Air Conditioning Flows," *Smart Mater. Struct*, Vol. 21, 2012, p. 045003.

[33] Zhu, D., et al., "A Novel Miniature Airflow Energy Harvester for Wireless Sensing Applications in Buildings," *IEEE Sensors J.*, Vol. 13, No. 2, 2013, pp. 691–700.

[34] Kinsler, L. E., *Fundamental of Acoustics*, 3rd ed., New York: Wiley, 1982.

[35] Matova, S. P., et al., "Harvesting Energy from Airflow with a Michromachined Piezoelectric Harvester Inside a Helmholtz Resonator," *J. Micromech. Microeng.*, Vol. 21, 2011, p. 104001.

[36] Kim, S.- H., et al., "An Electromagnetic Energy Scavenger from Direct Airflow," *J. Micromech. Microeng.*, Vol. 19, 2009, p. 094010.

[37] St. Clair, D., et al., "A Scalable Concept for Micropower Generation Using Flow-Induced Self-Excited Oscillations," *Appl. Phys. Lett.*, Vol. 96, 2010, p. 144103.

[38] He, X., and J. Gao, "Wind Energy Harvesting Based on Flow-Induced-Vibration and Impact," *Microelectronic. Eng.*, Vol. 111, 2013, pp. 82–86.

[39] He, X., et al., "A Micromachined Low-Frequency Piezoelectric Harvester for Vibration and Wind Energy Scavenging," *J. Micromech. Microeng.*, Vol. 23, No. 12, 2013, p. 125009.

[40] Liu, H., et al., "Development of Piezoelectric Microcantilever Flow Sensor with Wind-Driven Energy Harvesting Capability," *Appl. Phys. Lett.*, Vol. 100, 2012, p. 223905.

[41] Liu, H., et al., "Flow Sensing and Energy Harvesting Characteristics of a Wind-Driven Piezoelectric $Pb(Zr_{0.52}, Ti_{0.48})O_3$ Microcantilever," *Micro Nano Lett.*, Vol. 9, 2014, pp. 286–289.

[42] Blevins, R. D., *Flow-Induced Vibration*, Princeton, NJ: Van Nostrand-Reinhold, 1977.

9

MEMS Thermal Energy Harvesting

Apart from vibration energy, thermal energy is another exploitable energy reservoir. For example, burning fuels such as oil and natural gas releases thermal energy in a car or a furnace. In a similar way, food that you eat is digested and releases energy for your body's growth and activity. Some of that energy maintains your body's internal temperature and other thermal energy is released to the air and objects around you. Thermoelectric energy harvesters can convert a steady-state temperature difference to electricity by the Seebeck effect. Pyroelectric energy harvesters can scavenge the time-dependent temperature variations to electricity by a proper thermodynamic cycle. This chapter will introduce the working principle of these two working mechanisms of energy harvesters. To improve the output power, some novel structural designs and new materials will be demonstrated.

9.1 Thermoelectric Energy Harvesting

9.1.1 Working Principle

Thermoelectric energy harvesters work based on the principle of the Seebeck effect. When the junctions of two different metals are maintained at different temperatures, the electromotive force is produced in the circuit. As shown in Figure 9.1, a thermoelectric harvester consists of two thermoelectric semiconductors (n-type and p-type) subjected to a temperature difference, $T_H - T_C$, and electrically connected in series on the top and bottom. In the n-type semiconductor, most charge carriers are negatively charged electrons, whereas in the other one most of the carriers are positively charged holes. Electrons and holes

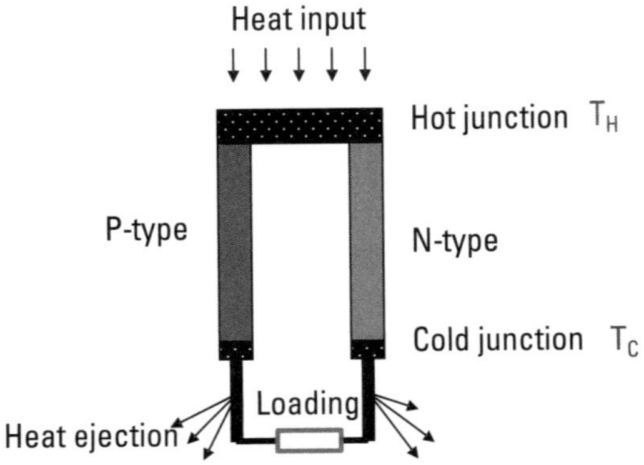

Figure 9.1 A schematic diagram of a single thermoelectric couple connected for thermoelectric power generation.

tend to accumulate on the cold side due to a temperature gradient. The voltages of the n- and p-type semiconductors add up and drive an electrical current through a loading resistance.

Currently, most commercial *thermoelectric power generators* (TPGs) [1] use Bi_2Te_3 as thermoelectric material due to its large thermoelectric figure of merit. Researchers at IMEC [2] presented the first prototype of wireless sensor nodes on human beings powered by thermoelectric harvesters with Bi_2Te_3 material. The first thermoelectric devices based on the V-VI-compounds Bi_2Te_3 and $(Bi,Sb)_2Te_3$ were manufactured by means of regular thin-film technology in combination with microsystem technology [3]. However, Bi_2Te_3 and its alloy are still difficult to obtain and more efforts are paid to make the material compatible with standard silicon chip fabrication processes. In addition, these thermoelectric materials are not CMOS compatible materials.

Large Seebeck effects are found in doped poly-Si, which makes thermoelectric devices fabricate by CMOS technology. With the development of microelectromechanical systems (MEMS) technology, the thin film of polysilicon is deposited and patterned to form thermocouples, and an array of thermocouples (i.e., called thermopile) could be arranged in a tiny area. The key challenge of thin-film material as thermopile is how to reduce the internal thermal resistance compared to the thermal contacts of the surrounding assembly, so that temperature difference between the two ends of thermocouple is small. Because the output voltage of the harvester is proportional to temperature difference between two junctions of thermocouples, appropriate heat flux path is necessary to be designed to obtain the largest temperature difference.

A schematic view of the planar test structure to measure the Seebeck coefficient is shown in Figure 9.2(a) [4, 5]. The polysilicon sample being characterized has a length of 670 μm and a width of 30 μm. It is contacted to

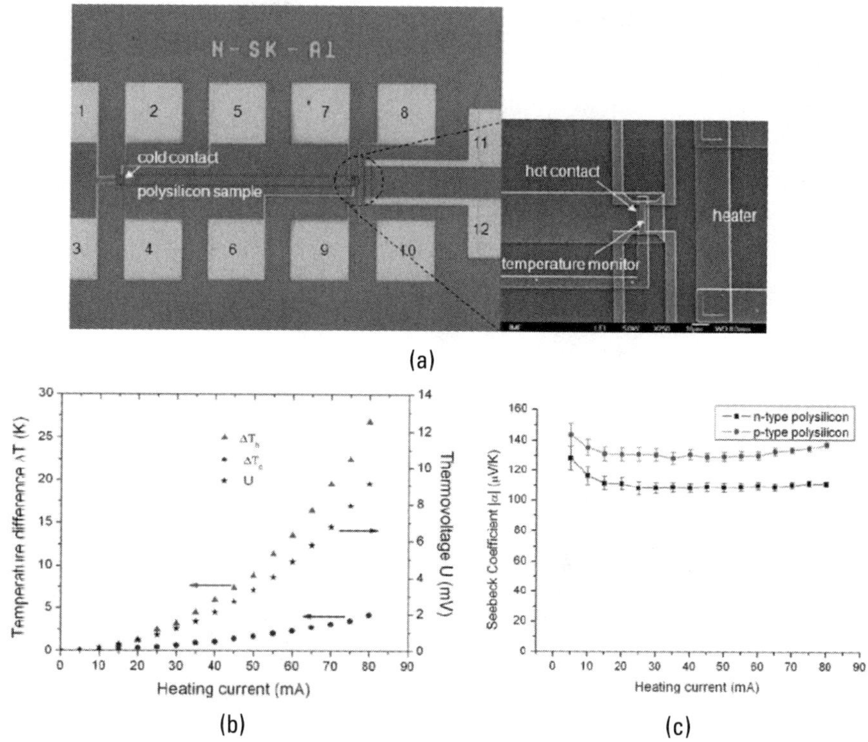

Figure 9.2 (a) Planar testing structure to measure the Seebeck coefficient; (b) the measured temperature difference and thermovoltage; and (c) the Seebeck coefficient of n-type and p-type polysilicon [4].

bottom aluminum at its two ends. A 110 μm × 20 μm large polysilicon heater of roughly 82 (pads 11 and 12) is placed 20 μm away from the hot junction. Two 5-μm-wide temperature monitors made of aluminum with a resistance of 0.16Ω at 300K are placed over these two contacts. The temperature monitors were integrated in a four-point measurement configuration (pads 1, 2, 3, 4 and pads 7, 8, 9, 10) enabling accurate measurements of small resistance changes and, thus, small temperature variations.

In the absence of heat dissipation in the integrated heating resistor, the temperature of the substrate and both samples contacts is T_0. When applying current to the heating resistor, the temperature of the hot and cold junctions of the polysilicon samples was increased from T_0 to T_H and T_C, respectively. The Seebeck coefficient is calculated as

$$\alpha\bigg|_{\frac{T_H+T_C}{2}} = \frac{U}{T_H - T_C} = \frac{U}{\Delta T_G} \tag{9.1}$$

where U is thermovoltage between the two ends of polysilicon and ΔT_G is the temperature difference between the hot and cold junctions.

For the p-type polysilicon, the measured thermovoltage and temperature at hot and cold junctions are shown in Figure 9.2(b). Based on (9.1), it is found that the Seebeck coefficient of p-type and n-type polysilicon is about 130 μV/K and −110 μV/K (here shown as the absolute value), respectively, as shown in Figure 9.2(c).

In thermoelectricity, efficiency is normally expressed as the ratio of the output electricity energy of loading resistance and the heat energy absorbed at the hot junction. It is also referred to as the dimensionless thermoelectric figure-of-merit ZT, and can be expressed by

$$ZT = \frac{\alpha^2 \sigma}{\lambda} \tag{9.2}$$

where σ is the electrical conductivity, and λ is the total thermal conductivity. The best thermoelectric materials Bi_2Te_3-Sb_2Te_3 alloys at room temperature have typical values of $\alpha = 225\mu$V/K, $\sigma = 10^5/\Omega$m, and $\lambda = 1.5$W/mK, which results in ZT ≈ 1 [6].

9.1.2 Design Consideration and Optimization

A thermocouple used in the macroworld comprises two electrical conductors of dissimilar metals or alloys that are joined at one end of a circuit. Thermopile refers to a large number of thermocouples that are electrically connected in series and arranged in meander lines. The maximum output power P is expressed as

$$P = UI = \frac{U^2}{4R_G} = \frac{m^2\alpha^2}{4R_G}\Delta T_G^2 \tag{9.3}$$

where m is the number of thermocouples, I is the electrical current, and R_G is the internal resistance of the harvester.

If the heat contributions by the Peltier effect and Joule heat are neglected, ΔT_G is simply expressed as

$$\Delta T_G = \frac{K_G}{K_G + K_C + K_H}\Delta T \tag{9.4}$$

where K_C and K_H are the thermal resistance at the cold side and hot side respectively, and ΔT is the temperature difference over the two sides of device. If the Peltier effect and Joule heat are included, the expression of ΔT_G is more complicated and has been given in [7]. To compare the performance of different thermoelectric harvesters, a figure φ, which is called a power efficiency factor [8], is defined as

$$\varphi = \frac{P}{A_G \Delta T^2} \tag{9.5}$$

where A_G is surface area. The power efficiency factor needs to be maximized by optimizing dimension of the thermal leg.

The device power factor p of a device expresses the electric energy yield for a certain temperature difference and device area A, where the heat is absorbed, and can be given by [9]

$$p = \frac{P}{\Delta T_G^2 A} \tag{9.6}$$

This value is used to compare the effectiveness of different energy-harvesting devices.

The thermoelectric harvester efficiency is expressed in terms of Carnot efficiency. Carnot efficiency is given by the following equation [10, 11]:

$$\eta_{\max} = \frac{T_H - T_C}{T_H} \frac{\sqrt{1 + ZT_M} - 1}{\sqrt{1 + ZT_M} + \frac{T_C}{T_H}} \tag{9.7}$$

where T_M is the average temperature $(T_H + T_C)/2$.

Infineon demonstrated micromachined TPGs using thermocouples made of poly-Si or poly-SiGe and with a cavity introduced into the substrate to force cross-plane heat flow mainly through the in-plane thermocouples [8, 12]. Because the thermal insulation on the front side of the device is not sufficient, the effective temperature difference between the two junctions of thermocouples is limited. However, IMEC group developed TPGs based on poly-SiGe materials and with improved energy conversion efficiency by creating trench under cold junctions and forming rim structure to match thermal resistance with ambient air [13]. To obtain the high temperature efficiency, the heat flow path is designed and perpendicular to the chip surface and guided by thermal connectors for a large thermal contact area [9]. Figure 9.3 shows the schematic of in-plane and cross-plane structure of the device. This structure is composed of three modules. Module A concerns the thermoelectric structure on a substrate. Module B is used to prevent an electrical short-circuit by the barrier layer of SiO_2. In Module C, the hot junctions are thermally insulated from the substrate by DRIE process, which results in the cold junctions being thermally connected with the bottom, cold side of the harvester. The bonded second wafer is to obtain a good thermal contact and to avoid cavity contamination. Lee et al. [14]

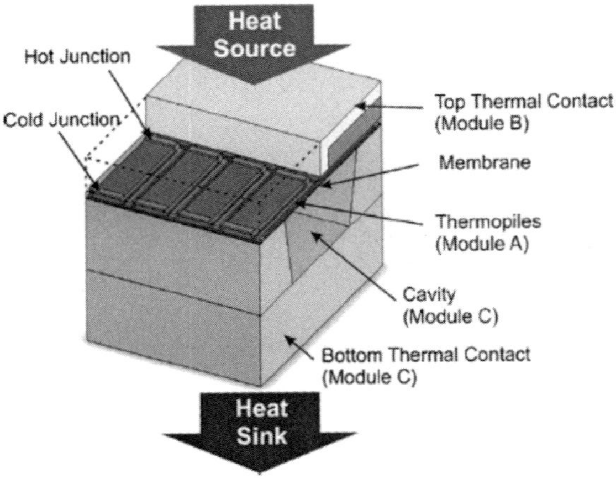

Figure 9.3 Schematic of the miconstructed thermoelectric generator with optimized heat flow path [9].

also investigated a design of CMOS MEMS-based TPGs with wafer-bonded vacuum cavities on both sides of suspended thermoelectric membrane. As we can see, vacuum cavity has become an effective approach to increase performance of TPGs.

Since heat transfers cross the device from bottom side to top side, the heat must be redirected to flow along the length of thermal legs to obtain large temperature difference. Designs of the bottom cavity to guide the heat flux have been seen in several reported TPGs [8–14]. However, in these devices, few efforts were made to reduce thermal insulation from the front side or prevent hot junctions easily cooled by ambient air. Heat loss by air convection from the top side is even heavier when the device surface has a high convection coefficient. For example, the harvester moves with the body during walking or sporting when powering wearable devices. Xie et al. at the Institute of Microelectronics, A*STAR, Singapore proposed a novel dual-cavity structure to optimize heat flux [5, 15]. The proposed power harvester is shown in Figure 9.4(a). Planar thermopiles are arranged on the substrate due to the limited thickness of thermoelectric material that is available in thin-film technology. Each thermal leg is embedded between the bottom and top cavities. The bottom cavities are created in the wafer level to guide the heat flux through thermal legs, as mentioned above, while the top cavities protect hot junctions from being cooled by air convection. In contrast, the area of the cold junctions is exposed to ambient air. Both the bottom and top cavities are sealed with a high vacuum level to minimize heat loss due to heat conduction and convection through air. With such a design, the planar thermopiles work like vertical configuration. A thin metal layer covers top surface of the device as a heat sink (i.e., a heat dissipation layer). This metal layer covers the whole device area such that the whole device area is

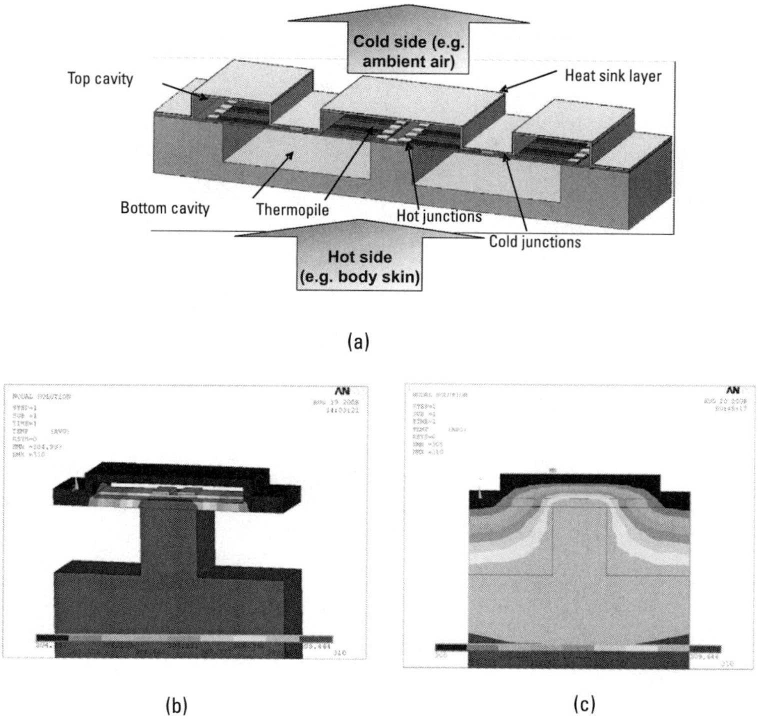

Figure 9.4 (a) Schematic view of the proposed thermoelectric power harvester; (b) FEA result for a design with top and bottom vacuum cavities = 3.75K; and (c) a design with top and bottom air cavities = 2.77K [5, 15].

assumed to be able to release heat to ambient air effectively. It makes that heat easily disperse from the cold sides to the air and then enlarge the temperature difference across the device.

A 3-D model is built with the software ANSYS for thermal finite element analysis (FEA). A fixed temperature difference of 5K between the top and bottom of the device is assumed as boundary conditions for thermal analysis. The optimized parameters of the harvester model have been determined by varying the length, width and thickness of the thermal legs. The variable to be maximized is the power efficiency factor, as shown in (9.5). For a layer thickness of 0.7 μm, an optimum is found for thermal legs of 16-μm effective length and 5-um width. According to this dimension, 31,536 thermocouples could be arranged in a chip with size of 5×5 mm^2. For the design with top and bottom vacuum cavities, the temperature difference between hot and cold junctions is 3.75K (i.e., ΔT_G) as shown in Figure 9.4(b). However, for the design with air cavities, the temperature difference ΔT_G is derived as only 2.77K when we consider heat loss due to air conduction and convection, as shown in Figure 9.4(c). Based on (9.3), the output power is a second-order function of ΔT_G, so the output power is improved about 83% based on the design with top and bottom vacuum cavities.

Based on (9.2), the increase in *ZT* results directly in the improvement of the cooling efficiency for Peltier modules [16] and of the electric generation efficiency of thermoelectric harvesters [17]. By superimposing the transient current pulse on the steady-state current, the effective *ZT* for the Peltier cooling was indeed increased to about two times of that under the steady-state condition [18]. Yamashita et al. investigated the effect of the periodically alternating temperature gradients on energy conversion efficiency [19]. The results show that the energy conversion efficiency of thermoelectric harvester operating under the temperature gradient (TG) alternating at an optimum period is several times as large as that of a harvester working under the conventional steady TG.

9.1.3 Microfabrication Process and Device Characterization

The first (standard configuration) thermoelectric harvester with over 100 elements is fabricated by the electrochemical MEMS process. It is easily scalable to batch-produce devices with thousands of thermoelectric elements at low cost [20]. Wafer-bonding technology for vacuum cavity is used to fabricate the thermoelectric harvester [8, 14]. Recently researchers proposed technologies for low-temperature wafer-level sealing technology for creating sealed vacuum cavities. For example, B. H. Stark [21] demonstrated vacuum sealing at a low temperature with a maximum of flexibility in package geometry by thick films of electroplated nickel; and P. Monajemi [22] presented a low-temperature packaging technique that utilizes decomposition of a sacrificial polymer through a photo-definable polymer overcoat. However, none of these reported low-temperature wafer-level sealing technologies can be exercised in a CMOS-compatible process. In other words, these technologies are post-CMOS processes and materials are not fully available in the standard CMOS manufacturing line.

A low-cost and low-temperature wafer level sealing technology based on CMOS materials and CMOS technology process is proposed to fabricate the thermoelectric harvester with two vacuum cavities. The detailed process steps are illustrated in Figure 9.5(a–g). The polysilicon layer is partially phosphorous-implanted with 180-keV energy to form the n-type thermoelectric legs and is subsequently boron-implanted under 80 keV in the other half-region to create the p-type thermoelectric legs. The doping dose of 10^{16} cm^{-2} is used in both steps. To obtain a better electrical contact between the Al and polysilicon, the wafer is annealed at 420°C for 30 minutes, and then is passivated by a Si_3N_4 layer. To optimize the heat flux direction within the harvester, bottom and top vacuum cavities are created. The bottom and top cavities are sealed by USG layer with low stress by PECVD at 400°C. Finally, 0.7-μm-thick aluminum is

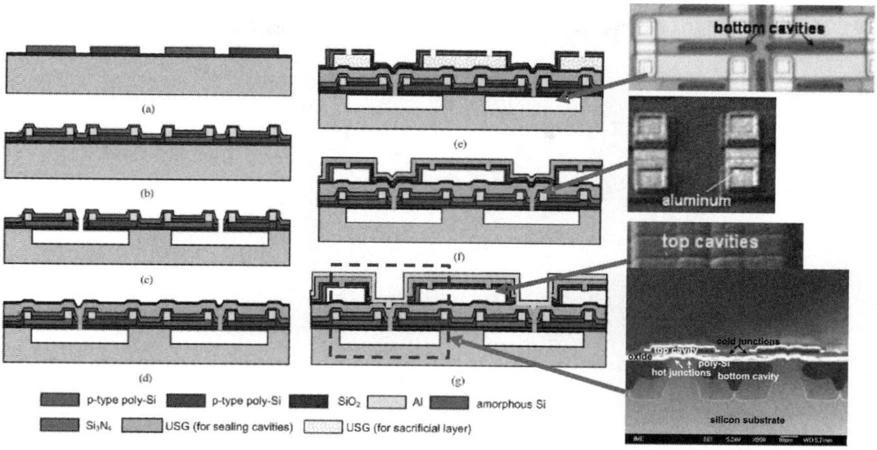

Figure 9.5 Fabrication process of the power generator. (a) Patterning and doping poly-Si; (b) metal junctions and Si_3N_4 passivation layer; (c) perforating holes, DRIE, and isotropic etching silicon; (d) sealing bottom cavities; (e) patterning USG sacrificial layer and opening holes; (f) removing USG sacrificial layer and sealing the top cavities; (g) partially etching oxide at area of cold junctions and coating metal on the whole device.

deposited to cover the whole device area as the heat sink layer [Figure 9.5(g)]. The sealed bottom cavity is about 30 μm in depth and 42 μm in width, while the encapsulated top cavity is 2 μm in height and 35 μm in width, and the lengths of both bottom and top cavities are about 670 μm.

The device chips under testing are diced into 1 cm × 1 cm size. They include 125,144 thermocouples that equal four harvester elements as shown in Figure 9.6(a). The measurement setup is depicted in Figure 9.6(b). The setup platform includes a hot plate, a brass heat sink, a commercial Peltier cooler, a CPU fan, two PT100 temperature sensors, two PID controller units, and a Hewlett Packard 34401A digital multimeter. The internal resistance of the harvester in the size of 1 cm × 1 cm is 52.8 MΩ, where 23 MΩ comes from the contact resistance of vias between poly-silicon and metal. Thus, the contact resistance occupies about 43.6% of the total internal resistance. The testing results show that the open-circuit voltage at 5K difference is 16.7V, and the derived output power is 1.3 μW/cm^2. The power efficiency factor φ of the fabricated harvester is calculated as 0.052 μWcm^{-2}K^{-2}. For the previous reported TPGs fabricated based on CMOS compatible process, the power efficiency factor with 0.0352 μWcm^{-2}K^{-2} and 0.016 μWcm^{-2}K^{-2} is for Infineon's device [7] and for the device reported by the University of Freiburg [9], respectively. The performance improvement is resulted from reducing the contact resistance by adding a nickel silicide layer between poly-Si and aluminum. It is expected that the contact resistance could be reduced to about one-quarter of the cur-

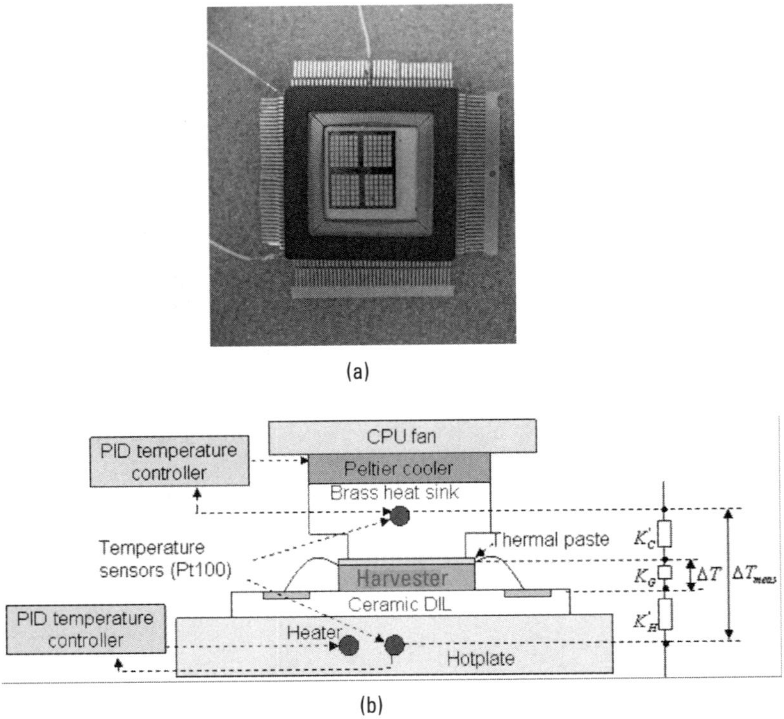

Figure 9.6 (a) The fabricated thermoelectric harvester; and (b) the schematic for testing the setup.

rently presented value. Therefore. the output power will increase about 48% (i.e., about 1.9 μW) for a chip in size of 1 cm^2 at a 5K temperature difference.

A thin-film batch process-based poly-SiGe is developed for mass production of 6- or 8-inch wafers [23], as shown in Figure 9.7. A 6-μm-thick SiO_2 layer is patterned to form blocks [Figure 9.7(a)]. A 0.7-μm undoped poly-SiGe layer is deposited and followed by implantation for forming n- and p-type legs. The thermocouples are released by using a pad etch [24].The harvester is placed between two aluminum blocks for testing purpose, which are supported by plastic spacers. The blocks are about 10 times larger than the devices to obtain larger thermal capacity than the devices. An open circuit voltage of 1.49V and an output power of 0.4 μW can be generated at a 3.5K temperature difference for a harvester with size of 1.0 mm × 2.5 mm. To reduce the microfabrication complexity compared to a free-standing membrane, the lateral heat conduction in the silicon overlayer is designed to establish temperature [25]. The output power of the SOI-based SbTe-TiW thermocouple is 0.56 μW/cm^2 at a temperature gradient of 18°K.

A single-layer grapheme (SLG) sheet has demonstrated excellent electrical and thermal conductivities [26, 27]. However, it will potentially become a new

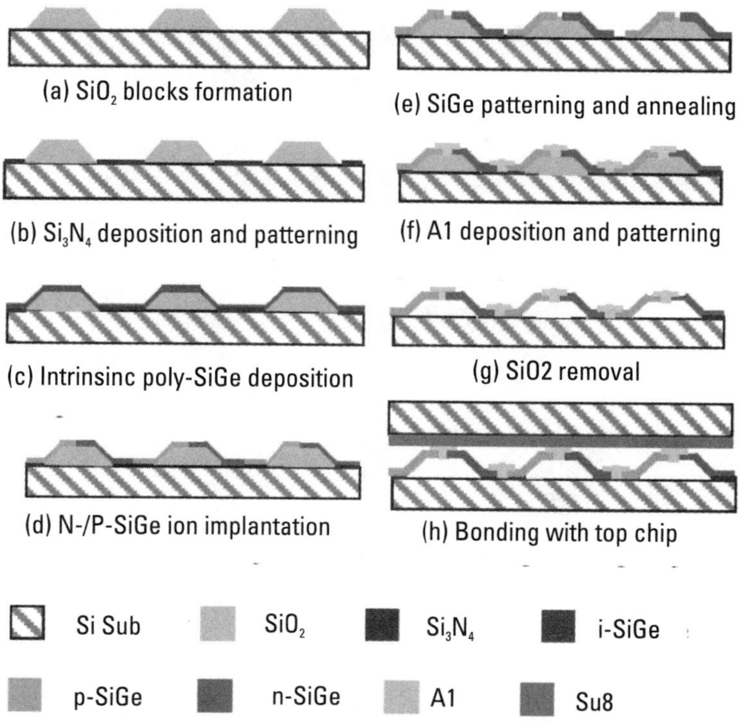

Figure 9.7 Process flow of the thermoelecrtric harvester [23].

thermoelectric material through analyzing the figure of merit (*ZT*) [28]. Based on the analytical model, *ZT* can reach a value of 0.17 to 1.02 for an isotope-doped SLG sheet under a sheet concentration range of 10^{16} to 10^{17} m^{-2} and a temperature range of 300K to 450K. To improve the Seebeck coefficient, the grapheme sheet would be appropriately patterned into nanoribbons [29, 30]. By introducing extended line defects (ELDs) in the length of the nanoribbons and edge roughness, the thermoelectric figure of merit can obtain a large value for zigzag grapheme nanoribbons [31].

Table 9.1 gives a performance comparison of Si-MEMS-based thermoelectric energy harvester among Infineon, IMEC, and IME (Singapore). Under a forced cooling 5K temperature difference, Infineon reported 1 $\mu W/cm^2$, while IME delivered 1.3 $\mu W/cm^2$, which increases 30%. Under naturally cooling, voltage from the IMEC's harvester was 96 mV, and 80 mV from IME. The internal resistance of IMEC's harvester is a little higher than that of IME, so the output power of the two harvesters is very close. However, in the current stage, power from both IME and IMEC is very small (in the nanowatt range) when harvesting energy from body heat.

Table 9.1
Performance Comparison of Si-MEMS-Based Thermoelectric Power Harvester

	Infincon (2004) [7]	**IMEC (2008) [13]**	**IME (2010) [15]**
Material	p/n—Poly Si	p/n-SiGe	p/nPoly-Si
Seebeck coefficient (μVK[1])	160	225	240
Electrical resistivity (mΩ cm)	1.5	3.6	2.3
Thermal conductivity ($Wm^{-1}K^{-1}$)	31	3.5	NA
Figure of merit of material	0.004	0.03	~0.005
Number of themocouple per cm^2	132,266	29,375	105,120
Contact resistance ($\Omega \mu m^2$)	5.8	$1.45*10^4$	415
Internal resistance per cm^2	31 MΩ	62.5 MΩ	52.8 MΩ
Open-circuit voltage (enforce cooling) (Vcm^2K^{-1})	2	2.5	3.2
Output power density (enforce cooling) ($\mu Wcm^{-2}K^{-2}$)	0.0426	0.026	0.052
Output power at 5K difference (enforce cooling) (μWcm^{-2})	1	0.65	1.3
Open-circuit voltage (on hotplate, naturally cooling) 5K difference, (mV)	NA	96	80
Output power (on body skin)	NA	0.3 nW	NA

9.2 Pyroelectric Energy Harvesting

9.2.1 Working Principle

Pyroelectric energy conversion can convert waste heat into electricity by alternatively heating and cooling a pyroelectric material resulting in electricity generations [32–35]. Pyroelectric harvesters depend on the property that the spontaneous polarization (and hence dielectric constant) of certain materials is temperature-dependent. The Ericsson thermal energy conversion cycle developed by Olsen is shown in Figure 9.8. The cycle consists of two isotherms (T_L and T_H) and two constant electric fields (V_1 and V_2). It starts by charging the pyroelectric element (PE) at a low temperature T_L by increasing the voltage from V_1 to V_2 (Point 1 to 2). Then the *PE* is discharged by heating the element from T_L to T_H at constant voltage V_2 where ideally $T_H > T_{Curie}$ (Point 2 to 3). Next, the material is further discharged by reducing the voltage from V_2 to V_1 at the constant temperature T_H (Point 3 to 4). Finally, the PE is recharged by cooling it from $T_L < T_{Curie}$ at a constant voltage V_1 (Point 4 to 1) [36]. The pyroelectric current I during the cycle is given by [37]

$$I = A_f \frac{dP_S}{dt} = A_f p \frac{dT}{dt} \tag{9.8}$$

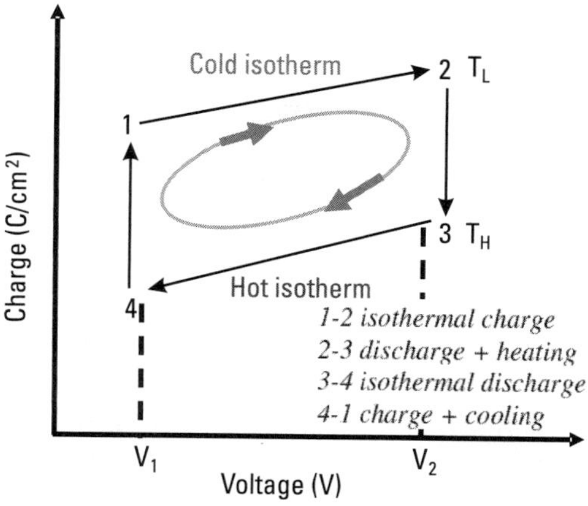

Figure 9.8 Change versus voltage at different temperatures. The Olsen power cycle is represented by the area between 1-2-3-4.

where A_f is the surface area of the pyroelectric thin film capacitor, P_S (C/m^2) is the pyroelectric thin-film polarization, T is the pyroelectric capacitor temperature, and p is the pyroelectric coefficient. The net output power N_p from the pyroelectric capacitor is:

$$N_p = V_{appl} I_p = V_{appl} p A_f \frac{dT}{dt} \tag{9.9}$$

where V_{appl} is the external applied voltage across the pyroelectric capacitor.

The surrounding area by points 1-2-3-4 corresponds to the electrical work output W, which is expressed as

$$W = \oint V_{appl} dq = \int N_p dt = \int V_{appl} p A_f \frac{dT}{dt} dt \tag{9.10}$$

Pyroelectric energy harvesters convert heat (Q_{in}) into electrical power (W) with efficiency [38]:

$$\eta = \frac{W}{Q_{in}} = \frac{W_E - W_P}{C_v \Delta T + Q_{Int} + Q_{Leak}} \tag{9.11}$$

where W_E is the generated electrical power, W_P is the power lost in the temperature cycle, C_v is the heat capacity of the pyroelectric device, Q_{Int} are the intrinsic heat losses in the thermal cycle, and Q_{Leak} are the heat leakages between the hot and cold sources.

The first proposed thermal energy conversion cycle was proposed by Clingman [39, 40], but energy conversion efficiencies were calculated as below

1% when the energy for temperature cycling was higher than that which could be extracted from the energy conversion cycle. Other theoretical modelings were created for studies, with overall predicted power conversion efficiencies of 10% to 15% [41] and 20% to 40% [42, 43], with Carnot efficiencies in the range of 50% to 80% [42–45].

9.2.2 Material Consideration

Pyroelectric materials possess a spontaneous polarization P_S without an applied electric field. P_S is defined as the electric dipole moment per unit volume averaged over the volume of the material and depends strongly on temperature due to the pyroelectric material's crystallographic structure [46]. Ferroelectricity materials have a spontaneous polarization $-P_S$ that can be reversed to $+P_S$ by applying an electric field. Based on mechanical properties of materials, the pyroelectric materials are divided into linear and nonlinear ones, as shown in Table 9.2 [47]. For nonlinear materials, the electrocaloric activity and the associated effective expected output power are given in this table. Because the electrocaloric effect has the ability to change the temperature (in the adiabatic case) or the entropy (in the isothermal case), the power of nonlinear pyroelectric harvesters is 10 to 600 times larger than that of linear pyroelectric energy harvesters due to their outstanding electrocaloric activity.

The (1-x)$Pb(Mg_{1/3}Nb_{2/3})O_3$–x$PbTiO_3$ ferroelectric ceramics have low Curie transition temperature and high pyroelectric activity [59, 60]. They may be obtained in ceramic or single crystal form, and exhibit very high electromechanical behavior [61, 62]. When $x = 0.10$, the Curie transition is about 40°C [63, 64], which results in high electrocaloric activity around room temperature. The $0.90Pb(Mg_{1/3}Nb_{2/3})O_3$–$0.10PbTiO_3$ ceramic was fabricated using a modified solid-state route [54]. Under a 50°C temperature variation and electric field cycle of 3.5 kV/mm, the harvested energy reached 186 mJ/cm^3.

PVDF and its co-polymer P(VDF-TrFE) as polymeric pyroelectric materials have been paid more attention due to their large pyroelectric coefficient [65]. Their repoling should be performed below the Curie temperature. The factors affecting the Curie temperature of P(VDF-TrFE) include the VDF/TrFE ratio, the applied electric field and polymer processing techniques [66]. Increasing the VDF/TrFE ratio and the applied electric field contributes to the increase of the Curie temperature [25, 65]. The Curie temperature for 60/40 P(VDF-TrFE) is approximately 66°C at the zero field [67] and increases to 120°C when the field is 527 kV/cm [68].

9.2.3 Performance Evaluation and Analysis

The MEMS-based pyroelectric harvester was schematically shown in Figure 9.9(a) [37, 38]. This device included a sandwich structure of P(VDF-TrFE)

Table 9.2
Comparison of Linear and Nonlinear Pyroelectric Materials

Linear materials	**Shape**	*P* (μC m–2K–1)	$\varepsilon_{33}^{o}(\varepsilon_0)$	**CE** (×106Jm^{-3}K^{-1})	P_{out} (μWcm^{-3})	**Ref.**
111 PMN-0.25PT single crystals	SC	1300–1790	961–1100	2.5	100–169	[48, 49]
PZT	C	533	1116	2.5	12.9	[50]
PMN-0.25PT ceramic	C	746	2100	2.5	13.4	[51]
PVDF	P	33	9	1.8	6.12	[52]
Nonlinear materials		Q_{ECE} (Jcm^{-3})	Applied electric amplitude (kV mm^{-1})	θ_{cold} (°C)	Pout (μWcm^{-3})	Ref.
$0.95PbSc_{0.5}Ta_{0.5}$–$0.05PbSc_{0.5}O_5$	C	4.2	2.5	–5	1570	[53]
$0.90Pb(Mg_{1/3}Nb_{2/3})\ O_3$–$0.10PbTiO_3$ ceramic	C	1.4	3.5	30	500	[54]
(110) 0.955 $Pb(Zn_{1/3}Nb_{2/3})\ O_3$–$0.10PbTiO_3$ crystal	SC	No measure	2	90	750	[55]
$0.90Pb(Mg_{1/3}Nb_{2/3})\ O_3$–$0.10PbTiO_3$ thin film	TF	15	90	75	4320	[56]
PZST75/20/5	C	9.8	3	160	2240	[57]
$PbZr_{0.95}Ti_{0.05}O_3$ thin film	TF	30	78	220	5960	[58]

Source: [47].

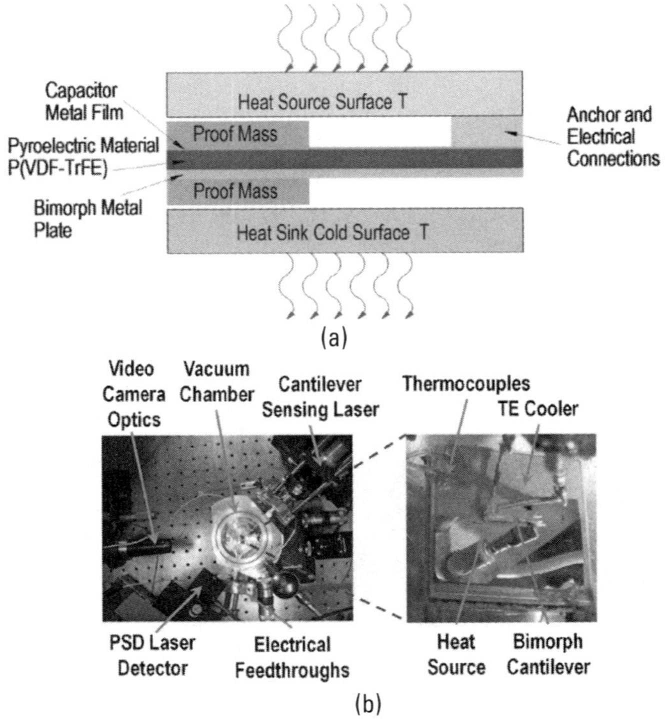

Figure 9.9 (a) Side view of the pyroelectric P(VDF-TrFE) harvester; and (b) vacuum test chamber and optical diagnostics [38].

dielectric layer and two metal films as capacitor electrodes, two additional small proof masses. The mass was located at the end of the cantilever for increasing the thermal mass and making good thermal contacts with the hot and cold surfaces. Because this cantilever structure initially heated through the anchor, the cantilever bended towards the cold surface. When contacting the cold surface, the cantilever bended towards the hot surface due to the heat loss, and vice versa. Good thermal contact with the hot and cold surfaces contributed to transfer substantial amounts of thermal energy to the pyroelectric capacitor on the cantilever. The testing platform was shown in Figure 9.9(b). A 20–50-mTorr vacuum pressure in the testing chamber was used to eliminate heat loss by gas conduction and convection. The motion of the tip of the cantilever was monitored by a position sensing laser and position sensitive detector. At a temperature difference of 70K and a temperature cycling frequency of 80 Hz, the output power of 1 mW with a power density of 400 W/L was obtained from this pyroelectric harvester.

By coupling the pyroelectric and semiconducting properties in ZnO, the pyroelectric nanoharvester was designed to use the anisotropic polarization in

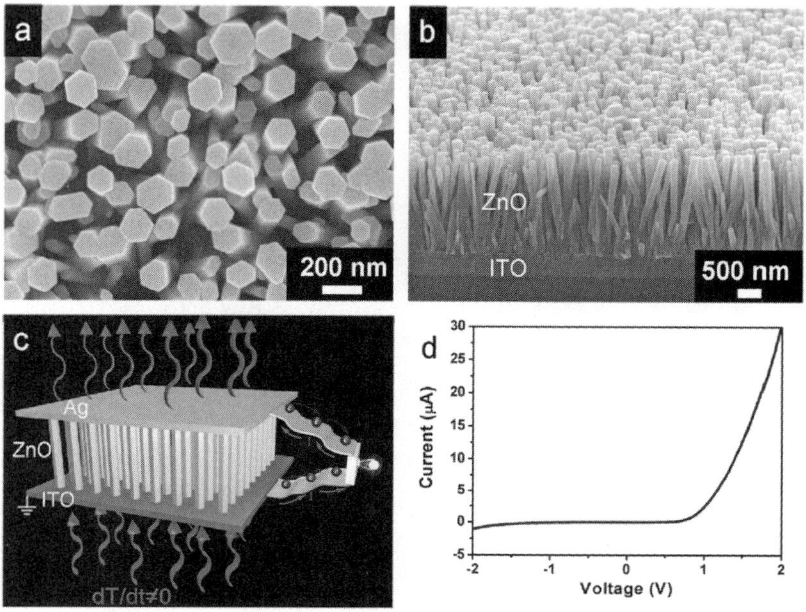

Figure 9.10 (a) SEM image of the as-grown ZnO nanowire array; (b) tilted cross sectional SEM image of the ZnO nanowire array; (c) schematic diagram of the pyroelectric nanogenerator; (d) I-V characteristics of the nanogenerator measured at room temperature [69].

ZnO nanowires created as a result of a time-dependent temperature fluctuation [69]. Figure 9.10(a, b) shows a diameter of about 200 nm and a length of 2 μm from ZnO nanowires, respectively. The experimental setup was shown schematically in Figure 9.10(c). The I-V curve in Figure 9.10(d) shows that the device exhibited a Schottky behavior at the interface. The characteristic coefficient of the heat flow conversion into electricity is about ~0.05 to 0.08Vm^2/W. The pyroelectric nanoharvester has the potential to power nanodevices, temperature imaging, radiation detection, X-ray generation, and space science [70–72].

In the future, enhancing the pyroelectric properties of the current materials will become the research hotspot through doping and material engineering [73]. The development of some new pyroelectric materials of nanowires and nanofibers or integrating multiple working mechanisms has a potential application of flexible textile and wearable electronics.

References

[1] Thermo Life Energy Corp., www.PowerByThermoLife.com.

[2] Leonov, V., et al., "Thermoelectric Converters of Human Warmth for Self-Powered Wireless Sensor Nodes," *IEEE Sensors Journal*, Vol. 7, No. 5, May 2007, pp. 650–657.

[3] Bottner, H., et al., "New Thermoelectric Components Using Microsystem Technologies," *Journal of Microelectromechanical Systems*, Vol. 13, No. 3, June 2004, pp. 414–420.

[4] Xie, J., et al., "Microstructures for Characterization of Seebeck Coefficient of Doped Polysilicon Films," *Microsyst. Technol.*, Vol. 17, 2011, pp. 77–83.

[5] Xie, J., et al., "Characterization of Heavily Doped Polysilicon Films for CMOS-MEMS Thermoelectric Power Generators," *J. Micromech. Microeng.*, Vol. 19, 2009, p. 125029.

[6] Sales, B. C., "Critical Review of Recent Approaches to Improved Thermoelectric Materials," *International Journal of Applied Ceramic Technology*, Vol. 4, 2007, pp. 291–296.

[7] Strasser, M., et al., "Micromachined CMOS Thermoelectric Generators as Onchip Power Supply," *Sens. Actuators A, Phys.*, Vol. 114, No. 2/3, September 2004, pp. 362–370.

[8] Glatz, W., S. Muntwyler, and C. Hierold, "Optimization and Fabrication of Thick Flexible Polymer Based Micro Thermoelectric Generator," *Sens. Actuators A, Phys.*, Vol. 132, No. 1, November 2006, pp. 337–345.

[9] Huesgen, T., P. Woias, and Kockmann, "Design and Fabrication of MEMS Thermoelectric Generators with High Temperature Efficiency," *Sensors and Actuators A*, Vol. 145, 2008, pp. 423–429.

[10] Yoshida, K., et al., "High–Energy Density Miniature Thermoelectric Generator Using Catalytic Combustion," *J. Microelectromech S.*, Vol. 15, 2006, pp. 195–203.

[11] Min, G., and D. M. Rowe, "Conversion Efficiency of Thermoelectric Combustion Systems," *IEEE Trans. on Energy Conver.*, Vol. 22, 2007, pp. 528–534.

[12] Strasser, M., et al., "Miniaturized Thermoelectric Generators Based on Poly-Si and Poly-SiGe Surface Micromachining," *Sensors and Actuators A: Physical*, Vol. 97-98, April 2002, pp. 535–542.

[13] Wang, Z., et al., "Realization of a Poly-SiGe Based Micromachined Thermopile," *Eurosensors 2008*, Dresden, Germany, 2008, pp. 1420–1423.

[14] Lee, C., and J. Xie, "Design and Optimization of Wafer Bonding Packaged Microelectromechanical Systems Thermoelectric Power Generators with Heat Dissipation Path," *Journal of Vacuum Science and Technology B: Microelectronics and Nanometer Structures*, Vol. 27, No. 3, May/June 2009.

[15] Xie, J., C. K. Lee, and H. H. Feng, "Design, Fabrication and Characterization of CMOS MEMS-Based Thermoelectric Power Generators," *Journal of Microelectromechanical Systems*, Vol. 19, No. 2, 2010, pp. 317–324.

[16] Goldshmid, H. J., *Thermoelectric Refrigeration*, New York: Plenum, 1964.

[17] Mahan, G. D., "Good Thermoelectronics" in *Solid State Physics*, Vol. 51, H. Ehrenreich and F. Spaepen, (eds.), New York: Academic Press, 1998, Vol. 51, p. 81.

[18] Thonhauser, T., et al., "Improved Supercooling in Transient Thermoelectronics," *Appl. Phys. Lett.*, Vol. 85, 2004, p. 3247.

[19] Yamashita, O., H. Odahara, and K. Satou, "Energy Conversion Efficiency of Thermoelectric Generator Under the Periodically Alternating Temperature Gradients," *Journal of Applied Physics*, Vol. 101, 2007, p. 023704.

[20] Snyder, G. J., et al., "Thermoelectric Microdevice Fabricated by a MEMS-Like Electrochemical Process," *Nature Materials*, Vol. 2, 2003, pp. 528–531.

[21] Stark, B. H., and K. Najafi, "A Low-Temperature Thin-Film Electroplated Metal Vacuum Package," *Journal of Microelectromechanical Systems*, Vol. 13, No. 2, April 2004, pp. 147–157.

[22] Monajemi, P., et al., "Wafer-Level MEMS Packaging Via Thermally Released Metal-Organic Membranes," *Journal of Micromechanics and Microengineering*, Vol. 16, 2006, pp. 742–750.

[23] Su, J., et al., "A Batch Process Micromachined Thermoelectric Energy Harvester: Fabrication and Characterization," *J. Micromech. Microeng.*, Vol. 20, 2010, p. 104005.

[24] Su, J., et al., "Process Development on Large-Topography Microstructures for Thermoelectric Energy Harvesters," *Proc. PowerMEMS 2008 + microEMS 2008*, 2008, pp. 365–368.

[25] Lee J., et al., "Thermoelectric Characterization and Power Generation Using a Silicon-On Insulator Substrate," *Journal of Microelectromechanical Systems*, Vol. 21, 2012, pp. 4–6.

[26] Balandin, A. A., "Thermal Properties of Graphene and Nanostructured Carbon Materials," *Nature Mater.*, Vol. 10, July 2011, pp. 569–581.

[27] Morozov, S. V., et al., "Giant Intrinsic Carrier Mobilities in Graphene and Its Bilayer," *Phys. Rev. Lett.*, Vol. 100, No. 1, 2008, pp. 016602-1–016602-4.

[28] Verma, R., S. Bhattacharya, and S. Mahapatra, "Thermoelectric Performance of a Single-Layer Grapheme Sheet for Energy Harvesting," *IEEE Trans. on Electron Devices*, Vol. 60, No. 6, 2013, pp. 2064–2070.

[29] Han, M., et al., "Energy Band-Gap Engineering of Graphene Nanoribbons," *Phys. Rev. Lett.*, Vol. 98, 2007, p. 206805.

[30] Zhang, A., et al., "Band-Gap Engineering in Graphene and Hexagonal BN Antidot Lattices: A First Principles Study," *Appl. Phys. Lett.*, Vol. 98, 2011, p. 023105.

[31] Karamitaheri, H., et al., "Engineering Enhanced Thermoelectric Properties in Zigzag Grapheme Nanoribbons," *Journal of Applied Physics*, Vol. 111, 2012, p. 054501.

[32] Olsen, R. B., and D. D. Brown, "High-Efficiency Direct Conversion of Heat to Electrical Energy- Related Pyroelectric Measurements," *Ferroelectrics*, Vol. 40, 1982, pp. 17–27.

[33] Olsen, R. B., et al., "A Pyroelectric Energy Converter Which Employs Regeneration," *Ferroelectrics*, Vol. 38, 1981, pp. 975–978.

[34] Olsen, R. B., "Ferroelectric Conversion of Heat to Electrical Energy: A Demonstration," *Journal of Energy*, Vol. 6, 1982, pp. 91–95.

[35] Olsen, R. B., D. A. Bruno, and J. M. Briscoe, "Pyroelectric Conversion Cycle of Vinylidene Fluoride-Trifluoroethylene Copolymer," *Journal of Applied Physics*, Vol. 57, 1985, pp. 5036–5042.

[36] Kouchachvili, L., and M. Ikura, "Pyroelectric Conversion-Effects of P(VDF-TrFE) Preconditioning on Power Conversion," *Journal of Electrostatics*, Vol. 65, 2006, pp. 182–188.

[37] Hunter, S. R., et al., "Review of Pyroelectric Thermal Energy Harvesting and New MEMs Based Resonant Energy Conversion Techniques, Energy Harvesting and Storage: Materials, Devices, and Applications III," *Proc. of SPIE*, Vol. 8377, 2012, pp. 83770D-1–14.

[38] Hunter, S. R., et al., "Development of MEMS Based Pyroelectric Thermal Energy Harvesters, Energy Harvesting and Storage: Materials, Devices, and Applications III," *Proc. of SPIE*, Vol. 8035, 2011, pp. 80350V-1–12.

[39] Clingman, W. H., and R. G. Moore, "Application of Ferroelectricity to Energy Conversion Process," *J. Appl. Phys.*, Vol. 32, 1961, pp. 675–681.

[40] Fattuzzo, E., H. Kiess, and R. Nitshe, "Theoretical Efficiency of Pyroelectric Power Converters," *J. Appl. Phys.*, Vol. 37, 1966, pp. 510–516.

[41] Gonzalo, A., "Ferroelectric Materials as Energy Converters," *Ferroelectrics*, Vol. 11, 1976, pp. 423–430.

[42] Drummond, A. D., "Dielectric Power Conversion," *Abstracts 10th Intersociety Energy Conversion Engineering Conf.*, Newark, DE, Vol. 950, 1975.

[43] Drummond, A. D., "Demonstration of a High Power Density Electrocaloric Heat Engine," *Ferroelectrics*, Vol. 27, 1980, pp. 213–218.

[44] Sebald, G., E. Lefeuvre, and D. Guyomar, "Pyroelectric Energy Conversion: Optimization Principles," *IEEE Trans. on Ultrasonics, Ferroelectrics, & Frequency Control*, Vol. 55, 2008, pp. 538–551.

[45] Guyomar, D., et al., "Toward Energy Harvesting Using Pyroelectric Material," *J. Intell. Mat. Sys. Struct.*, Vol. 20, 2009, pp. 265–271.

[46] Lang, S. B., "Pyroelectricity: From Ancient Curiosity to Modern Imaging Tool," *Physics Today*, Vol. 58, 2005, pp. 31–36.

[47] Sebald, G., D. Guyomar, and A. Agbossou, "On Thermoelectric and Pyroelectric Energy Harvesting," *Smart Mater. Struct.*, Vol. 18, 2009, p. 125006.

[48] Sebald, G., et al., "Electrocaloric and Pyroelectric Properties of 0.75Pb(Mg1/3Nb2/3)O3–0.25PbTiO3 Single Crystals," *J. Appl. Phys.*, Vol. 100, 2006, p. 124112.

[49] Davis, M., D. Damjanovic, and N. Setter, "Pyroelectric Properties of (1 – *x*)Pb(Mg1/3Nb2/3)O3–*x*PbTiO3 and (1 – *x*)Pb(Zn1/3Nb2/3)O3–*x*PbTiO3 Single Crystals Measured Using a Dynamic Method," *J. Appl. Phys.*, Vol. 96, 2004, pp. 2811–2815.

[50] Ng, W. Y., et al., "Pyroelectric properties of PZT/P(VDF-TrFE) 0–3 Composites," *ISAF 2000: Proc. 2000 12th IEEE Int. Symp. on Applications of Ferroelectrics*, Vol. 2, 2000, pp 767–770.

[51] Guyomar, D., et al., "Ferroelectric Electrocaloric Conversion in 0.75(PbMg1/3Nb2/3O3)–0.25(PbTiO3) Ceramics," *J. Phys. D: Appl. Phys.*, Vol. 39, 2006, pp. 4491–4496.

[52] Lang, S. B., and S. Muensit, "Review of Some Lesser-Known Applications of Piezoelectric and Pyroelectric Polymers," *Appl. Phys. A*, Vol. 25, 2006, pp. 125–134.

[53] Shebanov, L., and K. Borman, "On Lead–Scandium Tantalite Solid Solutions with High Electrocaloric Effect," *Ferroelectrics*, Vol. 127, 1992, pp. 143–148.

[54] Sebald, G., S. Pruvost, and D. Guyomar, "Energy Harvesting Based on Ericsson Pyroelectric Cycles in Relaxor Ferroelectric Ceramic," *Smart Mater. Struct.*, Vol. 17, 2008, p. 015012.

[55] Khodayari, A., et al., "Non Linear Pyroelectric Energy Harvesting from Relaxor Single Crystals," *IEEE Trans. on Ultrason. Ferroelectr. Freq. Control*, Vol. 56, 2009, pp. 693–699.

[56] Mischenko, A. S., et al., "Giant Electrocaloric Effect in the Thin Film Relaxor Ferroelectric 0.9PbMg1/3Nb2/3O3–0.1PbTiO3 Near Room Temperature," *Appl. Phys. Lett.*, Vol. 89, 2006, p. 242912.

[57] Tuttle, A., and D. A. Payne, "The Effects of Microstructure on the Electrocaloric Properties of Pb (Zr, Sn, Ti) O3 Ceramics," *Ferroelectrics*, Vol. 37 1981, p. 603.

[58] Mischenko, A. S., et al., "Giant Electrocaloric Effect in Thin-Film PbZr0.95Ti0.05O3," *Science*, Vol. 311, 2006, pp. 1270–1271.

[59] Guyomar, D., et al., "Ferroelecrtric Electrocaloric Conversion in 0.75 $(PbMg_{1/3}Nb_{2/3}O_3)$–0.25 $(PbTiO_3)$ Ceramics," *J. Phys. D: Appl. Phys.*, Vol. 39, 2006, p. 4491.

[60] Sebald, G., et al., "Electrocaloric Properties of High Dielectric Constant Ferroelectric Ceramics," *J. Eur. Ceram. Soc.*, Vol. 27, 2007, p. 4021

[61] Benayad, A., et al., "Segregation Study and Segregation Modeling of Ti in $Pb[(Mg_{1/3}Nb_{2/3})$ 0.60 $T_{i0.40}]O_3$ Single Crystal Grown by Bridgman Method," *Mater. Res. Bull.*, Vol. 41, 2006, p. 1069.

[62] Sebald, G., L. Lebrun, and D. Guyomar, "Stability of Morphotropic <110> Oriented 0.65 PMN-0.35PT Single Crystals," *IEEE Trans. on Ultrason. Ferroelectr. Freq. Control*, Vol. 51, 2004, p. 1491.

[63] Babooram, K., H. Tailor, and Z. -G. Ye, "Phase Formation and Dielectric Properties of $0.90Pb(Mg_{1/3}Nb_{2/3})O_3$-$0.10PbTiO_3$ Ceramics Prepared by a New Sol-Gel Method," *Ceram. Int.*, Vol. 30, 2004, p. 1411.

[64] Park, J. -H., et al., "Piezoelectric Properties in PMN-PT Relaxor Ferroelectronics with MnO_2 Addition," *J. Eur. Ceram. Soc.*, Vol. 21, 2001, p. 1383.

[65] Lang, S. B., and D. K. Das-Gupta, "Pyroelectricity: Fundamentals and Applications," in *Handbook of Advanced Electronic and Photonic Materials and Devices, Vol. 4*, San Diego, CA: Academic Press, 2001.

[66] Kim, K. M., et al., "Curie Transition, Ferroelectric Crystal Structure and Ferroelectricity of VDF/TrFE (75/25) Copolymer: 1. The Effect of the Consecutive Annealing in the Ferroelectric State on Curie Transition and Ferroelectric Crystal Structure," *Journal of Polymer Science*, Vol. 32, 1994, pp. 2435–2444.

[67] Navid, A., C. S. Lynch, and L. Pilon, "Purified and Porous Poly(Vinylidene Fluoridetrifluoroethylene) [P(VDF-TrFE)] Thin Films for Pyroelectric Infrared Sensing and Energy Harvesting," *Smart Materials and Structures*, Vol. 19, 2010, p. 055006.

[68] Olsen, R. B., and D. A. Bruno, "Pyroelectric Conversion Materials," *Proceedings of the 21st Intersociety Energy Conversion Engineering Conference*, American Chemical Society, Washington, D.C., 1986, pp. 89–93.

[69] Yang, Y., et al., "Pyroelectric Nanogenerators for Harvesting Thermoelectric Energy," *Nano Letters*, Vol. 12, 2012, pp. 2833–2838.

[70] Lang, S. B., "Pyroelectricity: From Ancient Curiosity to Modern Imaging Tool," Phys. Today, Vol. 58, 2005, pp. 31–36.

[71] Cooper, J., "A Fast Response Total-Radiation Detector," Nature, Vol. 194, 1962, pp. 269–271.

[72] Brownridge, J. D., "Pyroelectric X-Ray Generator," Nature, Vol. 358, 1992, pp. 287–288.

[73] Lingam, D., et al., "Nano/Microscale Pyroelectric Energy Harvesting: Challenges and Opportunities," International Journal of Smart and Nano Materials, Vol. 4, 2013, pp. 229–245.

10

Nano-Based Energy Harvesting

With the quick development of nanodevices and nanosystems, nanogenerator-based piezoelectric materials have attracted more attention because they can scavenge vibrational and triboelectric energy into electricity for powering these devices. Because the nanodevices and nanosystems normally operate at a very low power in the range from nanowatts to microwatts, the environmental energy scavenged by the nano-based harvesters could be enough to power the system. By coupling their semiconducting and piezoelectric properties, mechanical strains of different piezoelectric materials and nanowires can be converted into electricity, including ZnO, PVDF, and PZT. This chapter introduces the working principle of nano-based harvesters (nanogenerators), including the piezoelectric effect in nanowires or nanofibers, and energy conversion efficiency. Then some fabrication processes of three major piezoelectric nanowires (or nanofibers) are introduced in detail to demonstrate their advantages and disadvantages. Some new structural designs of nano-based energy harvesters and their applications are also shown.

10.1 Piezoelectric Effect in Nanowires and Nanofibers

The working principle of nano-based harvesters relies on the piezoelectric potential created in the nanowires or nanofibers by an external strain under a small force. A dynamic straining of the nanowire or nanofiber generates a transient flow of the electrons in the external load because of the driving force of the piezopotential [1, 2] and the Fermi level at the two contacts is the fundamental mechanism of the nanoharvesters. The advantage of using nanowires is that they can be excited by tiny physical motions under the frequency range from 1

Hz to thousands of hertz [3]. It is ideal for scavenging random energy such as body motion, triboelectricity, and air flow existing in the environment.

The physical principle for creating, separating, and preserving the piezoelectric charges in the nanowires or nanofibers is a coupling of piezoelectric and semiconducting properties [4]. Figure 10.1 shows the working principle of nanoharvesters excited by the AFM tip. When the deflection of the vertical nanowire (NW) by the AFM tip causes a strain field, the outer surface is tensile with a positive strain and the inner surface is compressive with a negative strain. A piezoelectric field along NWs is generated inside the NW volume by the piezoelectric effect, with the piezoelectric field is closely parallel to the NW at the outer surface and antiparallel to the NW at the inner surface. As a result, the electric potential distributed across the width of the NW at the top end from the compressed to the stretched side surface is approximately between the negative and positive potentials [5].

The energy conversion efficiency of nanogenerators is an important performance parameter for the application of energy scavenging. It is defined as the ratio between the electric energy that is delivered the load and the input mechanical energy from the applied strain. The electric energy delivered by the nanogenerator is equal to the Joule heating energy and can be expressed as [6]

$$E_{electric} = Q = \int_{t_1}^{t_2} I^2 \bullet R \bullet dt = R\int_{t_1}^{t_2} I^2 \bullet dt \tag{10.1}$$

where Q is the Joule heating energy, R is the loading resistance, and I is the instantaneous current.

The elastic strain energy can be estimated using [7]

$$E_{mech} = \frac{1}{2} EA\varepsilon^2 L_0 N \tag{10.2}$$

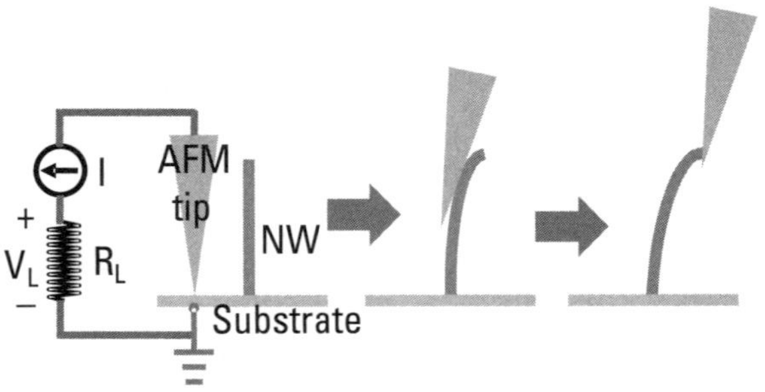

Figure 10.1 The converting mechanism of a piezoelectric nanogenerator.

where E is Young's modulus of the material, A is the cross-sectional area, ε is the strain applied on the material, L_0 is the length of piezoelectric nanowires, and N is the number of NWs.

Therefore, the energy conversion efficiency is calculated as

$$\eta = \frac{E_{electric}}{E_{mech}} \times 100\% = \frac{2R\int_{t_1}^{t_2} I^2 \bullet dt}{EA\varepsilon^2 L_0 N} \tag{10.3}$$

It is seen that the energy conversion efficiency is strongly related to the piezoelectricity in NWs or nanofibers. This piezoelectricity depends on the piezoelectric materials and their processing parameters and conditions.

10.2 ZnO Nanowire Harvesters

10.2.1 ZnO Nanowire Growth Approaches

One-dimensional (1-D) ZnO nanomaterial has three key advantages [8]. First, it exhibits both semiconducting and piezoelectric (PZ) properties that can form the basis for electromechanically coupled sensors and transducers. Its energy bandgap and large excitation binding energy are 3.37 eV and 60 meV at room temperature [9]. Second, ZnO is relatively biosafe and biocompatible, and it can be used for biomedical applications with little toxicity. Third, ZnO forms the diverse nanostructures, such as nanowires (NWs), nanobelts (NBs), nanosprings, nanorings, nanobows, and nanohelices. Moreover, the ZnO nanowires can be easily grown on a wide variety of low-cost substrates using inexpensive solution methods [10]. Various methods are used for the synthesis of ZnO nanostructures such as physical vapor deposition (PVD), molecular beam epitaxy (MBE), pulsed laser deposition (PLD), and hydrothermal and chemical methods. Among these methods, solution-based chemical methods are more advantageous because they are simple, low-cost, and compatible for flexible substrates, capable of large scaling up and growth occurs at relatively low temperatures [9]. N-type ZnO nanowires were grown onto p-type silicon substrates by the low-temperature solution method [11]. ZnO NWs vertically on the substrate consisted of two steps such as seed layer formation and growth of nanowires on the seeds. The spin-coated zinc acetate film was heated to 400°C in air for 1 hour to obtain the ZnO seed layer. Then the substrate with seed layers was dipped into a mixture of equimolar (25 mM) zinc nitrate hexahydrate (Zn $(NO_3)_2 \cdot 6H_2O$ and hexamethylenetetramine (HMTA) solution and it was kept for more than 10 hours at a constant temperature around 50°C. The SEM results show that the denser array of nanowires of a size of about 50 nm and a length of >500 nm is grown uniformly in the substrate [12].

Figure 10.2(a) shows the piezoelectric ZnO nanowires grown on the c-plane-oriented α-Al_2O_3 substrate by the vapor-liquid-solid (VLS) process, using Au particles as a catalyst [13, 14]. The growth is initiated and guided by the Au particle, and the expitaxial relationship between ZnO and Al_2O_3 leads to the alignment. Figure 10.2(b) shows the NW grown along the [0001] direction and has side surfaces of $\{01\bar{1}0\}$. To increase the growth rate and reduce the vaporization temperature from 1,300°C to 900°C, the epitaxial growth of ZnO on sapphire surface with carbon power is employed [13]. Figure 10.2(c) shows the typical morphology of the aligned ZnO NWs on the sapphire substrate. The successful alignment of ZnO NWs on sapphire substrates is attributed to the very small lattice mismatches between the substrates and ZnO. In the case of sapphire, the $\{11\bar{2}0\}$ (plane-orientated substrate is always used because the smallest lattice mismatch is along the c-axis of Al_2O_3 and the *a*-axis of ZnO [15]. The aligning quality is contributed by chamber pressure, oxygen partial pressure [16], and the thickness of the catalyst layer [17].

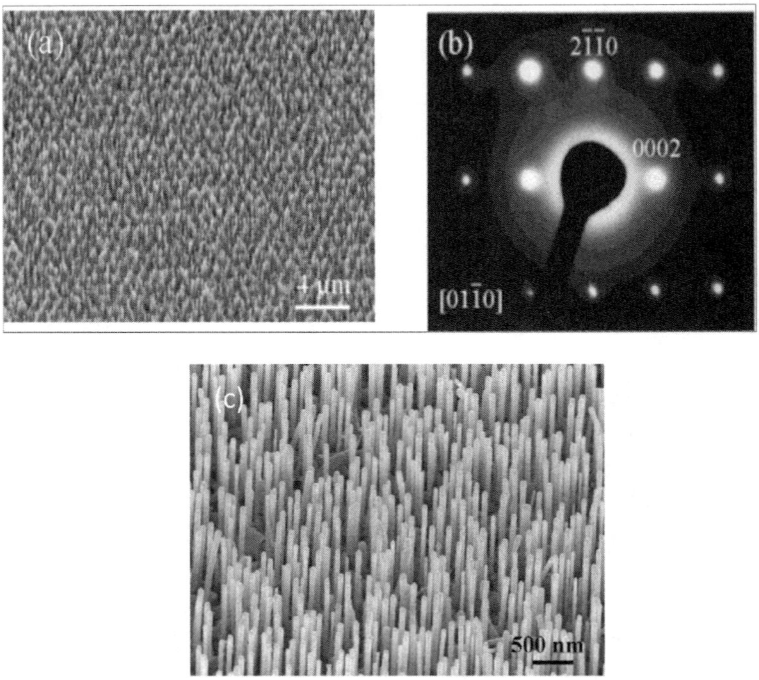

Figure 10.2 Experimental design for converting nanoscale mechanical energy into electrical energy by a vertical piezoelectric (PZ) ZnO NW: (a) scanning electron microscopy images of aligned ZnO NWs grown on an Al_2O_3 substrate; (b) transmission electron microscopy images of ZnO NWs; (c) ZnO NWs grown on sapphire substrate [13].

ZnO has its unique nanostructures and is formed owing to the three quasi-stable {0001}, $\{11\bar{2}0\}$, $\{01\bar{1}0\}$ surfaces and the corresponding growth directions. Zn can serve as catalyst that leads to the growth of ZnO nanostructure, or alternatively, the oxidation of Zn can also form ZnO nanostructures [18]. Figure 10.3 shows the cylindrically arranged NWs grown by a vapor-solid process [15]. The core of the cylinder fiber is likely made of Zn. Besides using silicon substrate, well aligned ZnO nanowires with an average length of ~5 μm had also been grown on GaN/AlN substrate through a vapor-solid process [19].

Recently, p-doped ZnO nanowires arrays for generator application had also been grown on the (001) silicon substrate using thermal vapor deposition [20, 21]. The solution containing 20-mM zinc acetate dehydrate $(Zn(C_3H_3O_2)\cdot 2H_2O)_2$ in ethanol is deployed for a thin aqueous film of zinc acetate on the substrate. ZnO thin films with a thickness of 100 to 200 nm were formed after annealing at 300°C for 30 minutes. This film is used as a seed layer for growing catalyst-free P-doped ZnO NW arrays vertically on the Si substrate. The growth temperature was controlled at 600°C. Compared to the low-cost and simple aqueous methods, the vapor-phase method usually employs vacuum, sophisticated equipment, and higher temperature, which restricts the type of substrate used. Among the aqueous grown methods, the hydrothermal method has emerged as a powerful method for the fabrication of one-dimensional nanostructures, offering significant advantages such as controllable structures and a cost-effective, low-temperature, substrate-independent technique that allows for easy integration. There are some reports of hydrothermal growth of ZnO nanostructures. Govender et al. [22] synthesized ZnO nanocolumns on gold-coated tin-oxide substrates by a low-temperature solution method. Greene et al. [23] reported the wafer-scale growth of ZnO nanorods and studied the temperature-dependent photoluminescence (PL) properties. Vayssieres et al.

Figure 10.3 Nanowire arrays grown by a vapor-solid process [16].

Figure 10.4 ZnO nanowire arrays by the hydrothermal method.

[24–26] reported the preparation of highly oriented ZnO microrods and microtubes and the diameter of the ZnO rods from 1–2 μm to 100–200 nm. Yang et al. [27] reported ZnO NW arrays with an average length of 2.5 μm and the diameter varied from 53 nm to 75 nm, as shown in Figure 10.4. A seed solution of 0.01-M zinc acetate (ZnAc) in 10-mL absolute ethanol was spin-coated on the substrate at 1,000 rpm for 1 minute and repeated three times. The growth precursor, which consisted of an aqueous 25-mM zinc nitrate hexahydrate and hexamethylenetetramine (HMT) solution, was added to 50 mL of the aqueous precursor.

In 2008, Wang et al. at Georgia Institute of Technology reported that the density-controlled growth of aligned ZnO NWs arrays was obtained without using ZnO seeds by the hydrothermal-based chemical approach [28]. By adjusting the precursor concentration, the density of ZnO NW arrays could be controlled within one order of magnitude (number of ZnO NWs per 100 μm^2) with one NW growing from one spot site. This novel method does not require ZnO seeds or external electrical field, and it can be grown at low temperatures and a large scale on any substrates regardless of being crystalline or amorphous. Moreover, the annealing process contributed to the oriented growth of aligned ZnO NWs and the chemistry of the growth is well documented [26, 29, 30]. To obtain high-quality patterned ZnO NW arrays, electron-beam lithography and a low-temperature hydrothermal method were used to achieve patterned and aligned growth of ZnO NWs at <100°C on general inorganic substrates, such as Si and GaN, without using a catalyst [31]. The laterally aligned NWs were realized by using different materials to activate or inhibit the growth of nanowires [32]. Two materials are used here: ZnO seeds for the growth, and the Cr layer for preventing local growth.

10.2.2 Design and Fabrication of ZnO Harvesters

Finite element method (FEM) calculations for analyzing the bending of a semiconducting piezoelectric ZnO nanowire for harvesters are demonstrated [33]. Figure 10.5 shows the results of the electric potential for a bent ZnO nanowire using the finite element method of COMSOL Multiphysics. The generated electrical potential is 0.3V based on theoretical calculations. A theoretical model that predicts the electrical power generated by an array of vertically aligned ZnO nanoribbons is proposed [34]. The theory suggests that an array of vertically aligned ZnO nanoribbons can produce as much as 1 nW mm^{-2} and 1–100 nW mm^{-3} electricity, which can power the device such as a 30-pW processor [35]. These values are significantly less than the power density of MEMS-based systems of comparable size (0.1–1 μWmm^{-3}) [36, 37].

In 2006, Professor Wang first proposed a piezoelectric nanogenerator based on ZnO nanowire arrays [8]. Because NW arrays were relatively less density and shorter length, AFM tip was employed to contact one NW without touching another. A constant normal force of 5 nN was applied between the tip and sample surface. The output voltage of 8 mV was measured as the tip scanned over the nanowires when the loading resistance of 500 MΩ was connected. The output power of the NW would be about 0.5 pW at the resonant frequency of around 10 MHz. To improve the power generation capabilities of the system, the ultrasonic waves driven the motion of the NWs are used to replace the AFM tip with a simpler source of mechanical energy, which results in the production of a continuous current [38, 39]. The output voltage of such nanowires generator has recently improved to 10 mV under the driving ultrasonic wave [38]. Figure 10.6 shows the working principle of nano-based

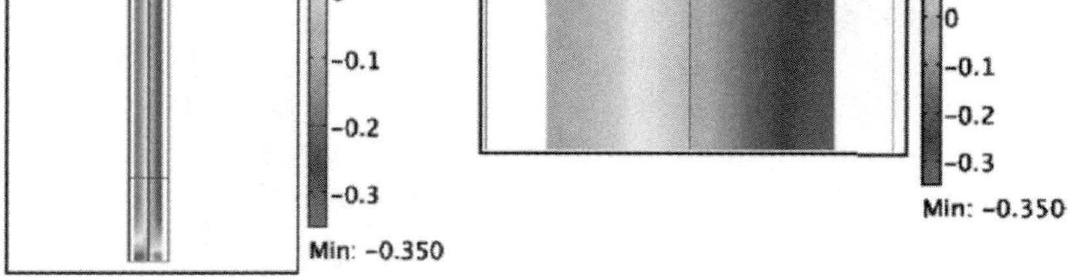

Figure 10.5 Color online FEM calculation of a ZnO NW bent by a lateral force of 80 nN: (a) the whole NW; and (b) the upper part of the NW [33].

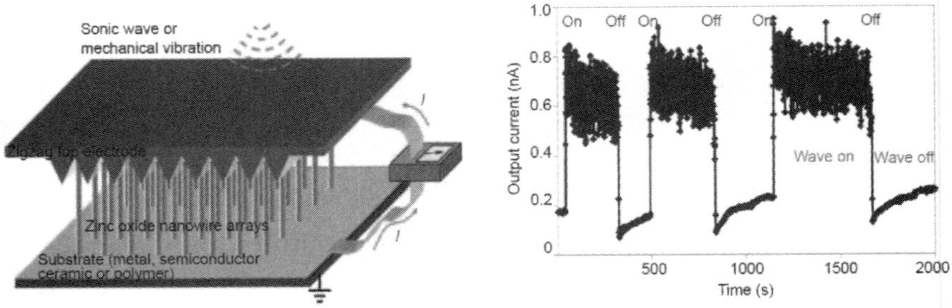

Figure 10.6 The mechanism of the nanogenerator driven by an ultrasonic wave [38].

harvester driven by ultrasonic wave, which includes four possible configurations of contact between a NW and the zigzag electrode, the current outputs. The output power of the nanogenerator fabricated with a substrate of area = 2 mm^2 is $W_{wave} = I_A V \approx 1$ pW.

Because the flexible polyester substrates hold one of the most promising form factors due to their wide degrees of freedom in shapes, they can be deployed for various nanogenerator applications, such as wallpaper and body patches [40–42]. The integration of ZnO nanowire-based energy harvesting devices into flexible polyesters or clothes was proposed for harvesting the mechanical energy from human motions [43]. The sound-driven piezoelectric device with Ag-doped ZnO nanowires generates the output power of 0.5 μW, which is almost 2.9 times that of a nanogenerator with undoped ZnO NWs. Various literature reported the fabrication of nanogenerator on different kinds of flexible substrate materials, such as Kapton [44], polythylene naphthalate (PEN) [45], and metal alloy (Phynox alloy) [46].

The low output power of ZnO nanogenerators hinders their practical applications. Different electrode architectures, including zig-zag, metal nanowire, and metal-coated ZnO nanowire electrodes, have been employed to improve the output voltages from nanowires. To scavenge effectively the energy from the human body movement, the ZnO nanogenerator integrates both piezoelectric and triboelectric effects [47]. Figure 10.7 shows ZnO NW growth and fabrication of the nanogenerator device. A low temperature hydrothermal method was used to grow a vertically aligned ZnO NW array on PET substrate. The double-sided ZnO NW substrate was sandwiched between the gold-coated ZnO nanowire substrates using polydimethylsiloxane (PDMS). This device was pressed using a 150-g weight and dried at 70°C overnight in the oven. The piezoelectric potential was generated by the strain applied to the nanogenerator such as bending and pressing. The electrostatic charge on the PDMS by the triboelectric effect is induced by the friction of the nanowires. The experimental

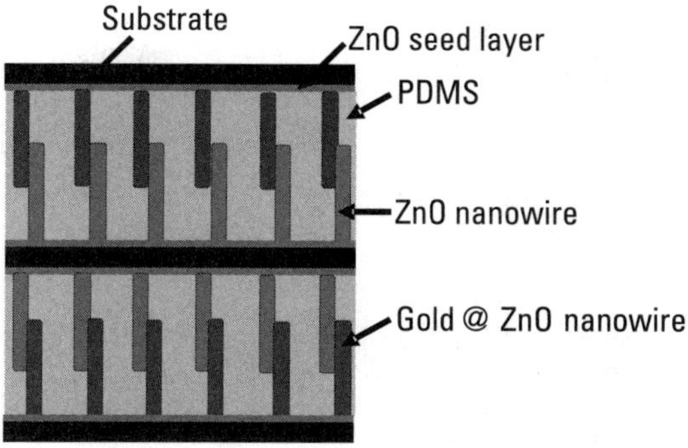

Figure 10.7 Schematic diagram of the ZnO nanowire growth and nanogenerator device fabrication process [47].

results show that the voltage and current values were 5V and 24V and 2 nA and 8 nA when stretching and pressing the fingers, respectively.

To increase the output voltage of the device, a novel design and process flow for fabricating an integrated nanogenerator based on position-controlled vertical ZnO NWs are developed to replace the Schottky contact, which prevents the current leakage through the internal structure [48]. Figure 10.8 shows the process flow of nanogenerator fabrication. The ITO layer was used not only as the bottom conductive electrode, but also the adhesion layer between the ZnO seed and the silicon substrate. PMMA layer deposited by spin-coating is applied cover the synthesized NWs. This layer as an insulating layer provides a potential barrier of infinite height, preventing the induced electrons in the electrodes from internal leaking through the ZnO/metal interface [49–51]. It replaces the Schottky contact. The second PMMA layer is used for packaging after the deposition of aluminum electrode layer. By scavenging biomechanical energy from a frog, the electric output from this device reached a record high of 58V and 134 μA, with a maximum power density of 0.78 W/cm^3.

Yang et al. [27] proposed a low-frequency, vibration-based energy harvester using ZnO cylindrical nanowires arrays (NWA) grown by the hydrothermal method. Figure 10.9(a) shows the cross-section drawing of the piezoelectric power generator, which includes the NWA deposited on silicon substrate, Cr/Au layer as the bottom electrode and silicon proof mass with the top electrode. The proof mass is deployed to scavenge kinetic energy from low-frequency vibrations, hence allowing the generator to achieve better performance. The generator is excited at two vibration resonant modes (i.e., normal mode A and shear mode B so as to maximize the output voltage generated). In normal mode A, the proof mass moves up and down along the longitudinal direction of the ZnO nanowires. In shear mode B, the direction of the proof mass movement is perpendicular to the length of the ZnO nanowires.

The total electrostatic energy stored in a ZnO nanowire is

$$E = \frac{\pi^5 \mu^2 \varepsilon_{11} Y^2 d^6}{32768 T^3} \left\{ e_{15} + \frac{\varepsilon_{11} C_{44}}{e_{15}} \right\}^{-2} = 0.0093 \mu^2 \frac{\varepsilon_{11} Y^2 d^6}{T^3} \left\{ e_{15} + \frac{\varepsilon_{11} C_{44}}{e_{15}} \right\}^{-2} \tag{10.4}$$

where μ is the coefficient of friction between the ZnO tip and silicon proof mass, $Y = C_{11} - C_{13}^2 / C_{33}$ is the Young's modulus for plane stress, C is the fourth-rank stiffness tensor, and d and T are the diameter and length of the ZnO nanowire, respectively. For ZnO, the values of C_{11}, C_{13}, and C_{33} are 210 GPa, 105 GPa, and 210 GPa, respectively. e is the third-rank stress constant tensor, and e is the second-rank electric permittivity tensor. For ZnO, $C_{44} = 42.5$ GPa, $e_{31} = -0.61$ C m^{-2}, $e_{33} = 1.14$ C m^{-2}, $e_{15} = -0.59$ C m^{-2}, and $\varepsilon_{11} = \varepsilon_{33} = 7.38 \times 10^{-11}$ F m^{-1} [52].

Figure 10.9(b, c) shows the output voltages and powers of the ZnO NWA from 10 Hz to 110 Hz at different accelerations under two modes, respectively. The open-circuit output voltage of the resonant generators at normal and shear mode are 1.1 mV and 3.59 mV, respectively. For an acceleration of 2.5g, the maximum output power obtained from normal and shear resonant modes are 0.2 pW and 2.39 pW, respectively.

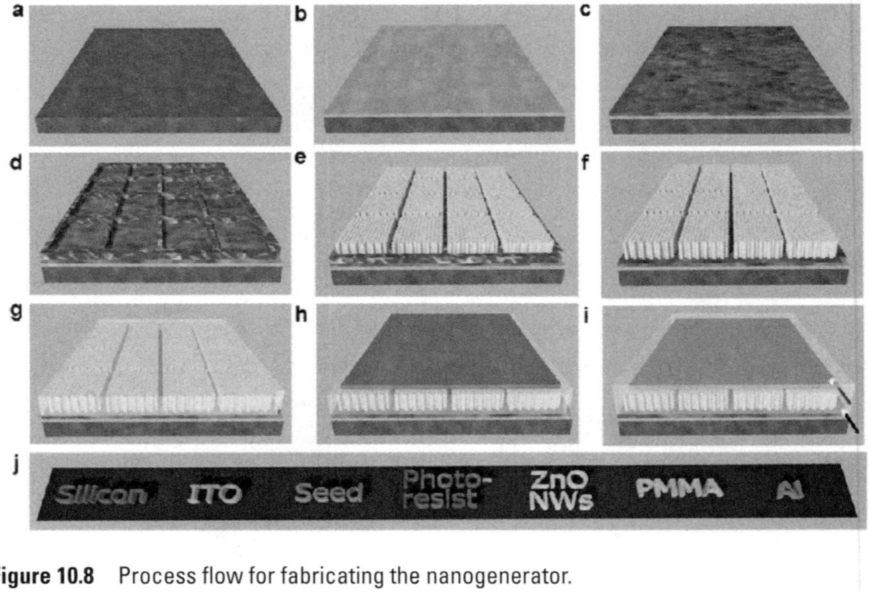

Figure 10.8 Process flow for fabricating the nanogenerator.

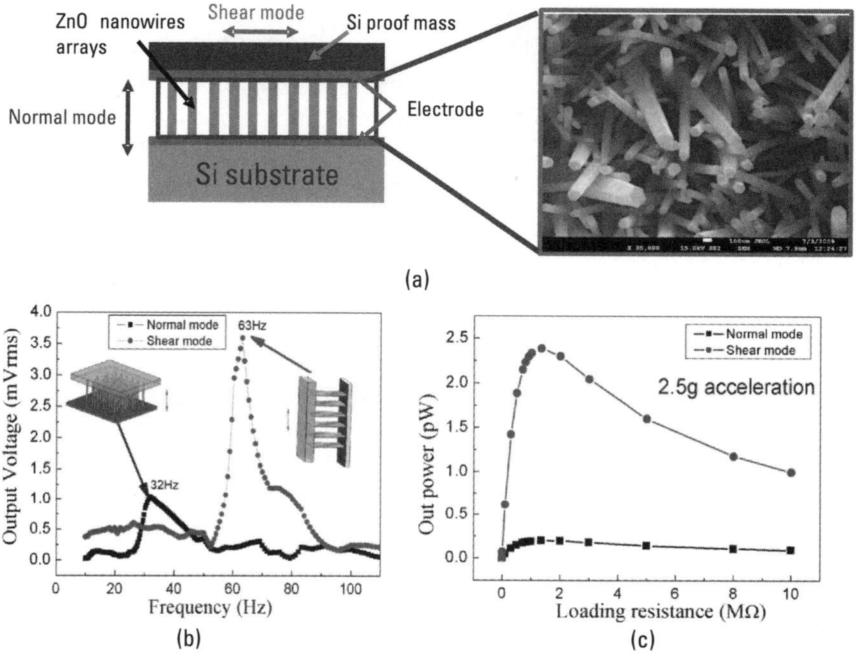

Figure 10.9 (a) Schematic diagram of the power generator which consists of ZnO nanowire arrays and Si proof mass for low-frequency vibration harvesting. (a) The harvester is excited and energy is scavenged by normal mode and shear mode; (b) the output voltages of the generator when it is excited under 2.5g acceleration during normal mode A and shear mode B; (c) the output power of the generator at 2.5g acceleration normal mde A and shear mode B.

Different circuits have been employed to convert the signals generated by the piezoelectric devices, such as rectifiers, voltage doublers [53, 54], integrated circuits (ICs) [55], and synchronized switch harvesting on inductor (SSHI) [56, 57]. Each circuit has its advantages and disadvantages. The rectifier circuits are easy to implement but inefficient. The ICs are smaller and more efficient, but less efficient than the SSHI circuits. But the SSHI circuits are difficult to implement and require a number of additional components. Heever et al. from Stellenbosch University, South Africa, investigated three different output circuit for ZnO nanogenerator, including a full-bridge rectifier, synchronized switch harvesting on inductor (SSHI) and the LTC3588 IC from Linear Technologies [58]. The full-bridge rectifier functioned extremely well under small current conditions. This circuit cannot store all the generated charge. SSHI circuit can solve this problem, but the current generated by the nanogenerators was too small to drive the SSHI circuits. The LTC3588 IC circuit can charges an input and output capacitor prior to supplying an output voltage, but it does not support continuous operations and works in a burst mode.

10.3 Organic PVDF-Based Nanofiber Harvesters

Polyvinylidene fluoride (PVDF) nanofibers have the unique good material properties in flexibility, lightweight, biocompatibility, and availability in ultralong lengths and various thicknesses and shapes. Therefore, they became an interesting candidate for energy-harvesting applications in wearable and/or implantable devices [7]. Meanwhile, PVDF has superior piezoelectric properties due to its polar crystalline structure. In nature, PVDF polymer is composed of at least five different structural forms depending on the chain conformation of trans (T) and Gauche (G) linkages. Figure 10.10 shows the crystalline structure of PVDF, including the α-phase as the most abundant form and β-phase as polar structure. β-phase has piezoelectric response due to its oriented hydrogen and fluoride (CH_2–CF_2) unit cells along with the carbon backbone, which is realized by electrical poling and mechanical stretching processes.

PVDF nanofibers fabricated by the conventional electrospinning process are under a high bias voltage (>10 kV), which could transform some nonpolar α-phase structures to polar β-phase structures for piezoelectricity [59–62].

Electrospinning can form fibers from solutions or melts and the diameters of fibers vary from tens of nanometers to micrometers. Due to high throughputs for polymers and its alloys and ceramics, it has been used to the fabrication of microdevices and nanodevices such as field effect transistors [63], sensors [64], actuators, and more [65].

The typical setup for conventional electrospinning is composed of four major components [66, 67]: a syringe pump to obtain a constant flow rate of the polymer solution, a dispense needle connected to a high voltage supply as a cathode, a high voltage power supply unit, and a collector electrode which

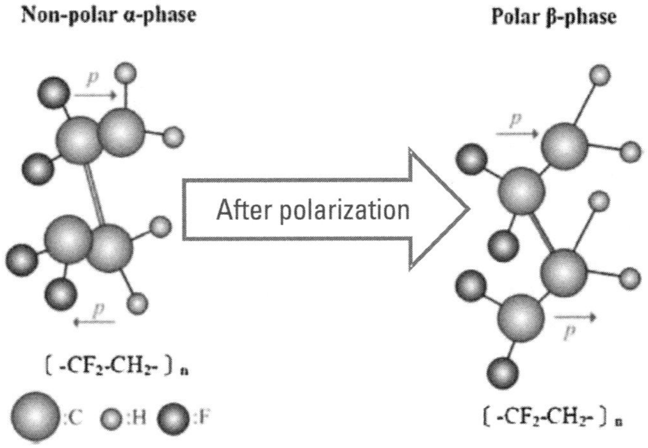

Figure 10.10 Schematic diagrams of PVDF crystalline structures, including nonpolar α- and β-phase.

collects electrospun nanofibers. When a high voltage is applied, a strong electrostatic field is excited between the needle tip and the collector electrode. The electrostatic force attracts the polymer melt out of the needle and is balanced by the surface tension force of the fluid. Hansen et al. demonstrated an energy harvesting system of a PVDF nanogenerator, which is fabricated by conventional far-field electrospinning process [68]. The 1.5-g PVDF power was dissolved in a mixture of 3 mL of DMF and 7 mL of acetone and heated at 60°C for 30 minutes. The voltage of 12 kV was applied to the syringe needle, and a feed rate of 50 μL/min was used. The electrospun fibers were collected onto two grounded copper pieces with a 2-cm gap, placed 15 cm away from the needle. Figure 10.11(a) shows the PVDF nanofiber network fabricated using a modified far-field electrospinning process. Figure 10.11(b) illustrates the integrated system with both piezoelectric PVDF nanofiber nanogenerator and a biofuel cell. Figure 10.11(c) shows the silver paste fixed at the ends of PVDF nanofibers. With a fixed strain rate of 1.67%/s, voltage and current outputs as high as 20 mV and 0.3 nA were recorded, respectively, as shown in Figure 10.11(d). The nanogenerator hybriding biochemical and biomechanical energy was also presented to drive the operation of a ZnO nanowire-based UV light sensor, as shown in Figure 10.12(b) [68]. When UV light was on the off status, the resistance of the ZnO nanowire was ~7 MΩ, and the corresponding peak voltage drop on the nanosensor was ~5 mV, as shown in Figure 10.12(a). When UV light was on the on status, the nanowire resistance dropped to ~800 kΩ and the

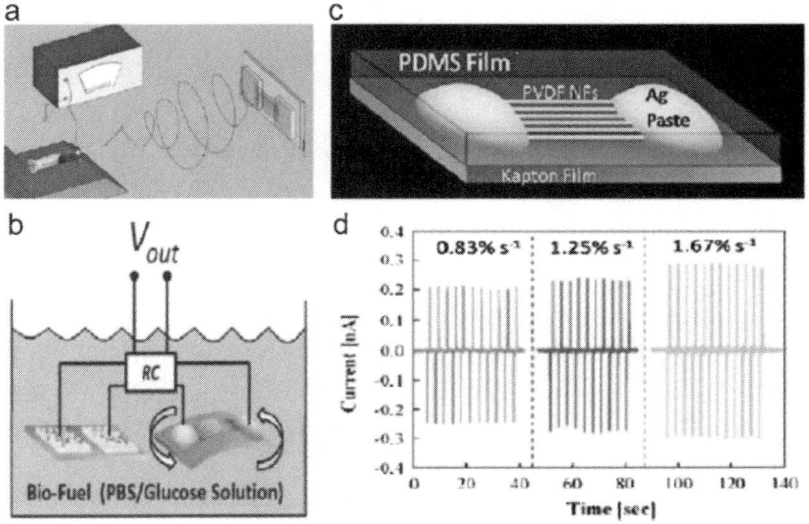

Figure 10.11 (a) Schematic diagram of the far-field electrospinning; (b) testing schematic; (c) with the PDMS and a post-electric poling process; and (d) testings results under different strain rates [68].

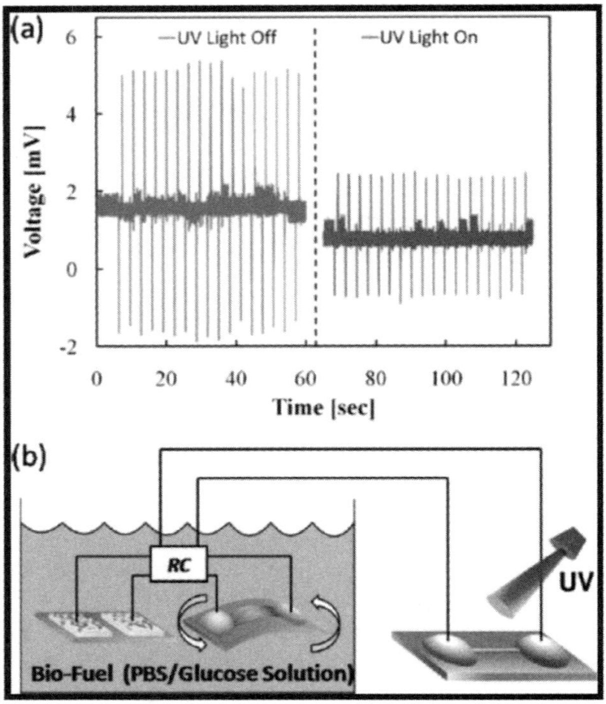

Figure 10.12 Integration of the hybrid BFC-NG device with a UV nonsensor to demonstrate a self-powered nanosystem: (a) voltage drop across the ZnO NW UV light sensor when the UV light is off and on. For illustration purposes only stabilized signals are displayed; (b) schematic illustration of the self-powered hybrid nanosystem [68].

peak voltage drop across the nanowire device was reduced to ~2.5 mV, which shows the potential of a fully self-powered nanosystem for in vivo applications.

Electrospinning processes can form PVDF fibers with simultaneous in situ mechanical stretch and electrical poling [69–71]. To improve piezoelectric properties of PVDF fibers, a direct-write electrospinning technique by means of near-field electrospinning (NFES) [72, 73] has been developed to produce orientation controllable depositions of fibers of various materials. The distance between needle and collector distance can be reduced to enhance the controllability of fiber deposition. Pu et al. from University of California, Berkeley, recently demonstrated the fabrication of PVDF fibers obtained by this method for the application of piezoelectric actuation [74]. The same research group reported nanogenerators based on electrospun PVDF nanofibers with high-energy conversion efficiency for potential wearable smart clothes to power handheld electronics through body movements [75]. The near field electrospinning (NFES) process, as shown in Figure 10.13(a), also possesses the inherent high electric field with in situ mechanical stretching for possible alignment of dipoles

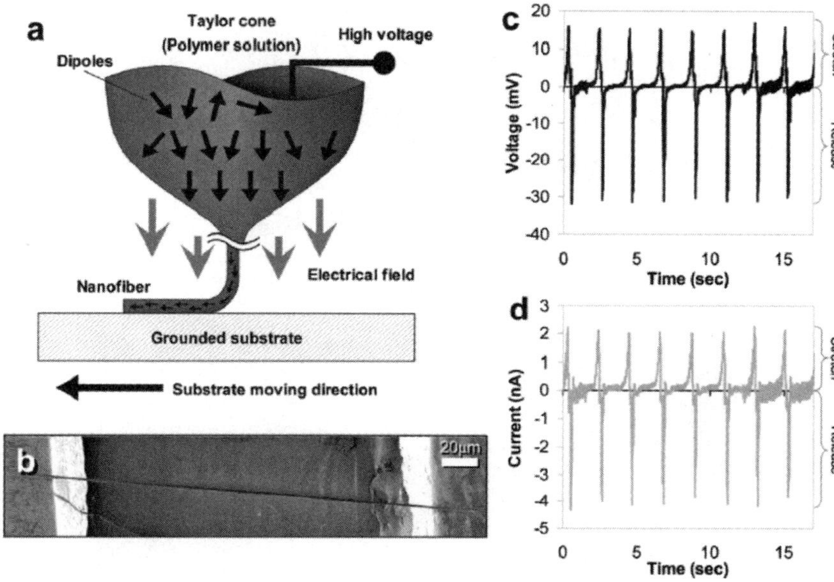

Figure 10.13 (a) Near-filed electrospinning (NFES) combining direct-write mechanical stretching and in situ electrical poling to create and place piezoelectric nanogenerators onto a substrate; (b) scanning electron microscope (SEM) image of a nanogenerator comprised of a single PVDF nanofiber, two contact electrodes, and a plastic substrate; (c) output voltage measured with respect to time under an applied strain at 2 Hz; and (d) output current measured with respect to time under applied strain at 2 Hz [69].

along the longitudinal direction of the nanofiber. The single PVDF nanofiber generates 0.5–3 nA of current and 5–30 mV of voltage under repeated long-term reliability tests without noticeable performance degradation, as shown in Figure 10.13(c, d).

To enhance the crystallinity of the β phase and piezoelectric properties, the modified carbon nanotubes (CNTs) were added into PVDF fibers [76]. Under the strong extensional force in NFES process, the unique crystalline structures of the PVDF/MWCNT were aligned in sequence along the fiber axis, which promoted the nucleation of highly oriented β-form extended-chain crystallites at the interface [77]. To improve the output power of the nanogenerators, Yu et al. attempted to increase the β-phase content and simultaneously raise the surface charge density of PVDF nanofiber mats by the addition of a suitable amount of MWCNTs to the spinning solution [78]. Figure 10.14(a) shows the schematic of the electrospinning apparatus. The generated energy in the capacitor for different concentration of MWCNTs dosages was shown in Figure 10.14(b). This nanogenerator can power a blue LED, as shown in Figure 10.14(c). Apart from CNTs, other materials, such as barium titanate

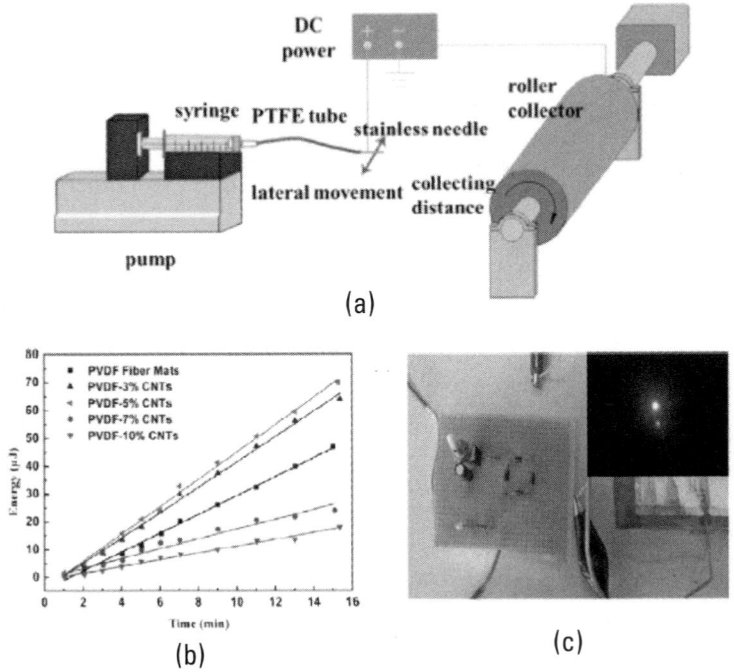

Figure 10.14 (a) Schematic of the homemade electrospinning apparatus; (b) the energy reserved in the capcitor versus charging time for different concentration of MWCNTs dosages; and (c) a digital image of the circuit for charging a capacitor and lighting a blue LED [78].

($BaTiO_3$) [79, 80], lead zirconate titanate (PZT) [81, 82], and grapheme [83], have added into PVDF fibers to improve their piezoelectric properties.

P(VDF-TrFE) generally exhibits good piezoelectric and ferroelectric behavior and a single all-transpolar crystalline phase (β-phase) that is stable at room temperature [84]. Persano et al. introduced a large area, flexible piezoelectric material that consists of sheets of electrospun fibers of the polymer poly (vinylidenefluoride-co-trifluoroethylene) [85]. Mandal et al. demonstrated a 43-μm-thick pressure type nanogenerator by far-field electrospinning of P(VDF-TrFE) with output voltage of about 400 mV under compression tests [86].

10.4 PZT Nanofiber Harvesters

PZT is a good piezoelectric material due to its higher voltage for energy-harvesting applications. As a ceramic bulk material, a thin-film PZT is more fragile in comparison to organic PVDF, but has demonstrated very good mechanical strength in nanowire form. PZT nanofibers prepared by an electrospin-

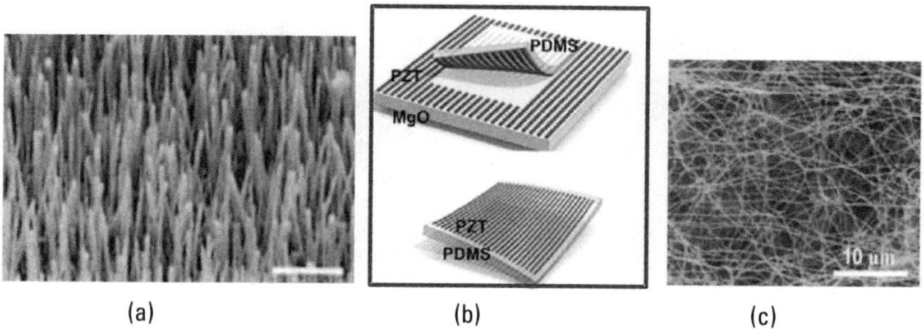

Figure 10.15 (a) SEM image of the epitaxially grown PZT nanowire arrays byt hydrothermal decomposition [88]; (b) a transfer process using lithography to define PZT ribbons and PDMS substrate to transfer the PZT fibers [89]; and (c) PZT nanofibers by electrospinning [90].

ning process exhibit an extremely high piezoelectric voltage constant (g_{33}, 0.079 Vm/N), high bending flexibility, and high mechanical strength [87]. Figure 10.15 shows the different fabrication processes of PZT nanowires, including epitaxially grown [Figure 10.15(a)] [88], transfer process [Figure 10.15(b)] [89], and electrospinning method [Figure 10.15(c)] [90]. To improve the output performance of nanoharvesters, the high piezoelectric coefficient of the fabricating material and integrated parallel and serial connection designs are two important factors. Ultralong aligned nanowires with length about 420 μm have been fabricated to obtain a high output [48]. The other effective method was stacked layers of shorter wires for nanogenerator, which can get rid of the superfluous hard substrates and keep nanowire's flexibility. Gu et al. [91] proposed a multilayer array of PZT nanowires for energy harvester to make a maximum output peak voltage of 209V. The electrospinning for the orientated nanofibers and suspending calcination techniques are used to prepare PZT nanofibers with an average crystal size of about 16 nm. The orientated PZT nanofiber film is cut and stacked layer by layer to form a multilayer film, as shown in Figure 10.16(a). Figure 10.16(f–h) shows that the synthesized VANA can be bent,

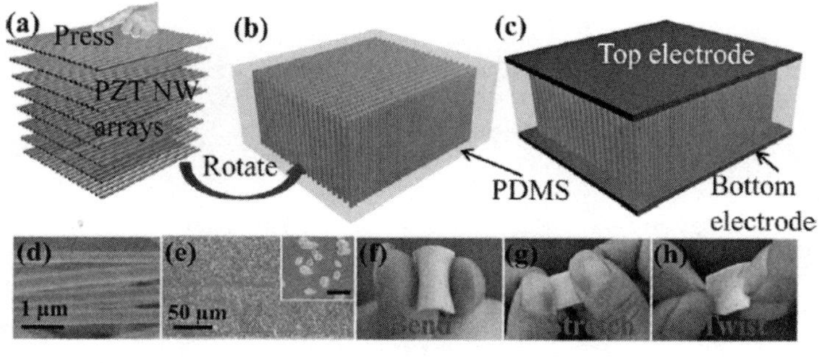

Figure 10.16 Fabrication process structure characterization of the NG.

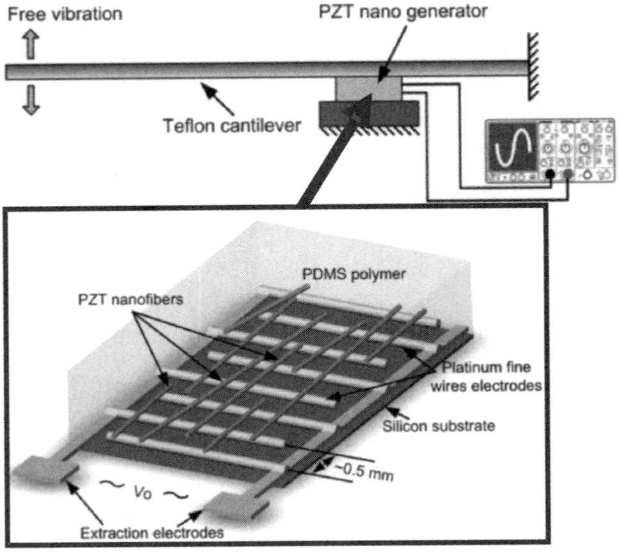

(a)

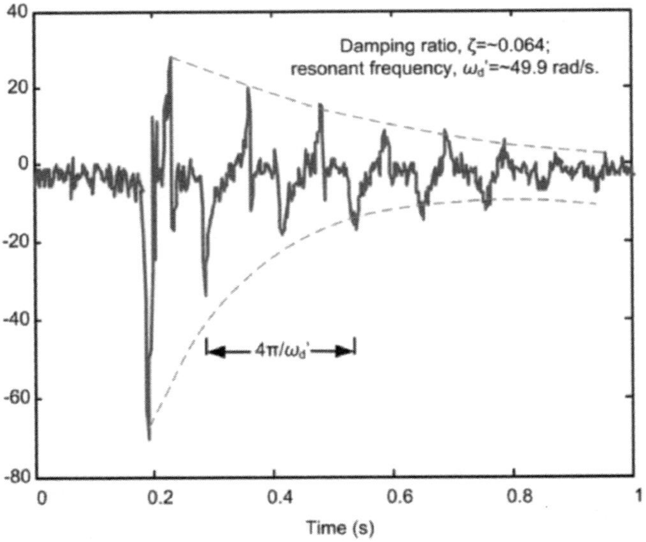

(b)

Figure 10.17 Energy harvetsed from the free vibration of a teflon cantilever: (a) schematic of the experimental setup; (b) the open circuit voltage output when the cantilever was under free vibration [90].

stretched, or twisted to a large degree without breaking its structure, which shows its potential application in flexible energy-harvesting and self-powered systems.

Because the PZT nanofibers could be woven into fabrics and made into composites, PZT nanofibers in energy-harvesting technology could make a portable, flexible, highly efficient device with a low-frequency vibration nature. Figure 10.17(a) shows the free vibration test using the PZT nanogenerator as a damper [90]. Electrospinning PZT nanofibers were deposited on the prepared interdigitated electrodes of platinum fine wire arrays. The diameters of PZT fibers were controlled to be about 60 nm by varying the concentration of polyvinylpyrrolidone (PVP) in the modified sol-gel solution. The output voltage from the nanogenerator was measured when a Teflon cantilever, placed on top of the nanogenerator, as shown in Figure 10.17(b). The measured result revealed that the amplitude of noise signal is only at about the 10-mV level. This confirmed that the power output from the PZT nanogenerator was in fact the energy harvested from mechanical vibration.

Lin et al. reviewed piezoelectric nanofibers for the application of energy harvesting, including PZT nanofibers and PVDF-based nanofibers [92]. Piezoelectric nanofibers, such as PVDF and PZT, are highly flexible and easy to fabricate for possible integration in implantable and/or flexible devices, as well as textile applications such as electric clothing. This tremendous variety in possible fields for fiber-based nanogenerators makes them a hot topic of research. Several groups demonstrated first prototypes of nanofiber nanogenerators. Further development of fiber-based nanogenerators focuses on the improvement of energy conversion efficiency.

References

[1] Yang, R. S., et al., "Power Generation with Laterally Packaged Piezoelectric Fine Wires," *Nature Nanotechnology*, Vol. 4, 2009, p. 34.

[2] Xu, S., et al., "Self-Powered Nanowire Devices," *Nature Nanotechnology*, Vol. 5, 2010, p. 366.

[3] Webster, J., "Piezoelectric Nanogenerators for Self-Powered Nanosensors and Nanosystems," in *Wiley Encyclopedia of Electrical and Electrical and Electronics Engineering*, 2012, pp. 1–20.

[4] Song, J. H., J. Zhou, and Z. Wang, "Piezoelectric and Semiconducting Coupled Power Generating Process of a Single ZnO Belt/Wire. A Technology for Harvesting Electricity from the Environment," *Nano Lett.*, Vol. 6, 2006, pp. 1656–1662.

[5] Wang, Z. L., "Energy Harvesting for Self-Powered Nanosystems," *Nano Res.*, Vol. 1, 2008, pp. 1–8.

[6] Zhu, G., et al., "Toward Large-Scale Energy Harvesting by a Nanoparticle-Enhanced Triboelectric Nanogenerator," *Nano Letters*, Vol. 13, 2013, pp. 847–853.

[7] Chang, J., et al., "Piezoelectric Nanofibers for Energy Scavenging Applications," *Nano Energy*, Vol. 1, 2012, pp. 356–371.

[8] Wang, Z. L., and J. H. Song, "Piezoelectric Nanogenerators Based on Zinc Oxide Nanowire Arrays," *Science*, Vol. 312, April 2006, pp. 242–246.

[9] Xu, S., and Z. L. Wang, "One-Dimensional ZnO Nanostructures: Solution Growth and Functional Properties," *Nano Res. Review Article*, 2011.

[10] Zeng, H., et al., "Electrochemical Deposition of ZnO Nanowire Arrays: Organization, Doping, and Properties," *Sci. Adv. Mater.*, Vol. 2010, 2, pp. 336–358.

[11] Kathalingam, A., and Jin-Koo Rhee, "Fabrication and Characterization of Solution Processed n-ZnO Nanowire/p-Si Heterojunction Device," *J. Nanosci. Nanotechnol.*, Vol. 12, 2012, pp. 6948–6954.

[12] Kathalingam, A., and J. -K. Rhee, "ZnO Nanowire Based Piezo and Photoelectric Effects Coupled Nanogenerator," *International Conference on Advanced Nanomaterials & Emerging Engineering Technologies (ICANMEET-2013)*, India, July 24–26, 2013, pp. 419–423.

[13] Wang, X. D., C. J. Summers, and Z. L. Wang, "Large-Scale Hexagonal-Patterned Growth of Aligned ZnO Nanorods for Nano-Optoelectronics and Nanosensor Arrays," *Nano Lett.*, Vol. 3, 2004, pp. 423–426.

[14] Wang, X. D., et al., "Growth of Uniformly Aligned ZnO Nanowire Heterojunction Arrays on GaN, AlN, and Al0.5Ga0.5N Substrates," *J. Am. Chem. Soc.*, Vol. 127, July 2005, pp. 7920–7923.

[15] Wang, Z. L., "ZnO Nanowire and Nanobelt Platform for Nanotechnology," *Materials Science and Engineering R*, Vol. 64, 2009, pp. 33–71.

[16] Song, J. H., et al., "Systematic Study on Experimental Conditions for Large-Scale Growth of Aligned ZnO Nanowires on Nitrides," *Phys. Chem. B*, Vol. 109, 2005, p. 9869.

[17] Wang, X. D., et al., "Density-Controlled Growth of Aligned ZnO Nanowires Sharing a Common Contact: A Simple Low-Cost, and Mask-Free Technique for Large-Scale Applications," *Phys. Chem. B*, Vol. 110, 2006, p. 7720.

[18] Gao, P. X., and Z. L. Wang, "Mesoporous Polyhedral Cages and Shells Formed by Textured Self-Assembly of ZnO Nanocrystals," *J. Am. Chem. Soc.*, Vol. 125, 2003, p. 11299.

[19] Liu, J. et al., "Toward High Output-Power Nanogenerator," *Appl. Phys. Lett.*, Vol. 92, April 2008, p. 173105.

[20] Lee, W. J., J. Kang, and K. J. Chang, "Defect Properties and p-Type Doping Efficiency in Phosphorous-Doped ZnO," *Phys. Rev. B*, Vol. 73, 2006, p. 024117.

[21] Lu, M. P., et al., "Piezoelectric Nanogenerator Using p-Type ZnO Nanowire Arrays," *Nano Lett.*, Vol. 9, No. 3, March 2009, pp. 1223–1227.

[22] Govender, K., et al., "Room-Temperature Lasing Observed from ZnO Nanocolumns Grown by Aqueous Solution Deposition," *Adv. Mater.*, Vol. 14, 2002, pp. 1221–1224.

[23] Greene, L. E., et al., "Low-Temperature Wafer-Scale Production of ZnO Nanowire Arrays," *Angew. Chem. Int. Edit.*, Vol. 42, 2003, pp. 3031–3034.

[24] Vayssieres, L., et al., "Purpose-Built Anisotropic Metal Oxide Material: 3D Highly Oriented Microrod Array of ZnO," *J. Phys. Chem. B*, Vol. 105, 2001, pp. 3350–3352.

[25] Vayssieres, L., et al., "Three-Dimensional Array of Highly Oriented Crystalline ZnO Microtubes," *Chem. Mater.*, 2001, pp. 4395–4398.

[26] Vayssieres, L., "Growth of Arrayed Nanorods and Nanowires of ZnO from Aqueous Solutions," *Adv. Mater.*, 2003, pp. 464–466.

[27] Yang, B., et al., "Modeling and Experimental Study of a Low Frequency Vibration Based Power Generator Using ZnO Nanowires Arrays," *J. Microelectromech. Syst.*, Vol. 21, 2012, pp. 776–778.

[28] Xu, S., et al., "Density-Controlled Growth of Aligned ZnO Nanowire Arrays by Seedless Chemical Approach on Smooth Surfaces," *Mater. Res.*, Vol. 23, 2008, p. 2072.

[29] Zhang, J., et al., "Shape Evolution of One-Dimensional Single-Crystalline ZnO Nanostructures in a Microemulsion System," *Cryst. Growth Des.*, Vol. 4, 2004, p. 309.

[30] Gupta, V., et al., "Optical Phonon Modes in ZnO Nanrods on Si Prepared by Pulsed Laser Deposition," *J. Cryst. Growth*, Vol. 287, 2006, p. 39.

[31] Xu, S., et al., "Patterned Growth of Vertically Aligned ZnO Nanowire Arrays on Inorganic Substrates at Low-Temperature Without Catalyst," *J. Am. Chem. Soc.*, Vol. 130, 2008, p. 14958.

[32] Qin, Y., R. S. Yang, and Z. L. Wang, "Growth by Horizontal ZnO Nanowire Arrays on Any Substrate," J. *Phys. Chem. C*, Vol. 112, 2008, p. 18734.

[33] Schubert, M. A., et al., "Finite Element Method Calculations of ZnO Nanowires for Nanogenerators," *Appl. Phys. Lett.*, Vol. 92, March 2008, p. 122904.

[34] Majidi, C., M. Haataja and D. J. Srolovitz, "Analysis and Design Principles for Shear-Mode Piezoelectric Energy Harvesting with ZnO Nanoribbons," *Smart Mater. Struct.*, Vol. 19, April 2010, p. 055027.

[35] Seok, M., et al., "The Phoenix Processor: A 30pW Platform for Sensor Applications," *IEEE Symp. on VLSI Circuits*, 2008, pp. 188–189.

[36] duToit, N. E., and B. L. Wardle, "Performance of Microfabricated Piezoelectric Vibration Energy Harvesters," *Integr. Ferroelectr.*, Vol. 83, 2007, pp. 13–23.

[37] Anton, S. R., and H. A. Sodano, "A Review of Power Harvesting Using Piezoelectric Materials (2003–2006)," *Smart Mater. Struct.*, Vol. 16, 2007, pp. R1–21.

[38] Wang, Z. L., "Energy Harvesting for Self-Powered Nanosystems," *Nano Res.*, Vol. 1, May 2008, pp. 1–8.

[39] Wang, X. D., et al., "Direct-Current Nanogenerator Driven by Ultrasonic Waves," *Science*, Vol. 316, 2007, pp. 102–105.

[40] Kim, H. J., et al., "Enhancement of Piezoelectricity via Electrostatic Effects on a Textile Platform," *Energy Environ. Sci.*, Vol. 5, 2012, p. 8932.

[41] Chang, C., et al., "Direct-Wire Piezoelectric Polymeric Nanogenerator with High Energy Conversion Efficiency," *Nano Lett.*, Vol. 10, 2010, p. 726.

[42] Lee, M. B., et al., "A Hybrid Piezoelectric Structure for Wearbale Nanogeneratos," *Adv. Mater.*, Vol. 24, 2012, p. 1759.

[43] Lee, S. H., et al., "Solution-Processed Ag-Doped ZnO Nanowires Grown on Flexible Polyester for Nanogenerator Applications," *Nanoscale*, Vol. 5, 2013, pp. 9609–9614.

[44] Zhu, G., et al., "Flexible High-Output Nanogenerator Based on Lateral ZnO Nanowire Array," *Nano Lett.*, Vol. 10, 2010, pp. 3151–3155.

[45] Kumar, B., and S. -W. Kim, "Energy Harvesting Based on Semiconducting Piezoelectric ZnO Nanostructures," *Nano Energy*, Vol. 1, 2012, pp. 342–355.

[46] Gaddam, V., et al., "A Novel Piezoelectric ZnO Nanogenerator on Flexible Metal Alloy Substrate," *IEEE Sensors Conference*, October 28-31, 2012, pp. 1–4.

[47] Saravanakumar, B., et al., "Fabrication of a ZnO Nanogenerator for Eco-Friendly Biomechanical Energy Harvesting," *RSC Advances*, Vol. 3, 2013, pp. 16646–16656.

[48] Zhu, G., et al., "Functional Electrical Stimulation by Nanogenerator with 58V Output Voltage," *Nano Letters*, Vol. 12, 2012, pp. 3086–3090.

[49] Zhu, G., et al., "Flexible High-Output Nanogenerator Based on Lateral ZnO Nanowire Array," *Nano Lett.*, Vol. 10, 2010, pp. 3151–3155.

[50] Hu, Y., et al., "Self-Powered System with Wireless Data Transmission," *Nano Lett.*, Vol. 11, 2011, pp. 2572–2577.

[51] Hu, Y., et al., "A Nanogenerator for Energy Harvesting from a Rotating Tire and Its Application as a Self-Powered Pressure/Speed Sensor," *Adv. Mater.*, Vol. 23, 2011, pp. 4068–4071.

[52] Ding, H. J., and W. Q. Chen, *Three-Dimensional Problems of Piezoelasticity*, Huntington, NY: Nova Science, 2001, p. 7.

[53] Ramadass, Y. K., and A. P. Chandrakasan, "An Efficient Piezoelectric Energy Harvesting Interface Circuit Using a Bias-Flip Rectifier and Shared Inductor," *IEEE J. Solid-State Circuits*, Vol. 45, 2010, p. 189.

[54] Ramadass, Y. K., "Energy Processing Circuits for Low-Power Applications," Ph.D. thesis, Massachusetts Institute of Technology, 2009.

[55] van Liempd, C., et al., "A 1 μW to 1 mW Energy Aware Interface IC for Piezoelectric Harvesters with 40 nA Quiescent Current and Zero-Bias Active Rectifiers," *IEEE Int. Solid-State Circuits Conf. Dig. Tech. Papers (ISSCC)*, 2013.

[56] Guyomar, D., et al., "Toward Energy Harvesting Using Active Materials and Conversion Improvement by Nonlinear Processing," *IEEE Trans. on Ultrason. Ferroelectr. Freq. Control*, Vol. 52, 2005, p. 584.

[57] Lallart, M., and D. Guyomar, "An Optimized Self-Powered Switching Circuit for Non-Linear Energy Harvesting with Low Voltage Output," *Smart Mater. Struct.*, Vol. 17, 2008, p. 035030.

[58] Heever, T. S. V. D., and W. J. Perold, "Comparing Three Different Energy Harvesting Circuits for a ZnO Nanowire Based Nanogenerator," *Smart Mater. Struct.*, Vol. 22, 2013, p. 105029.

[59] Farrar, D., et al., "Permanent Polarity and Piezoelectricity of Electrospun Helical Poly (Amino Acid) Fibers," *Adv. Mater.*, Vol. 23, 2011, pp. 3954–3958.

[60] Wang, Y. R., et al., "A Flexible Piezoelectric Force Sensor Based on PVDF Fabrics," *Smart Mater. Struct.*, Vol. 20, 2011, p. 045009.

[61] Zheng, J., et al., "Polymorphism Control of Poly(Vinylidene Fluoride) Through Electrospinning," *Macromol. Rapid Commun.*, Vol. 28, 2007, pp. 2159–2162.

[62] Ribeiro, C., et al., "Influence of Processing Conditions on Polymorphism and Nanofiber Morphology of Electroactive Poly(Vinylidene Fluoride) Electrospun Membranes," *Soft Mater.*, Vol. 8, 2010, pp. 274–287.

[63] Gonzalez, R., and N. Pinto, "Electrospun Poly(3-Hexylthiophene-2,5-Diyl) Fiber Field Effect Transistor," *Synthetic Metals*, Vol. 151, 2005, pp. 275–278.

[64] Ding, B., et al., "Electrospun Nanomaterials for Ultrasensitive Sensors," Materials Today, Vol. 13, 2010, pp. 16–27.

[65] Corres, J. M., et al., "Optical Fiber Humidity Sensors Using PVdF Electrospun Networks," *Sensors Journal*, Vol. 11, 2011, pp. 2383–2387.

[66] Yarin, A. L., S. Koombhongse, and D. H. Reneker, "Taylor Cone and Jetting Liquid Droplets in Electrospinning of Nanofibers," *Journal of Applied Physics*, Vol. 90, 2001, p. 4836.

[67] Reneker, D. H., et al., "Bending Instability of Electrically Charged Liquid Jets of Polymer Solutions in Electrospinning," *Journal of Applied Physics*, Vol. 87, 2000, p. 4531.

[68] Hansen, B. J., et al., "Hybrid Nanogenerator for Concurrently Harvesting Biomechanical and Biochemical Energy," *ACS Nano*, Vol. 4, 2010, pp. 3647–3652.

[69] Yee, W. A., et al., "Morphology, Polymorphism Behavior and Molecular Orientation of Electrospun Poly(Vinylidene Fluoride) Fibers," *Polymer*, Vol. 48, 2007, pp. 512–521.

[70] Zhao, Z., et al., "Preparation and Properties of Electrospun Poly(Vinylidene Fluoride) Membranes," *J. Appl. Polym. Sci.*, Vol. 97, 2005, pp. 466–474.

[71] Ren, X., and Y. Dzenis, "Novel Continuous Poly(Vinylidene Fluoride) Nanofibers," *Mater. Res. Soc. Symp. Proc.*, Vol. 920, 2006, pp. 55–61.

[72] Sun, D., et al., "Near-Field Electrospinning," *Nanoletters*, Vol. 6, 2006, pp. 839–842.

[73] Chang, C., K. Limkrailassiri, and L. Lin, "Continuous Near-Field Electrospinning for Large Area Deposition of Orderly Nanofiber Patterns," *Appl. Phys. Lett.*, Vol. 93, 2008, p. 123111.

[74] Pu, J., et al., "Piezoelectric Actuation of Direct-Write Electrospun Fibers," *Sens. Actuators A*, Vol. 164, 2010, pp. 131–136.

[75] Chang, C., et al., "Direct-Write Piezoelectric Polymeric Nanogenerator with High Energy Conversion Efficiency," Nanoletters, Vol. 10, 2010, pp. 726–731.

[76] Zhang, Y. J., Z. D. Zhao, and W. X. Yu, "Preparation and Characterizations of PVDF/MWCNT Nanocomposites," *Polymer Materials Science and Engineering*, Vol. 26, No. 6, 2010, pp. 141–144.

[77] Liu, Z. H., et al., "Piezoelectric Properties of PVDF/MWCNT Nanofiber Using Near-Field Electrospinning," *Sensors and Actuators A: Physical*, Vol. 193, 2013, pp. 13–24.

[78] Yu, H., et al., "Enhanced Power Output of an Electrospun PVDF/MWCNTs-Based Nanogenerator by Tuning Its Conductivity," *Nanotechnology*, Vol. 24, 2013, p. 405401.

[79] Banno, H., and K. Ogura, "Dielectric and Piezoelectric Properties of a Flexible Composite Consisting of Polymer and Mixed Ceramic Powder of PZT and PbTiO3," *Ferroelectrics*, Vol. 95, 1989, pp. 171–174.

[80] Chan, H. L. W., M. C. Cheung, and C. L. Choy, "Study on BaTiO3/P(VDF–TRFE) 0–3 Composite," *Ferroelectrics*, Vol. 224, 1999, pp. 541–548.

[81] Vatansever, D., et al., "An Investigation of Energy Harvesting from Renewable Sources with PVDF and PZT," *Smart Mater. Struct.*, Vol. 20, 2011, p. 055019.

[82] Wei, B., and Y. Daben, "Dielectric and Piezoelectric Properties of 0–3 Composite Film in PCM/PVDF and PZT/PVDF," *Ferroelectrics*, Vol. 157, 1994, pp. 427–430.

[83] Rahman, M. A., et al., "Fabrication and Characterization of Highly Efficient Flexible Energy Harvesters Using PVDF-Graphene Nanocomposites," *Smart Mater. Struct.*, Vol. 22, 2013, p. 085017.

[84] Jin, X. Y., K. J. Kim, and H. S. Lee, "Grazing Incidence Reflection Absorption Fourier Transform Infrared (GIRA-FTIR) Spectroscopic Studies on the Ferroelectric Behavior of Poly(Vinylidene Fluoride–Trifluoroethylene) Ultrathin Films," *Polymer*, Vol. 46, 2005, pp. 12410–12415.

[85] Persano, L., et al., "High Performance Piezoelectric Devices Based on Aligned Arrays of Nanofibers of Poly(Vinylidenefluoride-Co-Trifluoroethylene)," *Nature Communications*, Vol. 4, 2013, p. 1633.

[86] Mandal, D., S. Yoon, and K. J. Kim, "Origin of Piezoelectricity in an Electrospun Poly(Vinylidene Fluoride-Trifluoroethylene) Nanofiber Web-Based Nanogenerator and Nano-Pressure Sensor," *Macromol. Rapid Commun.*, Vol. 32, 2011, pp. 831–837.

[87] Chen, X., et al., "Potential Measurement from a Single Lead Ziroconate Titanate Nanofiber Using a Nanomanipulator," *Applied Physics Letters*, Vol. 94, 2009, p. 253113.

[88] Xu, S., B. J. Hansen, and Z. L. Wang, "Piezoelectric-Nanowire-Enabled Power Source for Driving Wireless Microelectronics," *Nature Communications*, Vol. 1, 2010, p. 93.

[89] Qi, Y., et al., "Piezoelectric Ribbons Printed onto Rubber for Flexible Energy Conversion," *Nano Lett.*, Vol. 10, 2010, pp. 524–528.

[90] Chen, X., et al., "1.6 V Nanogenerator for Mechanical Energy Harvesting Using PZT Nanofibers," *Nano Lett.*, Vol. 10, 2010, pp. 2133–2137.

[91] Gu, L., et al., "Flexible Fiber Nanogenerator with 209V Output Voltage Directly Powers a Light-Emitting Diode," *Nano Lett.*, Vol. 13, 2013, pp. 91–94.

[92] Chang, J., et al., "Piezoelectric Nanofibers for Energy Scavenging Applications," *Nano Energy*, Vol. 1, 2012, pp. 356–371.

11

Applications of Energy Harvesters

With the development of extremely low-power devices for wireless sensor networks, energy harvesters have the possibility of powering these devices in a wide array of potential applications. This chapter demonstrates some applications and requirements of energy harvesters for powering the implantable medical or wearable devices, tire pressure sensors in automobiles, and wireless sensors in structural health monitoring. Meanwhile, some application examples are also shown for the implement of energy-harvesting devices.

11.1 Bio-MEMS Applications

Implantable biotechnology has undergone rapid development during the past two decades. Some implantable devices either have become realities today or are currently being investigated [1]. Considering the feasibility of bio-implantable applications and the lifetime of many implantable devices, the methods of powering these devices have become the research hotspot. Normally power is obtained from either implanted batteries or from an external transcutaneous power source. Implanted batteries with an electronic device need to be replaced when they are depleted. Replacement of implanted batteries requires frequent, costly surgeries with increased risk of complications. Documented clinical experience with pacemakers and implanted defibrillators highlight the limitations of implanted batteries [2]. The mean time to when pacemaker battery replacement is needed is 8 years after the initial implantation [3]. Depletion of implanted batteries is the reason for more than 70% of pacemaker replacement surgeries and the complication rate after a replacement surgery is three times greater than that for the initial implant [4]. For implanted defibrillators, patients require battery replacement surgery 3 to 4 years after initial implantation with costs up to $10,000 [5]. Currently there are two methods to solve this problem, namely either the longevity of the power source or the ability to recharge the batteries [6]. One of the earliest and most-cited studies about scavenging biomechanical sources was by Starner at the MIT Media Laboratory [7]. Figure 11.1 shows a summary of output power from body-driven generation methods. It is seen that 67W of power is available for a 68-kg man walking using the fall of the heel through a 5-cm distance. This result is promising given the relatively large amount of available power compare to other biomechanical sources. A viable wearable computer was made with a PIC16C71 processor from Microchip Technology. This chip requires only 18 mW at 4 MHz and 0.1 mW at 32 kHz [8]. It is feasible to power these wearable devices by scavenging the biomechanical energy.

The energy harvesters for biomedical applications should meet the requirements of low-frequency, high response for small input and small volume. The operating frequency of body and organ motion normally locates the range from a few hertz up to a few tens of hertz, which would be the design goal of the optimized resonating frequencies of the energy harvesters [9]. Second, compared to some machine-mounted applications, a significant variation of the motion amplitude and frequency possibly exists, so the energy harvesters have a high response for wide range of input signals by structural design or dynamically tunable mechanisms. Finally, size should be unobtrusive for implanted applications, with a volume of 1 cm^3 being a reasonable limit [10]. Buren et al. [9] investigated and optimized the inertial micropower generators for human walking motion. The simulated results show that the maximum power densities

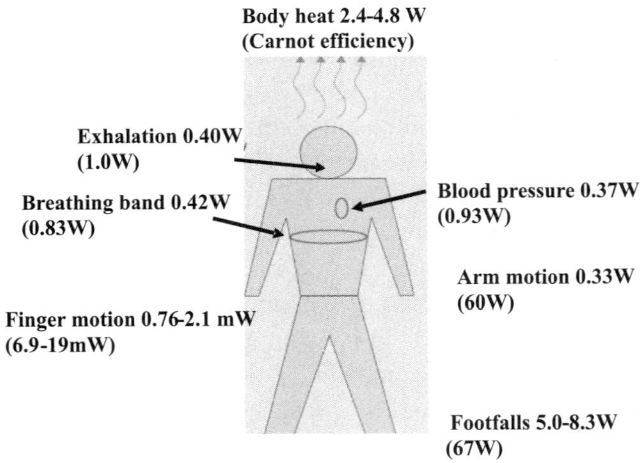

Figure 11.1 Power from body-driven sources [7].

for human powered motion vary between 8.7 and 2,100 μW/cm^3, depending on generator size and the location of the body on which it is mounted. The electrostatic energy harvester was fabricated using a three-wafer construction [10]. The central wafer contains a silicon proof mass, polyimide suspension, and metalization for electrical contact. The proof mass separated from the frame by deep reactive ion etching (DRIE) is about 0.12g and measures about 11 mm × 11 mm × 0.4 mm thick, as shown in Figure 11.2(a). The output power can be calculated as

$$P \approx \frac{1}{160} A_0 \omega \rho \alpha a^4 \tag{11.1}$$

where A_0 is the maximum external acceleration, ω is the frequency, ρ is the proof mass density, α is an aspect ratio <1, and $a \times a$ is the flat square of area. When 1g is typical of body motion, and an aspect ratio α = 0.2, the curve between the device size and output power is shown in Figure 11.2(b). For the device with a = 11 mm, for operation at 30 Hz, about 80 μW should be obtained, or 2.6 μJ/cycle.

Thin PVDF film in shell structure was used in energy harvester because of its flexibility in the form of polymer thin film [11]. The shell structure was obtained by attaching the PVDF film onto the precurved polyester thin film. When this structure is applies to wearable energy harvester, it is easily deformed and provides excellent wearable comfort. Figure 11.3 shows the photograph of a fabric with embedded piezoelectric shell structure. The maximum output voltage of general elbow motion is obtained around 25V. The proposed shell struc-

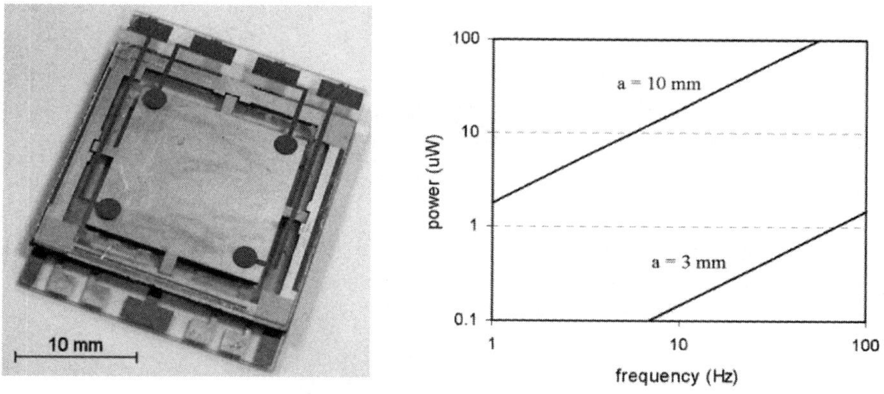

(a) (b)

Figure 11.2 (a) Prototype energy harvester; and (b) maximum output power versus frequency [10].

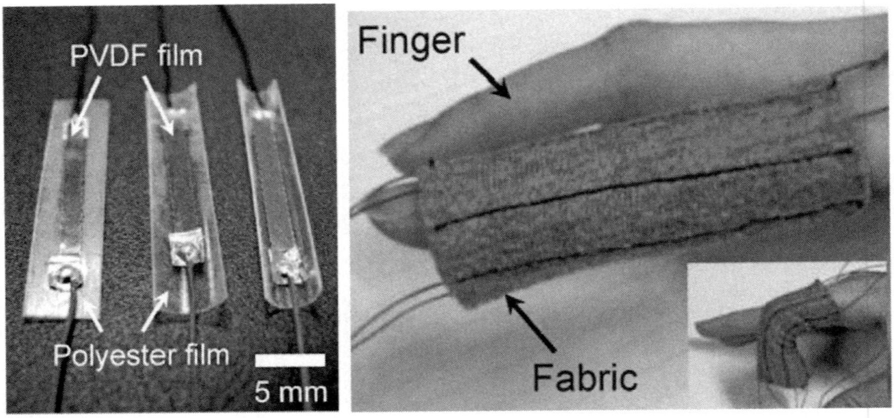

Figure 11.3 Fabric with embedded wearable piezoelectric shell structure [11].

ture device has some advantages such as high output power, a simple fabrication process, a low cost, a light weight, and a good wearing sensation.

Most of implanted energy harvesters scavenge the vibration energy from bone strain [12, 13], the motion of respiration [14] and knee [15], and heart contractions [16, 17]. Thermoelectric energy harvester driven by body heat was proposed in the biomedical application [18]. A thermoelectric harvester based on a surface micromachined poly-SiGe thermopile was realized by Wang et al. from IMEC in Belgium [19]. The fabricated thermopile chips with 4 mm × 4 mm were shown in Figure 11.4(a). Each chip contains either 2,350 or 4,700 thermocouples connected electrically in series. The thermopile chip was attached to a specially shaped metal plate by using thermally epoxy, as displayed in Figure 11.4(b). Next, a top radiator featured with an array of pins is fixed onto the metal spacer and the pillars, as shown in Figure 11.4(c). Because the internal electrical resistance was too high (i.e., about 20 ΩM), the output power was estimated to be about 0.3 nW at the optimized loading conditions. The power generated from these harvesters have been in the microwatt or submicrowatt range and the potential applications have been small-power consuming devices, such as telemetry of sensor data, drug pump delivery systems and pacemakers [2].

Currently, the power supply uses conventional thin-film batteries and power capacitors in drug delivery systems [20]. However, they are difficult to recharge within the body and the systems cannot work due to their short time. Allameh et al. from Princeton University have proposed using a piezoelectric energy harvester to recharge and/or power drug delivery systems by scavenging mechanical energy from moving fluids or solids within the body [21]. For example, the piezoelectric energy harvesters may be inserted between the coping of a dental crown and the metal abutment. The mechanical energy is provided

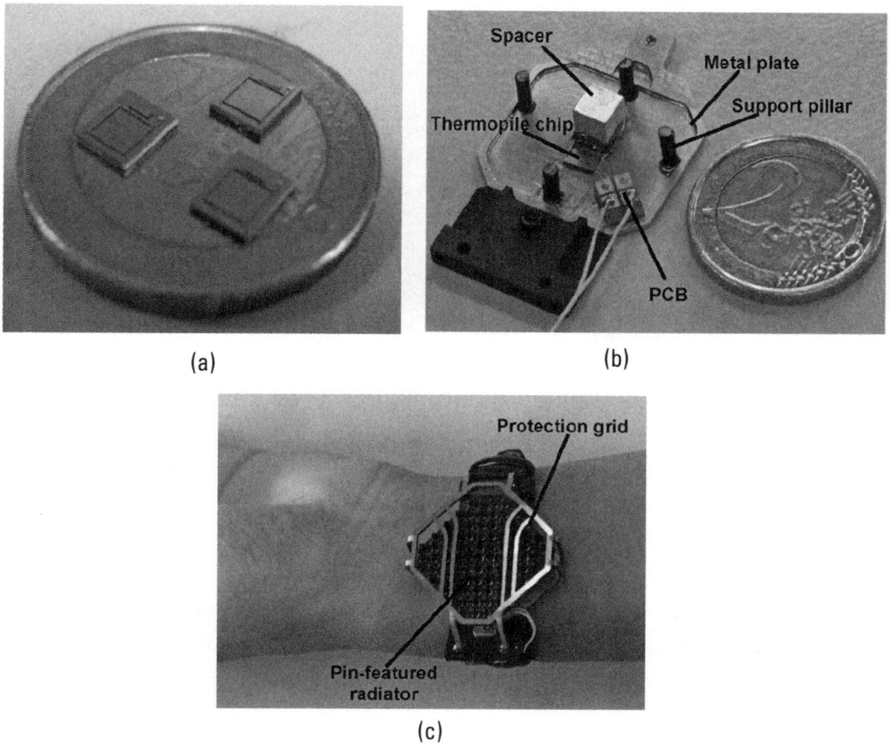

Figure 11.4 Photos of the wearable harvester for human body applications: (a) photograph of micromachined thermopile chips placed on one Euro coin; (b) TEG being assembled; and (c) TEG being worn on human body [19].

annually by 1 million contact cycles. The calculation results show that the kinetic energy from moving blood is insufficient to generate the required power levels in piezoelectric thin film generators with appropriate sizes for bio-MEMS structures [22]. Therefore, methods of enhancing this low energy output will become a research hotspot. Proper structural and electronic circuit designs contribute to improve the energy output.

11.2 Tire Pressure Monitoring in Automobiles

Over the last decade, many sensors for the application of automotive industry have been rapidly developed. The National Highway Traffic Safety Administration (NHTSA) has issued a mandatory regulation for tire pressure monitoring system (TPMS) [23]. Since September 1, 2007, all light vehicles sold in the United States must be equipped with a TPMS as a result of the TREAD Act. TPMS modules available on the market are usually powered by coin cell batteries. However, self-powered sensor devices have been paid more attention in the automotive industry due to the limitations of the available power and inaccessibility-related maintenance costs [24–26]. There are at least three motivations to replace the battery [25]: (1) 10 years is not sufficient in some cases, (2) car makers are concerned with warranty costs if batteries do not last as long as expected, and (3) in some usage scenarios, such as in the trucking industry, batteries last fewer than 10 years. Because rotating wheels can provide significant vibration amplitude, vibration energy harvesters may be a potential energy source for TPMS. When a vibration energy harvester mounted on the inside of a car tire at the inner liner position, which is subject to acceleration in three directions, this harvester can scavenge the vibration energy from different directions and convert to electrical energy due to the oscillation of the proof mass relative to the frame. For TPMS, the main vibration source is generated by gravity during rotation of wheels from the alternating tangential acceleration. This vibration frequency usually varies from 1 Hz to about 20 Hz, much lower than resonant frequencies of typical miniature energy harvesters. Masaki et al. developed a compact electrotret energy harvester for wireless sensor applications [27]. Figure 11.5(a) shows an external view of an actual lightweight harvester. Its size and weight are, respectively, $20 \times 20 \times 4$ mm^3 and 3.7g. The internal structure is composed of an electricity-generating part and a mechanical part fabricated by the MEMS process and a high-precision processing technology, respectively, as shown in Figure 11.5(b). Because these two parts are separated, and the vibrating stress does not apply to the electricity-generating part, high reliability is obtained. Meanwhile, the vibrating stress can be reduced by lightening the electricity part with a simple mechanism. Figure 11.5(c, d) are photographs of the CYTOP electrets substrate and of the collector substrate, respectively. The CYTOP film can be deposited by spin coating process, and its thickness could be about 15 μm by multiple spin-on. The charging was processed to these electret electrodes, and a surface potential of about –700V was charged [28]. The output power of this device was enhanced from 40 to 100 μW at an acceleration of 0.15g and a low resonance frequency of 30 Hz.

When automobiles accelerate or decelerate, the frequency of tires varies with time. In order to collect the vibration energy effectively at varied resonant frequencies, some wideband energy harvesters have been proposed, such as tunable resonance [29], cantilever arrays [30, 31], and nonresonant [32, 33] and frequency upconversion [34–36] mechanisms. Some examples are demonstrated in Chapter 5. For frequency-up mechanism energy harvesters, the mechanical contact for driving between the moving part and stoppers inevitably caused wear and energy loss. To address this problem, a noncontact frequency upconversion piezoelectric harvester based on magnetic repulsive force has been demonstrated [37, 38]. Figure 11.6(a) schematically shows the structure of the frequency upconversion harvester. The central sliding mass was constrained within an in-wall lubricated tube. Permanent magnets were fixed on the central

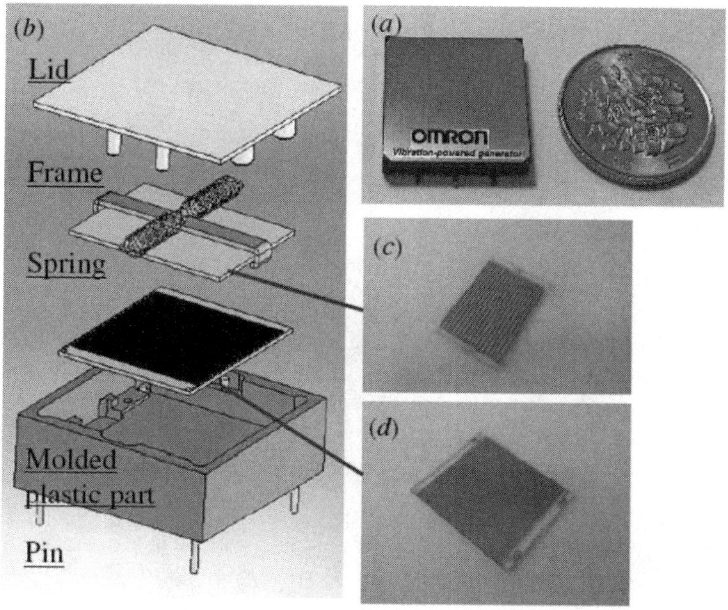

Figure 11.5 Photographs and schematic of electret harvester: (a) appearance, (b) schematic of internal structures, (c) electrets substrate with patterned electret electrodes, and (d) collector substrate in the electret harvester [27].

mass and both ends of the tube, forming a nonlinear oscillator. A pair of piezoelectric microcantilevers was placed at an end magnet on each. The central mass was made from a cylindrical NdFeB magnet and an additional Tungsten mass. The microcantilevers were composed of a 30-μm-thick polyvinylidene fluoride (PVDF) piezoelectric film and a 50-μm-thick copper foil with cyanoacrylate adhesive. The testing results demonstrated that over 10 μW of average power was generated within the frequency range of 10 Hz to 22 Hz at 1g acceleration.

In addition to the broadband energy harvester concept, the reliability of the harvester and the sensor system mounted in the tire is the main challenge due to withstanding very high accelerations. The maximum acceleration peaks of 2,900g must be taken into account for the design to fulfill the lifetime and automotive quality requirements [39]. A MEMS-based piezoelectric energy harvester for TPMS has developed for noise and shock excitations [40]. Figure 11.7(a) shows the high acceleration shocks generated in tires. The radial acceleration drops to a low value when the tire contacts the road (contact patch). For 60 km/h one observes a shock with amplitude of ~120g with a duration of ~10 ms. The measurement of the acceleration in the tire was performed with an acceleration measurement module of a three-axis accelerometer, as shown in Figure 11.7(b). The module measures accelerations up to 2,000g with a sample rate of 2 kHz. The data can be transmitted with a 2.4-GHz wireless link or

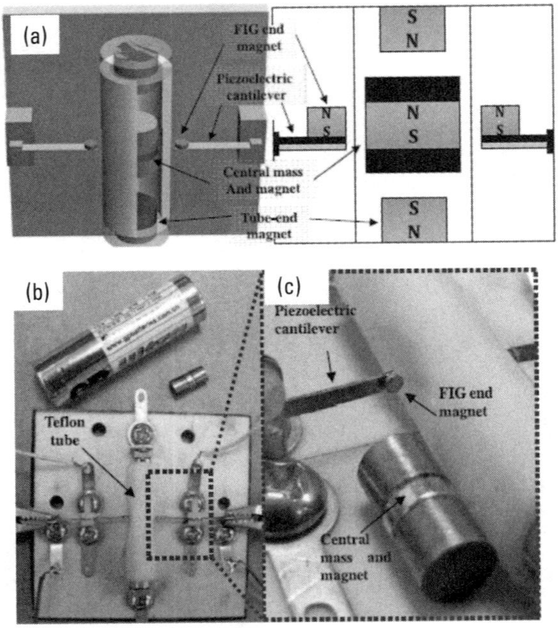

Figure 11.6 (a) The 3-D schematic (left) and structural layout (right) of the frequency upconversion piezoelectric energy harvester. (b) Fabricated harvester with its size compared with an AA battery. (c) Close-up view of the frequency increased harvester part [38].

stored in an 8-Gb small memory card. To obtain ultralow losses in the rectification of the output voltage of the (piezoelectric) vibration harvester, an optimized IMEC/Holst Centre active rectifier circuit has been added in the power management circuitry [41]. The average power consumption of this system is ~7 μW for pressure and temperature transmission every 30 seconds. By optimizing the system power consumption, the power generation of the piezoelectric energy harvester will make autonomy possible in a real TPMS product.

Apart from the reliability problem, the weight, size, and integration issues maybe are evaluated for TPMS applications. For the development of a MEMS device, a later integration of the harvester with the TPMS sensor will be allowed.

11.3 Structural Health Monitoring

Structural health monitoring (SHM) is the process of detecting damage in structures. The goal of SHM is to improve the safety and reliability of aerospace, civil, and mechanical infrastructure by detecting damage before it reaches a critical state [42]. Discenzo et al. [43] developed a prototype self-powered

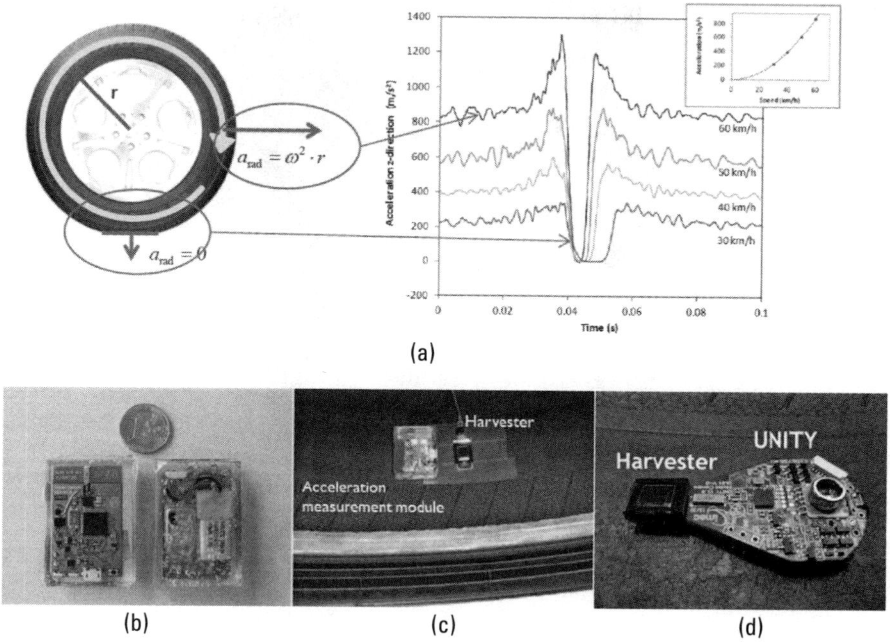

Figure 11.7 (a) The accelerations at the different locations of inner tire. (b) Acceleration measurement module with three-axis accelerometer with the possibility of wireless communication and data storage on small memory card. The power is delivered by a small Li-ion battery. (c) Mounting of the acceleration module in the tire together with a piezo-electric energy harvester. The harvester is not used during the measurements as power source; only the generated voltage is measured. (d) In-house developed low-power TPMS module (Unity) operating autonomous with the mounted piezo-electric energy harvester [40].

sensor node that performs sensing and local processing and telemeters for pump condition monitoring applications, shown in Figure 11.8. A wireless mote system was integrated with a piezoelectric energy-harvesting technique. This piezoelectric harvester was mounted to the bottom of the enclosure and worked at a nominal rotational frequency of 7,800 cps or 130 Hz. The generated power of this device was recorded interval by the mote as the rectified voltage. The maximum power output of 40 mW was obtained. However, the output power was only 10% of the amount expected under these operating conditions due to the slight discrepancy between the excitation frequency and the piezoelectric device's resonant frequency. The authors suggested the development of an adaptive-tuning harvester to accommodate uncertainties in the environment and in equipment operation. Moreover, the harvesters should be optimized for improving the reliability to operate in harsh, dynamic environment. For their application, the minimum setup and installation effort would be required.

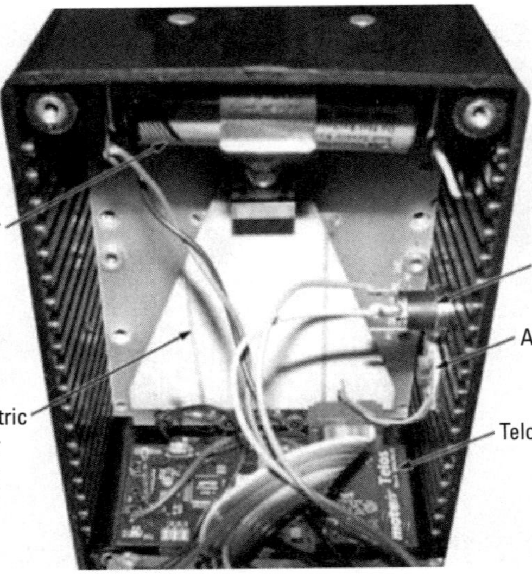

Figure 11.8 Self-powered sensor node [43].

For aerospace application, the implementation of SHM system at the design term could optimize the structures in safety critical areas which would reduce weight [44]. It would reduce fuel consumption and broad maximum range. As a result, the aircraft performance would be improved and the running costs of an aircraft would be reduced. There are two available energy sources, namely vibration and thermal energy, in the aircraft environment. Research has shown that power levels in the microwatt range can be obtained by piezoelectric and electromagnetic devices and in the milliwatt range by thermal electric generators (TEG) [45]. Vullers et al. at Cardiff University developed a series of high-tech, thin-film TEGs including the MPG-D751, which was selected for the investigation into the feasibility of powering an SHM system [46]. Figure 11.9 shows the locations of thermocouples to simulate output voltage of an individual TEG. It is possible to derive the output power for a particular temperature scenario when the internal resistance of the TEG is known. A novel TEG design for aircraft applications was able to generate the output voltage of 3.3V for a temperature gradient of 20°C [47]. The efficiency of the complete system can be increased by about 50% by increasing storage voltage and conversion efficiency.

U.S.-based MicroGen Systems, a spin-off from the University of Vermont, is a commercial company about developing energy-harvesting technologies. The MEMS-based device features an AlN piezoelectric microcantilever with a single end-proof mass. External vibration force bends the cantilever up and down, producing ac electricity. MicroGen's recently announced BOLT

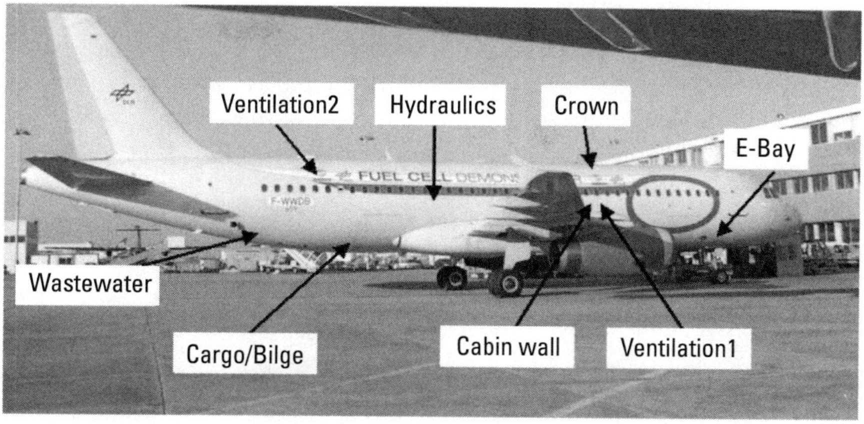

Figure 11.9 Locations of thermocouples used for the TEG power output simulation [46].

family (BOLT-M0600X shown in Figure 11.10) of micropower harvester enables an output of 25 to 500 μW under the resonant frequency of 600 Hz at 3.3 V_{DC}, when a 300-μF storage capacitor is connected. The output power of the harvesters is depended on configuration and frequency. The harvesters are specifically designed for the low-level operation and can be targeted for fixed-frequency industrial, structural monitoring, and commercial applications. For example, power cells can operate at harmonics of 50/60 Hz in the 100 to 1,500-Hz range. This device can be fabricated in volume on semiconductor production lines.

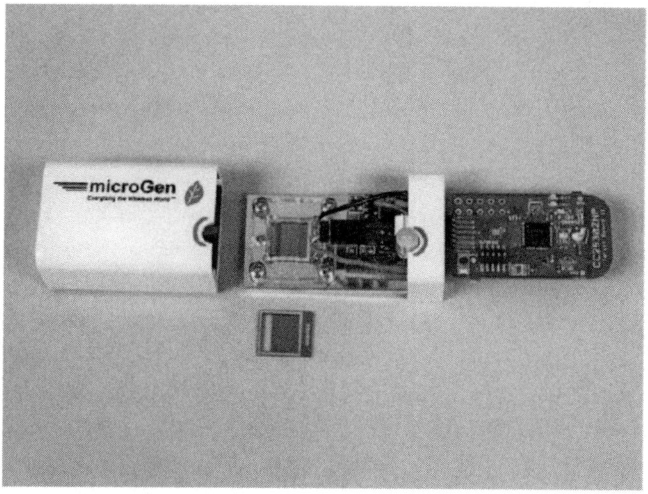

Figure 11.10 MicroGen BOLT-M0600X energy harvester [48].

Currently, in reality these energy harvesting devices would not be able to directly power an SHM system due to the varying nature of the sources. Therefore a further power management system would be developed to generate a stable power source from the dynamic nature of the energy sources for powering an SHM system [46].

References

[1] Pelletier, F., "Inner Space," *MPT Medicare Technology*, Wicklow, 2004, pp. 32–34.

[2] Lewandowski, B., "An Implantable, Stimulated Muscle Powered Piezoelectric Generator," Ph.D. dissertation, 2009.

[3] Kindermann, M., et al., "Longevity of Dual Chamber Pacemakers: Device and Patient Related Determinants," *Pacing Clin. Electrophysiol.*, Vol. 24, No. 5, May 2001, pp. 810–815.

[4] Deharo, J. C., and P. Djiane, "Pacemaker Longevity. Replacement of the Device," *Ann. Cardiol. Angeiol. (Paris)*, Vol. 54, No. 1, January 2005, pp. 26–31.

[5] Vorperian, V. R., S. Lawrence, and K. Chlebowski, "Replacing Abdominally Implanted Defibrillators: Effect of Procedure Setting on Cost," *Pacing Clin. Electrophysiol.*, Vol. 22, No. 5, May 1999, pp. 698–705.

[6] Soykan, O., "Power Sources for Implantable Medical Devices," *Business Briefing: Medical Device Manufacturing & Technology*, 2002, pp. 76–79.

[7] Starner, T., "Human-Powered Wearable Computing," *IBM Systems Journal*, Vol. 35, 1996, p. 618.

[8] Preliminary data sheet on PIC 16C71, Microchip Technology, 1992. http://ww1.microchip.com/downloads/en/DeviceDoc/30272a.pdf.

[9] Buren, T. V., et al., "Optimization of Inertial Micropower Generators for Human Walking Motion," *IEEE Sensors Journal*, Vol. 16, No. 1, 2006, pp. 28–38.

[10] Miao, P., et al., "MEMS Inertial Power Generators for Biomedical Applications," *Microsystem Technologies*, Vol. 12, 2006, pp. 1079–1083.

[11] Yang, B., and K. S. Yun, "Piezoelectric Shell Structures as Wearable Energy Harvesters for Effective Power Generation at Low-Frequency Movement," *Sensors and Actuators A: Physical*, Vol. 188, 2012, pp. 427–433.

[12] Elvin, N., A. A. Elvin, and M. Spector, "A Self-Powered Mechanical Strain Energy Sensor," *Smart Mater. Struct.*, Vol. 10, 2001, pp. 293–299.

[13] Platt, S. R., S. Farritor, and H. Haider, "On Low-Frequency Electric Power Generation with PZT Ceramics," *IEEE-ASME Transactions on Mechatronics*, Vol. 10, No. 2, April 2005, pp. 240–252.

[14] Hausler, E., and L. Stein, "Implantable Physiological Power Supply with PVDF Film," in *Medical Applications of Piezoelectric Polymers*, P. M. Galletti, D. E. De Rossi, and A. S. De Reggi, (eds.), New York: Gordon and Breach Science Publishers, 1988, pp. 259–264.

[15] Donelan, J. M., et al., "Biomechanical Energy Harvesting: Generating Electricity During Walking with Minimal User Effort," *Science*, Vol. 319, No. 5864, February 2008, pp. 807–810.

[16] Ko, W. H., "Piezoelectric Energy Converter for Electronic Implants," *19th Annual Conference of the Society for Engineering in Medicine and Biology*, 1966, p. 67.

[17] Ko, W. H., "Power Sources for Implant Telemetry and Stimulation Systems," in *A Handbook on Biotelemetry and Radio Tracking*, C. J. Amlaner and D. MacDonald, (eds.), Elmsford, NY: Pergamon Press, 1980, pp. 225–245.

[18] MacDonald, S., "Biothermal Power Source for Implantable Devices," U.S. Patent 6,640,137, October 28, 2003.

[19] Wang, Z. Y., et al., "Characterization and Optimization of Polycrystalline $Si_{70\%}$Ge30% for Surface Micromachined Thermopiles in Human Body Applications," *J. Micromech. Microeng.*, Vol. 19, 2009, p. 094011.

[20] Kovacs, T. A., *Micro Machined Transducers*, New York: McGraw-Hill, 1998.

[21] Allameh, S. M., et al., "Piezoelectric Generators for Biomedical and Dental Applications," *J. Mater. Sci.: Mater. Med.*, Vol. 18, 2007, pp. 39–45.

[22] Collinson, M., "Analysis of the Response Characteristics of Piezoelectric Cells in Cyclic Loading," Princeton University, 2005.

[23] www.nhtsa.dot.gov, FMVSS 138.

[24] Lohndorf, M., et al., "Evaluation of Energy Harvesting Concepts for Tire Pressure Monitoring Systems," *Technical Digest PowerMEMS 2007*, Freiburg, Germany, November 28–29, 2007, pp. 331–334.

[25] Roundy, S., "Energy Harvesting for Tire Pressure Monitoring Systems: Design Considerations," *Proceedings of PowerMEMS 2008+ microEMS 2008*, Sendai, Japan, November 9–12, 2008.

[26] Frey, A., et al., "Piezoelectric MEMS Energy Harvesting Module Based on Non-Resonant Excitation," *Solid-State Sensors, Actuators and Microsystems Conference (TRANSDUCERS)*, Beijing, China, June 5–9, 2011, pp. 683–686.

[27] Masaki, T., et al., "Power Output Enhancement of a Vibration-Driven Electrets Generator for Wireless Sensor Applications," *J. Micromech. Microeng.*, Vol. 21, 2011, p. 104004.

[28] Tsutsumino, T., et al., "Micro Seismic Electret Generator for Energy Harvesting," *Proc. 6th Int. Workshop Micro and Nanotechnology for Power Generation and Energy Conversion Applications (PowerMEMS2006)*, 2006, pp. 279–282.

[29] Leland, E. S., and P. K. Wright, "Resonance Tuning of Piezoelectric Vibration Energy Scavenging Generators Using Compressive Axial Preload," *Smart Mater. Struct.*, Vol. 15, 2006, pp. 1413–1420.

[30] Sari, I., T. Balkan, and H. Kulah, "An Electromagnetic Micro Power Generator for Wideband Environmental Vibrations," *Sensors and Actuators A.*, Vol. 145-146, 2008, pp. 405–413.

[31] Liu, J. Q., et al., "A MEMS-Based Piezoelectric Power Generator Array for Vibration Energy Harvesting," *Microelectronics Journal*, Vol. 39, 2008, pp. 802–806.

[32] Bowers, B. J., and D. P. Arnold, "Spherical, Rolling Magnet Generators for Passive Energy Harvesting from Human Motion," *J. Micromech. Microeng.*, Vol. 19, No. 9, 2009, p. 094008.

[33] Yang, B., and C. Lee, "Non-Resonant Electromagnetic Wideband Energy Harvesting Mechanism for Low Frequency Vibrations," *Microsyst. Technol.*, Vol. 16, 2010, pp. 961–966.

[34] Galchev, T., et al., "A Piezoelectric Frequency-Increased Power Generator for Scavenging Low-Frequency Ambient Vibration," *Micro Electro Mechanical Systems (MEMS)*, Hong Kong, China, January 2010, pp. 1203–1206.

[35] Gu, L., and C. Livermore, "Impact-Driven, Frequency Up-Converting Coupled Vibration Energy Harvesting Device for Low Frequency Operation," *Smart Mater. Struct.*, Vol. 20, 2011, p. 045004.

[36] Liu, H., et al., "Piezoelectric MEMS-Based Wideband Energy Harvesting Systems Using a Frequency-Up-Conversion Cantilever Stopper," *Sens. Actuators A: Phys.*, Vol. 186, 2012, pp. 242–248.

[37] Tang, Q. C., X. Y. Xia, and X. X. Li, "Non-Contact Frequency-Up-Conversion Energy Harvester for Durable & Broad-Band Automotive TPMS Application," *IEEE 25th International Conference on Micro Electro Mechanical Systems (MEMS)*, January 2–February 2, 2012, pp. 1273–1276.

[38] Tang, Q. C., Y. L. Yang, and X. X. Li, "Bi-Stable Frequency Up-Conversion Piezoelectric Energy Harvester Driven by Non-Contact Magnetic Repulsion," *Smart Mater. Struct.*, Vol. 20, 2011, p. 125011.

[39] Automotive Electronics Council (AEC), AEC-Q200, 2010.

[40] Schaijk, R. V., et al., "A MEMS Vibration Energy Harvester for Automotive Applications," *Proc. of SPIE Smart Sensors, Actuators and MEMS*, Vol. 8763, 2013, pp. 876305-1–5.

[41] van Liempd, C., et al., "A 1μW-to-1mW Energy-Aware Interface IC for Piezoelectric Harvesting with 40nA Quiescent Current and Zero-Bias Active Rectifiers," ISSCC Conference Proceedings, 2013, p. 21.

[42] Worden, K., and J. M. Dulieu-Barton, "An Overview of Intelligent Fault Detection in Systems and Structures," *International Journal of Structural Health Monitoring*, Vol. 3, No. 1, 2004, pp. 85–98.

[43] Discenzo, F. M., D. Chung, and K. A. Loparo, "Pump Condition Monitoring Using Self-Powered Wireless Sensors," *Sound and Vibration*, Vol. 40, No. 5, 2006, pp. 12–15.

[44] Chang, F. -K., and J. F. C. Markmiller, "A New Look in Design of Intelligent Structures with SHM. Structural Health Monitoring 2006," *Proceedings of the Third European Workshop on Structural Health Monitoring*, 2006, pp. 5–20.

[45] Vullers, R. J. M., et al., "Micro Power Energy Harvesting," *Solid-State Electronics*, Vol. 53, No. 7, 2009.

[46] Pearson, M. R., et al., "Energy Harvesting for Aerospace Structural Health Monitoring Systems," *Journal of Physics: Conference Series*, Vol. 382, 2012, p. 012025.

[47] Samson, D., et al., "Wireless Sensor Node Powered by Aircraft Specific Thermoelectric Energy Harvesting," *Sensors and Actuators A*, No. 172, 2011, pp. 240–244.

[48] http://www.microgensystems.co/content/MicroGen_M0600X-datasheet_Jun2013.pdf.

About the Authors

Bin Yang received an M.Sc. degree from the Department of Material Science and Engineering at Jilin University, Changchun, China, in 2003, and a PhD degree from the Department of Electrical Science and Technology at Shanghai Jiao Tong University, Shanghai, China, 2006. He was a research fellow with the Department of Mechanical Engineering at the National University of Singapore and a senior research engineer with the Institute of Microelectronics, the Agency for Science, Technology and Research, Singapore, from November 2007 to April 2009. From 2009 to April 2011, he was a senior research and development engineer with AAC Acoustic Technologies in Singapore. Currently, he is an associate professor in the Department of Micro/Nano Electronics, School of Electronic Information and Electrical Engineering, Shanghai Jiao Tong University, China. He wrote Chapter 3 of *Nanofabrication and Its Application in Renewable Energy* (Royal Society of Chemistry, Thomas Graham House, U.K.). His current research interests include micro/nano energy harvesting, bio-MEMS, piezoelectric actuators, and sensors.

Huicong Liu received B.Eng. and M.Sc. degrees from the Department of Mechanical Engineering, University of Science and Technology Beijing, Beijing, China, in 2006 and 2008, respectively, and a Ph.D. degree from the Department of Mechanical Engineering, National University of Singapore (NUS), Singapore, in 2013. She was a research fellow in the Department of Electrical and Computer Engineering, NUS, from August 2012 to August 2013. Currently, she is an associate professor in the Robotics and Microsystems Center and the Collaborative Innovation Center of Suzhou Nano Science and Technology, Jiangsu Provincial Key Laboratory of Advanced Robotics, Soochow University, Suzhou, China. Her research interests are MEMS/NEMS energy harvesters and self-powered MEMS devices and systems.

Jingquan Liu received B.Eng., M.Sc., and Ph.D. degrees from the Department of Mechanical Engineering, Jilin University, Changchun, China, in 1993, 1996, and 2000, respectively. He was a postdoctoral research associate in electrical engineering at Shanghai JiaoTong University, Shanghai, China, from 2000 to 2002. From 1998 to 1999, he was a researcher with Yamagata University, Yamagata, Japan. From February 2006 to January 2007, he was a visiting scholar at the Department of Electrical Engineering, the California Institute of Technology, Pasadena, California. Currently, he is a professor in the Department of Micro/Nano Electronics, School of Electronic Information and Electrical Engineering, Shanghai Jiao Tong University, China. His current research interests include MEMS, biomedical microsystems, and micro and nano fabrication technology.

Chengkuo Lee received an M.S.degree in materials science and engineering from National Tsing Hua University, Hsinchu, Taiwan, in 1991, an M.S. degree in industrial and system engineering from Rutgers University, New Brunswick, New Jersey, in 1993, and a Ph.D. degree in precision engineering from The University of Tokyo, Tokyo, Japan, in 1996. He worked as a foreign researcher in the Nanometer Scale Manufacturing Science Laboratory of the Research Center for Advanced Science and Technology, The University of Tokyo, from 1993 to 1996. He also worked in the Mechanical Engineering Laboratory, AIST, MITI of Japan as a JST Research Fellow in 1996. Thereafter, he became a senior research staff member of the Microsystems Laboratory, Industrial Technology Research Institute, Hsinchu, Taiwan. In September 1997, he joined Metrodyne Microsystem Corporation, Hsinchu, Taiwan, and established the MEMS device division and the first micromachining fabrication for commercial purposes in Taiwan. He was the manager of the MEMS Device Division between 1997 and 2000. He was an Adjunct Assistant Professor in the Electrophysics Department of National Chiao Tung University, Hsinchu, Taiwan, in 1998, and an adjunct assistant professor in the Institute of Precision Engineering of National Chung Hsing University, Taichung, Taiwan, from 2001 to 2005. In August 2001, he cofounded Asia Pacific Microsystems (APM), where he first became a vice president of R&D, before becoming a vice president of the Optical Communication Business Unit and the special assistant to the chief executive officer in charge of international business and technical marketing for the MEMS foundry service. From 2006 to 2009, he was a senior member of the technical staff at the Institute of Microelectronics, A-STAR, Singapore. Currently, he is an associate professor in the Department of Electrical and Computer Engineering, National University of Singapore, Singapore. He is the coauthor of the book *Advanced MEMS Packaging* (McGraw-Hill, 2010). He has contributed to more than 200 international conference papers and extended abstracts and 150 peer-reviewed international journal articles in the fields of sensors, actuators, energy harvesting, MEMS, NEMS, nanophotonics, and nanotechnology. He is also the holder of nine U.S. patents.

Index

Recent Titles in the Artech House Microelectromechanical Systems (MEMS) Series

Fundamentals and Applications of Microfluidics, Second Edition, Nam-Trung Nguyen and Steven T. Wereley

Hermeticity Testing of MEMS and Microelectronic Packages, Suzanne Costello, and Marc P. Y. Desmulliez

Introduction to Microelectromechanical (MEM) Microwave Systems, Héctor J. De Los Santos

An Introduction to Microelectromechanical Systems Engineering, Nadim Maluf

MEMS Mechanical Sensors, Stephen Beeby et al.

Micro and Nano Energy Harvesting Technologies, Bin Yang, Huicong Liu, Jingquan Liu, and Chengkuo Lee

Microfluidics for Biotechnology, Jean Berthier and Pascal Silberzan

Organic and Inorganic Nanostructures, Alexei Nabok

Post-Processing Techniques for Integrated MEMS, Sherif Sedky

RF MEMS Circuit Design for Wireless Communications, Héctor J. De Los Santos

Wireless Sensor Network, Nirupama Bulusu and Sanjay Jha